Springer Series in Optical Sciences   Volume 68

Springer Series in Optical Sciences
Editorial Board: A. L. Schawlow K. Shimoda A. E. Siegman T. Tamir

Managing Editor: H. K. V. Lotsch

42 **Principles of Phase Conjugation**
By B. Ya. Zel'dovich, N. F. Pilipetsky, and V. V. Shkunov

43 **X-Ray Microscopy**
Editors: G. Schmahl and D. Rudolph

44 **Introduction to Laser Physics**
By K. Shimoda 2nd Edition

45 **Scanning Electron Microscopy**
Physics of Image Formation and Microanalysis
By L. Reimer

46 **Holography and Deformation Analysis**
By W. Schumann, J.-P. Zürcher, and D. Cuche

47 **Tunable Solid State Lasers**
Editors: P. Hammerling, A. B. Budgor, and A. Pinto

48 **Integrated Optics**
Editors: H. P. Nolting and R. Ulrich

49 **Laser Spectroscopy VII**
Editors: T. W. Hänsch and Y. R. Shen

50 **Laser-Induced Dynamic Gratings**
By H. J. Eichler, P. Günter, and D. W. Pohl

51 **Tunable Solid State Lasers for Remote Sensing**
Editors: R. L. Byer, E. K. Gustafson, and R. Trebino

52 **Tunable Solid-State Lasers II**
Editors: A. B. Budgor, L. Esterowitz, and L. G. DeShazer

53 **The CO_2 Laser** By W.J. Witteman

54 **Lasers, Spectroscopy and New Ideas**
A Tribute to Arthur L. Schawlow
Editors: W. M. Yen and M. D. Levenson

55 **Laser Spectroscopy VIII**
Editors: W. Persson and S. Svanberg

56 **X-Ray Microscopy II**
Editors: D. Sayre, M. Howells, J. Kirz, and H. Rarback

57 **Single-Mode Fibers** Fundamentals
By E.-G. Neumann

58 **Photoacoustic and Photothermal Phenomena**
Editors: P. Hess and J. Pelzl

59 **Photorefractive Crystals in Coherent Optical Systems**
By M. P. Petrov, S. I. Stepanov, and A. V. Khomenko

60 **Holographic Interferometry in Experimental Mechanics**
By Yu. I. Ostrovsky, V. P. Shchepinov, and V. V. Yakovlev

61 **Millimetre and Submillimetre Wavelength Lasers**
By N. G. Douglas

62 **Photoacoustic and Photothermal Phenomena II**
Editors: J. C. Murphy, J. W. Maclachlan Spicer, L.C. Aamodt, and B. S. H. Royce

63 **Electron Energy Loss Spectrometers**
The Technology of High Performance
By H. Ibach

64 **Handbook of Nonlinear Optical Crystals**
By V. G. Dmitriev, G. G. Gurzadyan, and D. N. Nikogosyan

65 **High-Power Dye Lasers**
Editor: F. J. Duarte

66 **Silver Halide Recording Materials for Holography and Their Processing**
By H. I. Bjelkhagen

67 **X-Ray Microscopy III**
Editors: A. G. Michette, G. R. Morrison, and C. J. Buckley

68 **Holographic Interferometry**
Editor: P. K. Rastogi

69 **Photoacoustic and Photothermal Phenomena III**
Editor: D. Bićanić

70 **Electron Holography**
By A. Tonomura

Volumes 1–41 are listed on the back inside cover

P.K. Rastogi (Ed.)

Holographic Interferometry

Principles and Methods

With 178 Figures, Some in Colour

Springer-Verlag Berlin Heidelberg GmbH

PRAMOD K. RASTOGI, Sr. Res. Engineer
Laboratory of Stress Analysis, Swiss Federal Institute of Technology
CH-1015 Lausanne, Switzerland

Editorial Board

ARTHUR L. SCHAWLOW, Ph. D.
Department of Physics, Stanford University
Stanford, CA 94305, USA

Professor KOICHI SHIMODA, Ph.D.
Faculty of Science and Technology
Keio University, 3-14-1 Hiyoshi,
Kohoku-ku
Yokohama 223, Japan

Professor ANTHONY E. SIEGMAN, Ph. D.
Electrical Engineering
E.L. Ginzton Laboratory, Stanford University
Stanford, CA 94305, USA

THEODOR TAMIR, Ph. D.
Polytechnic University
333 Jay Street, Brooklyn, NY 11201, USA

Managing Editor: Dr. HELMUT K.V. LOTSCH
Springer-Verlag, Tiergartenstrasse 17, D-69121 Heidelberg, Germany

ISBN 978-3-662-13990-5

Library of Congress Cataloging-in-Publication Data. Rastogi, Pramod. Holographic interferometry/Pramod Rastogi. p. cm.--(Springer series in optical sciences; v. 68) Includes bibliographical references and index.
ISBN 978-3-662-13990-5   ISBN 978-3-540-48078-5 (eBook)
DOI 10.1007/978-3-540-48078-5
1. Holographic inter- ferometry. I. Title. II. Series. TA 1555.R37   1994   621.36'75--dc20 93-39992

This work is subject to copyright. All rights are reserved, whether the whole or part of the material is concerned, specifically the rights of translation, reprinting, reuse of illustrations, recitation, broadcasting, reproduction on microfilm or in any other way, and storage in data banks. Duplication of this publication or parts thereof is permitted only under the provisions of the German Copyright Law of September 9, 1965, in its current version, and permission for use must always be obtained from Springer-Verlag Berlin Heidelberg GmbH.
Violations are liable for prosecution under the German Copyright Law.

© Springer-Verlag Berlin Heidelberg 1994
Originally published by Springer-Verlag Berlin Heidelberg New York in 1994
Softcover reprint of the hardcover 1st edition 1994

The use of general descriptive names, registered names, trademarks, etc. in this publication does not imply, even in the absence of a specific statement, that such names are exempt from the relevant protective laws and regulations and therefore free for general use.

Typesetting: Macmillan India Ltd., Bangalore-25

SPIN: 10068856   54/3140/SPS – 5 4 3 2 1 0 – Printed on acid-free paper

Preface

Holographic interferometry is an important and widely used technique in the measurement of variations in certain important physical quantities such as displacements, strains, densities, etc. in solid mechanics and flow problems. The fundamentals of the technique stand on a firm theoretical foundation. The applications of the method have developed at a rapid pace and a stage has reached where it would seem hazardous, if not impossible, to single out outright a field of research not yet penetrated by the technique. A survey of literature shows that a significant amount of research and development work is being carried out in holographic interferometry. The shortcomings of the technique are being addressed and the interest in the field is growing substantially.

The aim of the present volume is to provide a valuable and up-to-date source of information in this fast expanding field. It is organized in eight chapters, with each chapter dealing with a particular aspect of holographic interferometry. The book has been written within a framework covering the principles and methods currently in use in holographic interferometry. The scope of the book has been limited to the study of opaque objects. Ample space has been devoted to a comprehensive treatment of the phenomena of fringe formation, with a particular emphasis on the quantitative evaluation of the holographic interference fringe patterns. The emergence of computer-aided fringe analysis and phase-shifting techniques have simplified considerably the quantitative real-time measurement of object shapes and deformations. The last two chapters provide a reasonably detailed overview of full field holographic methods for the measurement of shapes, displacements, derivatives, difference displacements and vibrations. Apart from reviewing the current state-of-the-art, the book also aims at stimulating research activities which are likely to expand further the horizons of holographic interferometry.

The book addresses to both researchers and practising engineers alike. To the researcher it should provide invaluable insights, an objective appraisal of the holographic methods and above all, directions to the present and emerging lines of research. To those contemplating the use of holographic interferometry in their respective disciplines, it presents a useful and pertinent information on available techniques and systems needed to solve their specific measurement problems. The outlook projected by the present volume when compared with those published earlier in the field is distinctly different in style and in content, with particular emphasis on evolving user requirements.

The state-of-the-art reports are meant to provide fairly detailed summaries of different facets to anyone who needs to apply holographic interferometry in nondestructive testing and metrological investigations in a wide range of scientific and engineering disciplines. Furthermore, the set of contributions contained in the book should be of value as a source of reference to engineers and research workers interested or active in the field.

I wish to express my sincere gratitude to Professor L. Pflug for his encouragement and support. I want to thank Dr. A. Lahee for her many helpful and friendly discussions. My special thanks go to Dr. H. Lotsch who in spite of his ill health worked tirelessly to get the project through in a timely manner. Finally, I wish to thank E. Crombie for her cheerful assistance.

Lausanne
January 1994

PRAMOD K. RASTOGI

Contents

1 **Introduction**
P.K. Rastogi 1
    References 6

2 **Basic Principles**
P. Hariharan. With 18 Figures 7

  2.1 The Development of Holography 7
  2.2 The Off-Axis Hologram 9
      2.2.1 Reflection Holograms 11
      2.2.2 Image Holograms 11
  2.3 The Reconstructed Image 12
  2.4 Image Speckle 13
      2.4.1 Signal-to-Noise Ratio 14
  2.5 Types of Holograms 14
      2.5.1 Thin Amplitude and Phase Gratings 14
      2.5.2 Volume Gratings 15
      2.5.3 Holograms of Diffusely Reflecting Objects 18
      2.5.4 Multiply Exposed Holograms 18
  2.6 Light Sources and Optical Systems 19
      2.6.1 Coherence Requirements 19
      2.6.2 Laser Beam Expansion 20
      2.6.3 Beam Polarization 21
      2.6.4 Optical Systems for Holography 21
      2.6.5 Holography with Pulsed Lasers 22
      2.6.6 Laser Safety 22
  2.7 The Recording Medium 22
      2.7.1 Effects of Nonlinearity 24
  2.8 Recording Materials 24
      2.8.1 Photographic Emulsions 25
      2.8.2 Photothermoplastics 25
      2.8.3 Photorefractive Crystals 26
  2.9 Holographic Interferometry 27
      2.9.1 Real-Time Holographic Interferometry 27
      2.9.2 Double-Exposure Holographic Interferometry 28
      2.9.3 Phase Difference in the Interference Pattern 29

	2.9.4	The Holodiagram	30
	2.9.5	Localization of the Interference Fringes	31
	References	31

3 Quantitative Determination of Displacements and Strains from Holograms
R.J. Pryputniewicz. With 19 Figures 33

3.1	Projection matrices: Definition and properties.		33
	3.1.1	Normal Projection	33
	3.1.2	Oblique Projection	35
3.2	Illumination, Observation, and Sensitivity Vectors		38
3.3	Determination of Displacements		39
	3.3.1	Determination of Displacements when Fringe Order is Known	41
	3.3.2	Determination of Displacements when Fringe Order is Unknown	47
	3.3.3	Determination of Displacements from Multiple Holograms	51
3.4	Determination of Strains and Rotations		53
	3.4.1	Spatial Dependence of the Fringe-Locus Function	56
	3.4.2	Rigid-Body Rotations and Homogeneous Strains	57
3.5	Electro-Optic Holography		59
	3.5.1	Fundamentals of EOH	59
	3.5.2	Electronic Processing of Holograms	60
		3.5.2.1 Static Measurements	60
		3.5.2.2 Dynamic Measurements	64
	3.5.3	Representative Applications of EOH	68
3.6	Conclusions		70
	References		72

4 Two-Reference-Beam Holographic Interferometry
R. Dändliker. With 22 Figures. 75

4.1	Electronic Interference Phase Measurement		75
	4.1.1	Phase-Shifting and Heterodyne Detection	76
	4.1.2	Double-Exposure with Two Reference Beams	78
4.2	Interferometry with Speckle Fields (Diffusely Scattering Objects)		82
	4.2.1	Speckle Statistics for Coherent Imaging	83
	4.2.2	Measuring Intensity with a Detector of Finite Size	84
	4.2.3	Interference Fringe Formation	86

4.3	Sources of Errors in Interference Phase Measurement		88
	4.3.1	Systematic Errors from Holographic Interferometry with Two Reference Sources	88
		4.3.1.1 Misalignment of Hologram and Reconstructing Reference Waves	88
		4.3.1.2 Cross-Talk from Holographic Recording	89
		4.3.1.3 Spurious Fringes Due to Overlapping Reconstructions	90
	4.3.2	Statistical Errors Due to Speckle Noise	90
		4.3.2.1 Non-Overlapping Reconstructions	91
		4.3.2.2 Overlapping Reconstructions	94
4.4	Applications		95
	4.4.1	Double-Exposure Holography with Phase-Shift Fringe Evaluation	95
	4.4.2	Multiple-Exposure Holographic Interferometry	98
	4.4.3	Double-Pulse Holography for 3-D Displacement Measurement	99
	4.4.4	Strain Measurement by Heterodyne Holographic Interferometry	102
4.5	Conclusions		106
	References		107

5 Phase-Shifting Holographic Interferometry
K. Creath. With 23 Figures 109

5.1	Background		109
	5.1.1	Real-Time Holographic Secondary Interference Fringes	109
	5.1.2	From Wavefront to Object Displacement	110
5.2	Phase-Measurement Basics		111
	5.2.1	General Phase Measurement Theory	112
	5.2.2	Phase-Modulation Techniques	114
	5.2.3	Phase Unwrapping	116
	5.2.4	Sampling Requirements	117
	5.2.5	Intensity-Modulation Requirements	119
	5.2.6	Measurement Examples	121
5.3	Temporal Phase-Measurement (Phase-Shifting) Algorithms		122
	5.3.1	Three Frames (90°, 120°, and General Phase Shifts)	122
	5.3.2	Four Frames (90° Phase Shifts)	123
	5.3.3	Five Frames (90° Phase Shifts)	124
	5.3.4	Carré (Four Frames with General Phase Shifts)	125
	5.3.5	Synchronous Detection ($2\pi/N$ Phase Shifts)	125
	5.3.6	$(N+1)$ Frames ($2\pi/N$ Phase Shifts)	126

	5.3.7	2 + 1 (90° Phase Shifts)	126
	5.3.8	Scanning Phase Shift (Random Phase Shifts)	127
5.4	Phase-Shifter Calibration		128
5.5	Major Error Sources		130
	5.5.1	Phase-Shifter Errors	132
	5.5.2	Detection (Intensity) Nonlinearities	134
	5.5.3	Quantization	136
	5.5.4	Vibration and Air Turbulence	137
5.6	Equipment and Experimental Consideration		139
	5.6.1	Cameras and Frame Grabbers	139
	5.6.2	Phase Shifters: Ramping Versus Stepping	141
	5.6.3	Alignment Considerations	142
5.7	Specialized Techniques		142
	5.7.1	Electronic Holography	142
	5.7.2	Time-Average Vibration Analysis	144
	5.7.3	Phase-Shifting Holographic Moiré	145
5.8	Emerging Trends		147
	References		148

6 Computer-Aided Evaluation of Holographic Interferograms
T. Kreis. With 27 Figures 151

6.1	Background		151
	6.1.1	Holographic Interferometry	151
	6.1.2	Digital Image Processing	153
6.2	Fringe Formation in Holographic Interferometry		155
	6.2.1	Double Exposure and Real-Time Holographic Interferometry	155
	6.2.2	Vibration Analysis by Holographic Interferometry	156
	6.2.3	Interference Phase Variation due to Deformation	157
	6.2.4	Interference Phase Variation due to Refractive-Index Variations	159
	6.2.5	Ambiguity of Fringe Patterns	160
	6.2.6	Distortions of Holographic Interferograms	163
	6.2.7	Simulation of Holographic Interference Patterns	163
6.3	Evaluation of Holographic Interferograms		164
	6.3.1	Qualitative Evaluation	164
	6.3.2	Quantitative Evaluation	165
	6.3.3	Condition of the Evaluation Matrix	168
	6.3.4	Determination of Sensitivity Vectors	169
	6.3.5	Correction of Perspective Distortion	170
6.4	Interference Phase Determination		173
	6.4.1	Fringe Skeletonizing	173
	6.4.2	Evaluation by Temporal Heterodyning	175

		6.4.3	Phase-Sampling Evaluation	177
			6.4.3.1 Phase Shifting and Phase Stepping............	178
			6.4.3.2 Solution of Phase Sampling Equation.........	179
			6.4.3.3 Phase Sampling with Unknown Phase Steps...	180
		6.4.4	Fourier-Transform Evaluation........................	184
			6.4.4.1 Spatial Heterodyning	184
			6.4.4.2 Fourier-Transform Evaluation Without Carrier....................................	185
			6.4.4.3 Fourier-Transform Evaluation of Phase Shifted Interferograms	192
		6.4.5	Further Methods for Interference Phase Determination......................................	193
		6.4.6	Interference Phase Demodulation.....................	195
		6.4.7	Comparison of Interference Phase Determination Methods	197
			6.4.7.1 Experimental Requirements	197
			6.4.7.2 Resolution and Precision.....................	199
			6.4.7.3 Errors and Distortions.......................	200
			6.4.7.4 Comparison Results	200
	6.5	Processing of Evaluated Data		201
		6.5.1	Curve Fitting by Gaussian Least Squares	201
		6.5.2	Strain and Stress Analysis............................	204
		6.5.3	Finite-Element Methods	205
		6.5.4	Boundary-Element Methods	206
		6.5.5	Fracture Mechanics	206
		6.5.6	Computer Tomography..............................	207
	6.6	Conclusions and Future Trends.............................		208
		References		209

7 Techniques to Measure Displacements, Derivatives and Surface Shapes. Extension to Comparative Holography
P.K. Rastogi. With 62 Figures 213

	7.1	Measurement of Out-of-Plane Displacements		214
		7.1.1	Basic Configurations................................	214
			7.1.1.1 Remarks on the Removal of Phase Ambiguities and Measurement of Small Deformations....	217
		7.1.2	Speckle Decorrelation and Fringe Visibility	218
		7.1.3	Compensation of Rigid-Body Movements and Fringe Control...	220
	7.2	Measurement of In-Plane Displacements		221
		7.2.1	Holographic Moiré.................................	222
			7.2.1.1 Basic Relations	222
			7.2.1.2 Generation of Auxiliary Fringes	224

		7.2.1.3	Localization	226
		7.2.1.4	Some Remarks Concerning Imaging Parameters	229
	7.2.2	Reconstructed Reference Wave Holographic Interferometry		230
	7.2.3	Conjugate-Wave Holographic Interferometry		232
	7.2.4	Phase-Shifted Holographic Moiré		233
7.3	Measurement of the Derivatives of Displacements			236
	7.3.1	Measurement of Slope Change and Curvature		238
		7.3.1.1	Holographic Moiré for Slope-Change Measurement	238
		7.3.1.2	Measurement of Curvature	243
		7.3.1.3	Measurement of Slope and Curvature using Modulated Diffraction Gratings	249
		7.3.1.4	Relationship between the Second Derivatives, Moments and Stresses	251
	7.3.2	Measurement of In-Plane Strains		252
		7.3.2.1	Measurement of In-Plane Strains using Modulated Diffraction Gratings	252
7.4	Contouring of Three-Dimensional Objects			254
	7.4.1	Wavelength-Difference Contouring		255
	7.4.2	Immersion Method		257
	7.4.3	Holographic Moiré Multiple-Sources Contouring		258
		7.4.3.1	Basic Technique	258
		7.4.3.2	Phase Management Solution	260
		7.4.3.3	Mechanism of Fringe Formation	263
		7.4.3.4	Performance Evaluation	264
	7.4.4	Holographic Moiré Object Tilt Contouring		266
	7.4.5	Multiple-Sources Contouring		268
		7.4.5.1	Basic Techniques	268
		7.4.5.2	Mechanism of Fringe Formation	270
		7.4.5.3	Enhancement of Contouring Sensitivity	271
		7.4.5.4	Orientation of the Contour Planes	274
7.5	Comparative Holographic Interferometry			274
	7.5.1	Equation of the Difference Displacement Vector of Two Surface Elements		275
	7.5.2	Difference Holographic Interferometry		277
	7.5.3	Comparative Holographic Moiré		280
	7.5.4	Formation of the Modulated Pattern		282
	7.5.5	Phase-Shifted Comparative Holography		285
7.6	Conclusions			288
	References			288

8 Study of Vibrations
C.S. Vikram. With 9 Figures 293

8.1 Time-Average Holographic Interferometry of Sinusoidal
    Vibration and Separable Motions.......................... 293
8.2 Real-Time Interferometry of Vibration 295
8.3 Time-Average Holography of Nonseparable Motions and
    Multiple Modes.. 296
8.4 Stroboscopic Holographic Interferometry................... 298
8.5 Temporally Modulated Holography.......................... 301
    8.5.1 Frequency Translation............................... 302
    8.5.2 Amplitude Modulation 302
    8.5.3 Laser Irradiance Modulation......................... 303
    8.5.4 Phase Modulation 303
    8.5.5 Holographic Subtraction............................. 305
    8.5.6 Other Modulation Techniques 306
8.6 Fringe-Shifting and Quantitative Analysis Techniques 306
8.7 Rotating Objects 311
    8.7.1 Synchronized Triggering............................. 312
    8.7.2 Rotating Holographic Plate and Reference Beam 312
    8.7.3 Use of Image-Derotator 312
8.8 Other Techniques and Applications 312
    References 316

Subject Index... 319

1. Introduction

P.K. Rastogi

Laboratory of Stress Analysis, Swiss Federal Institute of Technology, CH-1015 Lausanne, Switzerland

One of the remarkable contributions of holography to metrology has been to extend the services of optical interferometry to study objects of arbitrary microstructures. The development of holographic interferometry has come around on the one hand by the ability of a hologram to record the complex wavefield scattered by a laser lit diffuse object (test piece) and on the other hand by the extraordinary characteristic of the same hologram to release the captured wavefield with high fidelity during the reconstruction process. Several basic approaches (double exposure, real time, etc.) have been developed around this phenomena to furnish interference fringe patterns. Basically, two holographic recordings, one corresponding to any initially recorded state of the object and the other corresponding to the slightly changed position of the same object, are needed to obtain an interference pattern which is indicative of the small movement undergone by the object. The resulting fringes are contours of constant displacement in the direction of the sensitivity vector, an optical geometry dependent parameter.

At this stage it may be fruitful to orient the reader to the developments achieved during the first decade and a half of holographic interferometry. The events leading to the discovery of holographic interferometry have been eloquently described by *Stetson* [1.1]. *Powell* and *Stetson* [1.2], *Stetson* and *Powell* [1.3], *Brooks* et al. [1.4], *Collier* et al. [1.5], *Burch* [1.6], *Haines* and *Hilderbrand* [1.7], and *Heflinger* [1.8] were among the first to demonstrate the interference of the holographic reconstructed wave fields. Static, real-time static and dynamic measurements were reported on diffusely reflecting surfaces. Over the next few years, a number of investigators concentrated their efforts on developing the conceptual foundations for holographic interferometry. Notable among these have been the work of *Aleksandrov* and *Bonch-Bruevich* [1.9], *Froehly* et al. [1.10], *Tsuruta* et al. [1.11], *Stetson* [1.12–14], *Sollid* [1.15], *Welford* [1.16], *Abramson* [1.17], *Walles* [1.18], *Prikryl* [1.19], and *Dubas* and *Schumann* [1.20]. The list is certainly not exhaustive. These investigators have, for the most part, attempted to analyse the formation and localization of fringes in holographic interferometry. Mathematical expressions were derived to relate the holographic interference fringes to the displacements and deformations of a diffusely reflecting object. The different approaches published in literature have contributed enormously to the general understanding of holographic interferometry. A self contained treatment of the subject can be found in the works of *Vest*

[1.21], and *Ostrovsky* et al. [1.22]. A thorough treatment of the fundamental concepts has been given by *Schumann* and *Dubas* [1.23].

Holographic interferometry has found an extensive use in experimental mechanics in the measurement of displacements and strains of rough objects. The method has been used with considerable amount of success in the non-destructive testing of materials and engineering components. The presence of flaws such as microcracks, voids, delaminations and material nonhomogeneities have all been unveiled by the method. An illustrative example is that of car tyres subjected to inspection for identifying possible flaws in their casings. In the presence of a flaw, the strength of the defective region is weaker than that of the region surrounding it. Under internal pressure change, the defective region in the car tyre shows more pronounced local deformation, manifesting itself on the surface as a pocket of fringe concentration. The perturbed region on the interferogram clearly pinpoints the presence of a problem area on the tyre.

Ever since it first appeared a little over quarter of century ago in the scientific literature, holographic interferometry has never ceased to grow. After having proved itself in the confines of the laboratories, the technique responded in a relatively successful manner to the calls of scientific and industrial communities to interact in their studies and research projects in sometimes less than ideal circumstances. Here was at issue the technique's potential to address real world problems, and at stake lay its future. The underlying advantages and shortcomings were exposed as efforts continued to explore and harness the potentials inherent in the method. The highly promising and coveted technique of the mid 60's had lost much of its shine by the late 70's as its limitations and weaknesses became better recognized. Although the instrumentation had improved dramatically over the years making the recording of holograms faster and easier, the quantitative analysis of fringe patterns for example remained time consuming and tedious. The technique's capacity to stage a come-back results largely from its abilities to create new options in its folds, enhance its potentialities and, above all, adapt itself relatively easily to the problem at hand.

The technique is still developing, it is as lively these days as in the first years of its history. If in the early years of its existence the flurry of activity in the field was more focused towards the understanding of the basic phenomena underlying the technique, the later years have seen this activity become more directed towards expanding its horizons, rendering it more user friendly and practical to handle, and developing automatic computer based systems for acquisition and analysis of the displayed fringe patterns. These and related developments during the last decade have contributed to a profound revitalization of the field.

The most striking thing about holographic interferometry has been the sweep of its reach. From the inspection of aircraft components to the study of crystal growth, from the studies of biological changes of bones to the measurement of wear in prosthesis, from the study of the embryonic behaviour in intact incubating eggs to the evaluation of the frost resistance of porous materials, from the study of the mechanical behaviour of reeds to the vibration analysis of tympanic membranes, from the measurement of time dependent shrinkage in concrete structures to the studies of orthodontic tooth movements, the applications of holographic interferometry have been far and wide touching upon fields as diverse as medicine, dentistry, biomechanics, heat transfer, plasma diagnostics, fracture mechanics, material testing, biology and environment. The cited fields are to be considered more on an indicative than on a comprehensive basis. The widespread use of holographic interferometry is thus evident. These various disciplines share a common interest in the all-round development and growth of holographic interferometry.

Holographic interferometry has provided a new dimension to the validation and development of analytical models. Extremely heterogeneous nature of composite materials for example makes writing down their constitutive relationships very complicated. Moreover, the global analytical model must reflect the effect of refined constitutive relations on the overall structural behaviour. The complex interaction between the level of refinement in these two domains calls for a non-destructive testing tool in order to allow one to control the extent of refinement in the numerical and constitutive models. With the numerical techniques becoming increasingly refined, equally high performance experimental procedures are needed to address the task of evaluation. Holographic interferometry is a highly qualified technique for this job and is destined to play an increasingly important role as the materials modelling becomes more and more complex.

Recent advances in digital image acquisition and processing technologies have been changing the nature of fringe pattern analysis towards an increasing reliance on computer based and optoelectronic approaches. The tedium and imprecision associated with the manual extraction of the quantitative data from the interferograms has given way to the automatic evaluation of phase encoded in the fringe pattern. The high-speed automatic reduction of fringe data not only provides a versatile service to practitioners but also, at the same time, makes the technique more economical and efficient to use.

The present work fills a gap which is characteristic of a rapidly growing field. Each chapter is tailored to provide an up-to-date overview of the work achieved in that particular topic. The outlook projected by the present volume when compared with those published recently in the field [1.24, 25] is distinctly different in style, in content and in the depth of presentation of the individual topics.

The quasi-monograph is organized in eight chapters. Broadly speaking, the following two chapters provide the background and basic principles governing the fringe formation in holographic interferometry. The next three chapters discuss the fringe analysis techniques, such as electronic phase measurement, phase-shifting and digital fringe processing, which have opened the way to an on-line analysis of fringe patterns. The last two chapters deal with the development of methods suitable for the measurement of specific displacement and deformation components, shapes and vibrations. The topic of comparative holography is also considered in some detail. The content of the book are summarized below.

Chapter 2 by P. Hariharan provides a clear understanding of the imaging properties of thin transmission holograms as applied to holographic interferometry. Types of holograms, light sources and basic optical systems used for holography, and major classes of recording media popularly employed in the practice of holographic interferometry are all briefly described. Finally, the schemes to generate fringe patterns and some early approaches used for their interpretation are succinctly presented.

Chapter 3 by R. Pryputniewicz considers the governing equations for quantitative determination of displacements and strains directly from the fringes of holographic interferometry. Starting from the concept of projection matrices, the author presents a mathematical analysis leading to the solutions for the determination of three displacement components for the cases when the fringe order is known, when the fringe order is unknown, and when multiple holograms are used. This is followed by a comprehensive presentation of the fringe-vector method of strain analysis. This method is of considerable importance in experimental mechanics as the quantity one is in many cases interested in measuring is strain and not displacement. A description of electro-optic holographic interferometry is presented. The chapter concludes with a discussion on automated fringe readout systems and instrumentation, thereof preparing the readers for the presentations to follow in the next three chapters.

The classes of techniques outlined in chaps. 4–6 may not be mutually exclusive in as far as their goals are concerned, but they do represent major distinctions in the current state-of-the-art of the fringe evaluation techniques. These distinctions are a reflection of different trends that have emerged for addressing automation in the realms of holographic interferometry. These techniques have made and promise to make significant contributions in the automatic readout and processing of fringe patterns.

Chapter 4 by R. Dändliker introduces the basics of the electronic interference phase measurement and the recording of double exposure holograms with two reference beams. Fringe formation of diffusely reflecting objects is described in some detail. The treatment emphasizes the statistical aspects of interferometry with speckle fields. Sources of errors are explicitly presented to help one in assessing the accuracy of measurements. The usefulness of the two reference beam holographic interferometry is impressively illustrated by means of some practical examples.

Phase shifting holographic interferometry is a powerful and versatile method in the quantitative measurement of deformations and shapes of diffuse object surfaces. Chapter 5 by K. Creath summarizes the present status of phase shifting holographic interferometry. After providing the background material, the author outlines the various algorithms which can be used with real-time holographic interferometry to measure the wavefront phase. A miscalibration of the phase shifter can be an important source of error in the measurement. Major sources of errors are elucidated and wherever possible means of reducing these errors are given. The chapter concludes with a brief discussion on some specialized techniques.

Chapter 6 by T. Kreis provides a comprehensive summary of the use of the computer based and optoelectronic approaches in the evaluation of holographic interferograms. The presentation emphasizes the measurements by Fourier transform and spatial-carrier phase shifting techniques. One of the major driving forces in the research on computer aided evaluation of interferograms results from the substantial storage and processing capabilities that are today becoming available in the end-user equipment.

Measurement of specific displacement components and their derivatives has been a field of active investigation in holographic interferometry. Chapter 7 by P.K. Rastogi outlines a wide range of configurations based on holographic interferometry that have been employed to determine the displacement, strain, slope change, and curvature contours of a deformed structure. This is followed by a presentation of procedures used for shape measurements of three-dimensional surfaces using holographic interferometry. Emphasis is placed more on the description of recently developed techniques. One of the most relevant successes obtained in the last decade concerns the development of comparative holography. The procedure enables one to compare wavefields emerging from two identical but nominally different object surfaces. The last part of the chapter is devoted to a discussion on comparative holography. Most of the optical configurations examined have been illustrated with experimentally obtained interference patterns.

Chapter 8 by C. Vikram is devoted to the study of vibrations using holographic interferometry. Holography provides a wholefield modal map of a vibrating object yielding, at the same time, quantitative measurement of the object's deformation. Of particular interest is the account given on the use of fringe-shifting techniques in the quantitative analysis of time-average concomitant fringe patterns. A description of holographic vibration analysis on high-speed rotating objects is also included.

Time to wind up this chapter. High sensitivity, measurement information spread up over the whole tested surface, noninvasive and real-time observation, and high-speed quantitative analysis are some of the excellent features around which holographic interferometry has built up its reputation slowly over the years. It can be said with all fairness that at present no other method can even approach, not to say match, the impact, wide acceptance and versatility of holographic interferometry in metrology and non-destructive testing. As for the future developments in store for holographic interferometry, what more can one modestly add to what has already been reported in the state-of-the-art reports contained in these volumes, except to quote from Shakespeare's Macbeth

> If you can look into the seeds of time,
> And say which grain will grow and which will not,
> Speak then to me, who neither beg nor fear
> Your favours nor your hate.

References

1.1 K.A. Stetson: Exp. Tech. **15**, 15 (1991)
1.2 R.L. Powell, K.A. Stetson: J. Opt. Soc. Am. **55**, 1593 (1965)
1.3 K.A. Stetson, R.L. Powell: J. Opt. Soc. Am. **55**, 1694 (1965)
1.4 R.E. Brooks, L.O. Heflinger, R.F. Wuerker: Appl. Phys. Lett. **7**, 248 (1965)
1.5 R.J. Collier, E.T. Doherty, K.S. Pennington: Appl. Phys. Lett. **7**, 223 (1965)
1.6 J.M. Burch: Prod. Eng. **44**, 431 (1965)
1.7 K.A. Haines, B.P. Hildebrand: Appl. Opt. **5**, 595 (1966)
1.8 L.O. Heflinger, R.F. Wuerker, R.E. Brooks: J. Appl. Phys. **37**, 642 (1966)
1.9 E.B. Aleksandrov, A.M. Bonch-Bruevich: Sov. Phys. -Tech. Phys. **12**, 258 (1967)
1.10 C. Froehly, J. Monneret, J. Pasteur, J.Ch. Vienot: Opt. Acta **16**, 343 (1969)
1.11 T. Tsuruta, N. Shiotake, Y. Itoh: Opt. Acta **16**, 723 (1969)
1.12 K.A. Stetson: Optik **29**, 386 (1969)
1.13 K.A. Stetson: Optik **31**, 576 (1970)
1.14 K.A. Stetson: Appl. Opt. **14**, 272 (1975)
1.15 J.E. Sollid: Appl. Opt. **8**, 1587 (1969)
1.16 W.T. Welford: Opt. Commun. **1**, 123 (1969)
1.17 N. Abramson: Appl. Opt. **9**, 97 (1970)
1.18 S. Walles: Opt. Acta **17**, 899 (1970)
1.19 I. Prikryl: Opt. Acta **21**, 675 (1974)
1.20 M. Dubas, W. Schumann: Opt. Acta **22**, 807 (1975)
1.21 CM Vest: *Holographic Interferometry* (Interscience, New York 1979)
1.22 Y.I. Ostrovsky, M.M. Butusov, G.V. Ostrovskaya: *Interferometry by Holography*, Springer Ser. Opt. Sci. Vol. 20 (Springer, Berlin, Heidelberg 1980)
1.23 W. Schumann, M. Dubas: *Holographic Interferometry*, Springer Ser. Opt. Sci. Vol. 16 (Springer, Berlin, Heidelberg 1979)
1.24 W. Schumann, J.P. Zürcher, D. Cuche: *Holography and Deformation analysis*, Springer Ser. Opt. Sci. Vol. 46 (Springer, Berlin, Heidelberg 1985)
1.25 Y.I. Ostrovsky, V.P. Schepinov, V.V. Yakovlev: *Holographic Interferometry in Experimental Mechanics*, Springer Ser. Opt. Sci. Vol. 60 (Springer, Berlin, Heidelberg 1991)

2. Basic Principles

P. Hariharan

CSIRO Division of Applied Physics, PO Box 218, Lindfield, NSW 2070, Australia

This chapter summarizes the basic principles of holographic imaging, as applied to holographic interferometry. In it, after a brief historical introduction, we will review image formation by a hologram, the basic types of holograms, light sources and optical systems used for holography, and hologram recording media and their characteristics, before discussing some of the basic techniques of holographic interferometry.

2.1 The Development of Holography

The unique characteristic of holography is its ability to record both the phase and the amplitude of a light wave. This is not possible with conventional imaging techniques, since all recording materials respond only to the intensity. Holography gets around this problem by using coherent illumination and introducing, as shown in Fig. 2.1, a reference beam derived from the same source. The photographic film records the interference pattern produced by this reference beam and the light waves scattered by the object.

The resulting recording (the hologram) contains information on the phase as well as the amplitude of the object wave, since the intensity at any point in this interference pattern depends on the phase as well as the amplitude of the object wave. If the hologram is illuminated once again with the original reference wave, as shown in Fig. 2.2, it reconstructs the object wave. An observer looking through the hologram sees a perfect three-dimensional image of the object.

In *Gabor's* [2.1] historical demonstration of holographic imaging, a transparency consisting of opaque lines on a clear background was illuminated with a collimated beam of monochromatic light, and the interference pattern produced by the directly transmitted beam (the reference wave) and the light scattered by the lines on the transparency was recorded on a photographic plate. When the hologram (a positive transparency made from this plate) was illuminated with the original collimated beam, it produced two diffracted waves, one reconstructing an image of the object in its original location, and the other, with the same amplitude but opposite phase, forming a second, conjugate image. However, a major drawback was that the conjugate image, as well as scattered light from the directly transmitted beam, seriously degraded the reconstructed image.

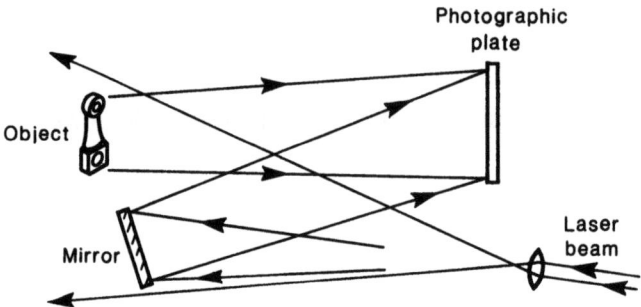

Fig. 2.1. Hologram recording: the interference pattern produced by the reference wave and the object wave is recorded

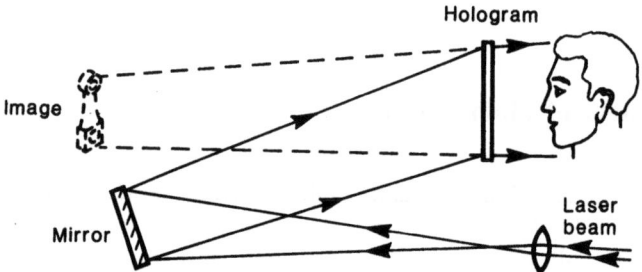

Fig. 2.2. Image reconstruction: light diffracted by the hologram recreates the object wave

The twin-image problem was finally solved when *Leith* and *Upatnieks* [2.2–4] developed the off-axis reference beam technique, already shown schematically in Figs. 2.1 and 2. This technique used a separate reference wave incident on the photographic plate at an appreciable angle to the object wave. As a result, when the hologram was illuminated with the original reference beam, the two reconstructed images were separated by large enough angles from the directly transmitted beam, and from each other, to ensure that they did not overlap.

The development of the off-axis technique resulted in a surge in activity in holography. One of its most important applications, which was discovered almost simultaneously by several groups, was holographic interferometry [2.5–9].

In holographic interferometry, at least one of the interfering waves is reconstructed by a hologram. Its major attraction is the fact that holography makes it possible to store a wave front and reconstruct it at a later time. As a result interferometric techniques can be used to compare two wave fronts which were originally separated in time or space, or even wave fronts of different wavelengths. In addition, since a hologram reconstructs the shape of an object with a rough surface faithfully, down to its smallest details, large-scale changes in its shape can be measured with interferometric precision.

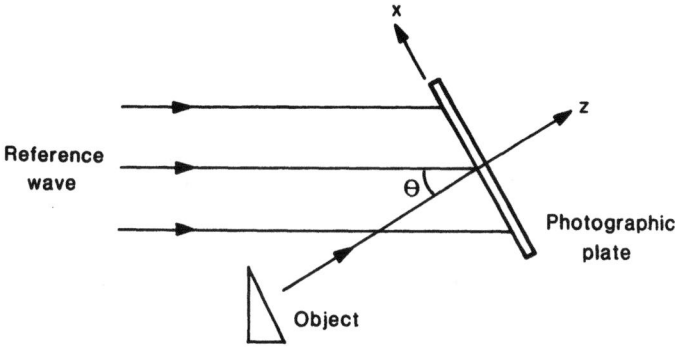

Fig. 2.3. The off-axis hologram: recording setup

2.2 The Off-Axis Hologram

We will now discuss the formation of an image by an off-axis hologram in more detail. We consider, for simplicity, the recording arrangement shown in Fig. 2.3, in which the reference beam is a collimated beam of uniform intensity, derived from the same source as that used to illuminate the object.

The complex amplitude at any point (x, y) on the photographic plate due to the reference beam can then be written as

$$r(x, y) = r \exp(i2\pi\xi x), \tag{2.1}$$

where $\xi = (\sin\theta)/\lambda$, since only its phase varies across the photographic plate, while that due to the object beam, for which both the amplitude and phase vary, can be written as

$$o(x, y) = |o(x, y)| \exp[-i\varphi(x, y)]. \tag{2.2}$$

The resultant intensity is, therefore,

$$\begin{aligned}I(x, y) &= |r(x, y) + o(x, y)|^2 \\ &= |r(x, y)|^2 + |o(x, y)|^2 + r|o(x, y)| \exp[-i\varphi(x, y)] \exp(-i2\pi\xi x) \\ &\quad + r|o(x, y)| \exp[i\varphi(x, y)] \exp(i2\pi\xi x) \\ &= r^2 + |o(x, y)|^2 + 2r|o(x, y)| \cos[2\pi\xi x + \varphi(x, y)]. \end{aligned} \tag{2.3}$$

For simplicity, we can assume that the amplitude transmittance of the photographic plate is a linear function of the intensity and is given by the relation

$$t = t_0 + \beta TI, \tag{2.4}$$

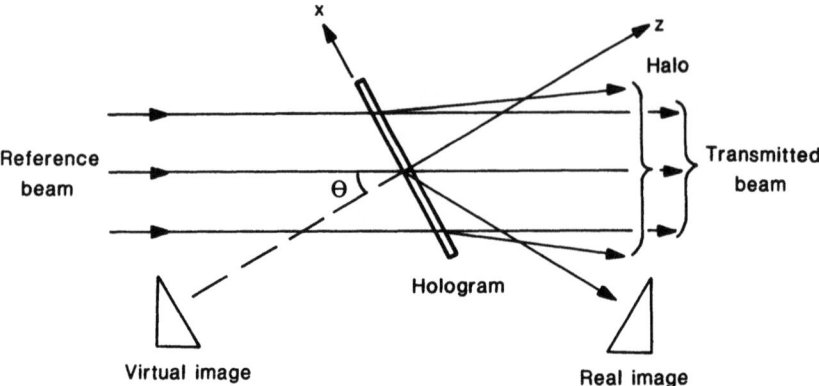

Fig. 2.4. The off-axis hologram: image reconstruction

where t_0 is a constant background transmittance, T is the exposure time and β is the slope (negative) of the amplitude transmittance vs. exposure characteristic of the photographic plate. The resultant amplitude transmittance of the hologram is then

$$t(x, y) = t_0 + \beta T\{|o(x, y)|^2 + r|o(x, y)| \exp[-i\varphi(x, y)] \exp(-i2\pi\xi x)$$
$$+ r|o(x, y)| \exp[i\varphi(x, y)] \exp(i2\pi\xi x)\}. \qquad (2.5)$$

When the hologram is illuminated once again with the same reference beam, as shown schematically in Fig. 2.4, the complex amplitude of the transmitted wave can be written as

$$u(x, y) = r(x, y)t(x, y)$$
$$= t_0 r \exp(i2\pi\xi x) + \beta T r |o(x, y)|^2 \exp(i2\pi\xi x)$$
$$+ \beta T r^2 o(x, y) + \beta T r^2 o^*(x, y) \exp(i4\pi\xi x). \qquad (2.6)$$

The first term on the right-hand side of (2.6) is the directly transmitted beam, while the second term yields a halo surrounding it, with approximately twice the angular spread of the object. The third term is identical to the original object wave, except for a constant factor $\beta T r^2$, and produces a virtual image of the object in its original position. The fourth term corresponds to the conjugate image, which, in this case is a real image. If the offset angle of the reference beam is made large enough, the virtual image can be separated from the directly transmitted beam and the conjugate image.

In this arrangement, corresponding points on the real and virtual images are located at equal distances from the hologram, but on opposite sides of it. Since the depth of the real image is reversed, it is called a *pseudoscopic image*, as opposed to the normal, or *orthoscopic virtual image*. It should also be noted that the sign of β only affects the phase of the reconstructed image, so that a "positive" image is always obtained, even if the hologram recording is a photographic negative.

2.2.1 Reflection Holograms

If the object and reference beams are incident on the photographic plate from opposite sides, the interference fringes produced are actually layers within the thickness of the emulsion layer, about half a wavelength apart. A hologram recorded in this manner, when illuminated with a point source of white light, can reconstruct a monochromatic image in reflected light. A simple way of recording a reflection hologram of a flat object is to attach a photographic plate to it and illuminate the object, through the plate, with a laser source.

2.2.2 Image Holograms

For some applications, there are advantages in recording a hologram of an image of the object formed by a lens. As shown in Fig. 2.5, the hologram plate is set in the central plane of the image, and a hologram is recorded in the normal fashion with an off-axis reference beam. When the hologram is illuminated with the original reference beam, part of the image lies in front of the hologram, and part of the image lies behind it. Since the image is very close to the hologram plane, it is possible to relax the spatial and temporal coherence requirements for the illumination used at the reconstruction stage [2.10].

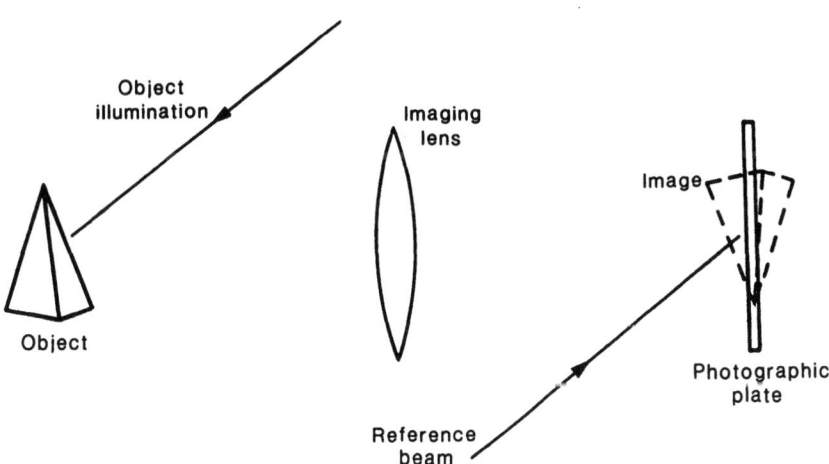

Fig. 2.5. Recording an image hologram

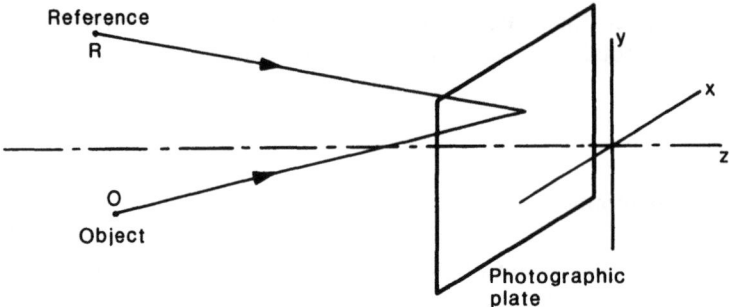

Fig. 2.6. Formation of the image of a point object

2.3 The Reconstructed Image

To study the characteristics of the reconstructed image and their dependence on the optical system, we consider, as shown in Fig. 2.6, the hologram of a point object $O(x_O, y_O, z_O)$, recorded with a reference wave from a point source $R(x_R, y_R, z_R)$, using light of wavelength λ_1. If the hologram is illuminated with monochromatic light of wavelength λ_2 from a point source $P(x_P, y_P, x_P)$, it can be shown that the coordinates of the image of O are [2.11]

$$x_I = \frac{x_P z_O z_R + \mu x_O z_P z_R - \mu x_R z_P z_O}{z_O z_R + \mu z_P z_R - \mu z_P z_O}, \qquad (2.7)$$

$$y_I = \frac{y_P z_O z_R + \mu y_O z_P z_R - \mu y_R z_P z_O}{z_O z_R + \mu z_P z_R - \mu z_P z_O}, \qquad (2.8)$$

$$z_I = \frac{z_P z_O z_R}{z_O z_R + \mu z_P z_R - \mu z_P z_O}, \qquad (2.9)$$

where $\mu = (\lambda_2/\lambda_1)$. The lateral magnification of the image can be defined as

$$M_{\text{lat}} = (dx_I/dx_O) = (dy_I/dy_O) = 1 \bigg/ \left[1 + z_O \left(\frac{1}{\mu z_P} - \frac{1}{z_R} \right) \right]. \qquad (2.10)$$

If the hologram is illuminated with the same reference wave used to record it, the image has the same size as the original object and coincides with it. However, any change in the position or wavelength of the point source used for reconstruction results in a change in the position and magnification of the reconstructed image.

2.4 Image Speckle

When a diffusely reflecting object is illuminated, each element on its surface produces a diffracted wave. With coherent light, these diffracted waves can interfere with each other. Since the optical paths to neighboring elements may differ by several wavelengths, local fluctuations are seen in the intensity in the far field. As a result, the image exhibits a speckled appearance. With polarized light, the intensity in the speckle pattern has, as shown in Fig. 2.7, the negative exponential distribution

$$p(I) = (1/2\sigma^2) \exp(-I/2\sigma^2), \tag{2.11}$$

where $2\sigma^2$ is the mean intensity [2.12]. The contrast of the speckle pattern is unity, and its appearance is almost independent of the nature of the surface, but the size of the speckles increases with the viewing distance and the f-number of the imaging system. With a circular pupil of radius ρ, the average size of the speckles in the image is

$$\Delta x = \Delta y = 0.61 \lambda f/\rho. \tag{2.12}$$

Speckle is a serious problem in holographic imaging. While a number of methods have been described to reduce speckle in the reconstructed image [2.13], the most common method is to record a number of holograms with the object illuminated from slightly different directions. Each of these holograms reconstructs the same image, but a different speckle pattern. If the images

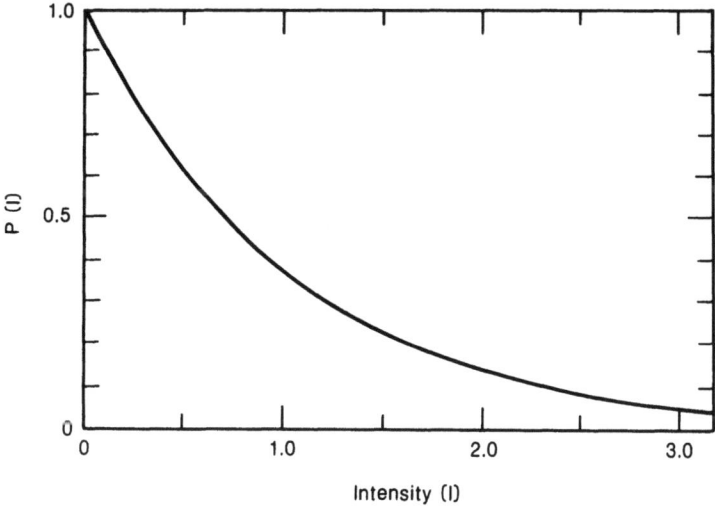

Fig. 2.7. Intensity distribution in a speckle pattern

produced by N such holograms are superposed, the contrast of the speckle pattern is reduced by a factor equal to $\sqrt{N}$.

2.4.1 Signal-to-Noise Ratio

Random spatial variations in the intensity of the reconstructed image, commonly caused by scattered light, are referred to as noise. However, when calculating the signal-to-noise ratio, the amplitudes of the signal and the noise must be added, since they are both encoded on a common carrier [2.14].

We consider the reconstructed image of a uniform bright patch on a dark background, and assume that the intensity due to the nominally uniform signal is I_S, while that of the randomly varying background is I_N. The noise N in the bright area is given by the variance of the resulting fluctuations of the intensity. It can then be shown that when $I_S \gg \langle I_N \rangle$, as is usually the case, the signal-to-noise ratio is

$$I_S/N = (I_S/2\langle I_N \rangle)^{1/2}. \tag{2.13}$$

Even a small amount of scattered light can result in relatively large fluctuations in intensity in the bright areas of the image.

2.5 Types of Holograms

So far, we have treated a hologram recorded on a photographic film as equivalent, to a first approximation, to a grating of negligible thickness with a spatially varying transmittance. However, with modified processing techniques, or with other recording materials, it is possible to reproduce the variations in the intensity in the interference pattern produced by the object and reference beams as variations in the refractive index, or the thickness, of the hologram. Accordingly, holograms can be classified, in the first instance, as amplitude and phase holograms.

In addition, if the thickness of the recording medium is much larger than the average spacing of the fringes, volume effects cannot be neglected. It is even possible to produce holograms in which the interference pattern consists of planes running almost parallel to the surface of the recording material, and which reconstruct an image in reflected light. Holograms recorded in thick media can therefore be subdivided into volume transmission holograms and volume reflection holograms.

2.5.1 Thin Amplitude and Phase Gratings

The amplitude transmittance of a thin amplitude grating can be written as

$$t(x) = t_0 + \Delta t \cos Kx, \tag{2.14}$$

where t_0 is the average amplitude transmittance, Δt is the amplitude of the spatial variations of $t(x)$, and $K = 2\pi/\Lambda$, where Λ is the average spacing of the fringes. The maximum amplitude in each of the two diffracted orders is obtained when $t_0 = \Delta t = \frac{1}{2}$, and is equal to a fourth of that in the incident wave, so that the maximum diffraction efficiency is

$$\eta_{max} = 0.0625. \tag{2.15}$$

If the phase shift produced by the recording medium is proportional to the intensity in the interference pattern, the complex amplitude transmittance of a thin phase grating can be written as

$$t(x) = \exp(-i\varphi_0)\exp[-i\Delta\varphi\cos(Kx)], \tag{2.16}$$

where φ_0 is a constant phase factor, and $\Delta\varphi$ is the amplitude of the phase variations. If we neglect this constant phase factor, the right hand side of (2.16) can be expanded to obtain the relation [2.15]

$$t(x) = \sum_{n=-\infty}^{\infty} i^n J_n(\Delta\varphi)\exp(inKx), \tag{2.17}$$

where J_n is the Bessel function of the first kind, of order n.

Equation (2.17) demonstrates that the incident beam is diffracted into a number of orders, with the diffracted amplitude in the nth order proportional to the value of the Bessel function $J_n(\Delta\varphi)$. Only the first order contributes to the image reconstructed by a hologram. The diffraction efficiency of the phase grating can therefore be written as

$$\eta = J_1^2(\Delta\varphi). \tag{2.18}$$

As shown in Fig. 2.8, the diffraction efficiency of a thin phase grating increases initially with the phase modulation and then decreases; its maximum value is

$$\eta_{max} = 0.339. \tag{2.19}$$

2.5.2 Volume Gratings

With a thick recording medium, the hologram is made up of layers corresponding to a periodic variation of transmittance or refractive index. If the two interfering wave fronts are incident on the recording medium from the same side, these layers are approximately perpendicular to its surface, and the hologram produces an image by transmission. However, it is also possible, as mentioned in Sect. 2.2.1, to have the two interfering wave fronts incident on the recording medium from opposite sides, in which case the interference surfaces run approximately parallel to the surface of the recording medium. In this case, the reconstructed image is produced by the light reflected from the hologram. In both cases, the diffracted amplitude is a maximum only when the Bragg condition is satisfied. With volume reflection holograms, the angular and wavelength selectivity can be high enough to produce an image with white light.

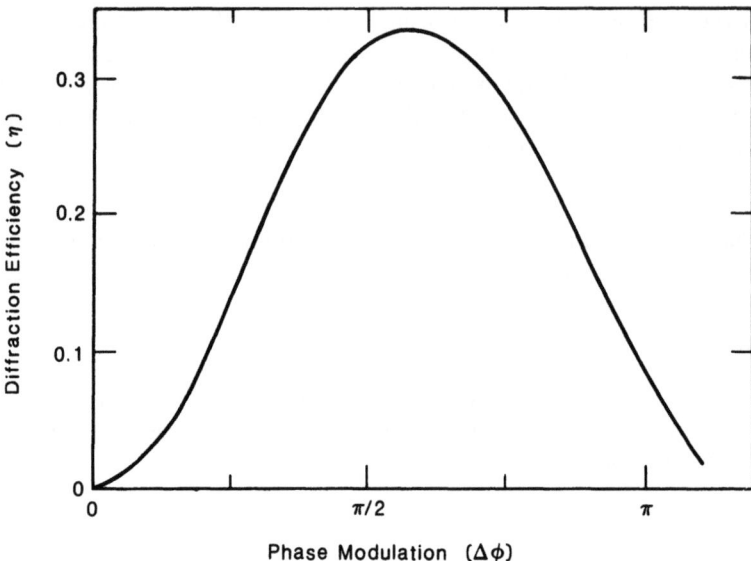

Fig. 2.8. Diffraction efficiency of a thin phase grating as a function of the phase modulation

When analyzing the diffraction of light by volume gratings, it is necessary to take into account the fact that the amplitude of the diffracted wave increases progressively, while that of the incident wave decreases, as they propagate through the grating. This problem was solved by the development of a coupled-wave theory [2.15, 16]. Some of the most important results for volume transmission gratings are summarized below.

We consider, in the first instance, a lossless, volume transmission phase grating of thickness d, with the grating planes running normal to its surface. If we assume that the refractive index varies sinusoidally, with an amplitude Δn, about a mean value n, the diffraction efficiency of the grating at the Bragg angle θ_B is

$$\eta_B = \sin^2 \phi, \tag{2.20}$$

where $\phi = \pi \Delta n d / \lambda \cos \theta_B$ is known as the modulation parameter. Initially, the diffraction efficiency increases as the modulation parameter ϕ is increased, until, when $\phi = \pi/2$, $\eta_B = 1$. Beyond this point, the diffraction efficiency decreases.

For a deviation $\Delta \theta$ in the angle of incidence from the Bragg angle, the diffraction efficiency drops to

$$\eta = \frac{\sin^2(\phi^2 + \chi^2)^{1/2}}{(1 + \chi^2/\phi^2)}, \tag{2.21}$$

where

$$\chi = \Delta\theta K d/2. \tag{2.22}$$

2.5 Types of Holograms

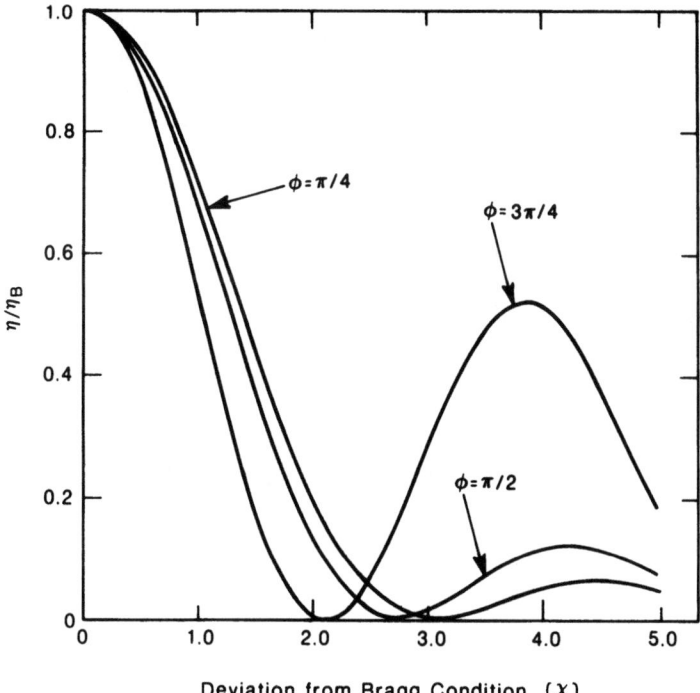

Fig. 2.9. Normalized diffraction efficiency of a volume phase transmission grating as a function of the deviation from the Bragg condition [2.15]

Figure 2.9 exhibits the normalized diffraction efficiency, as a function of the parameter χ, for three values of the modulation parameter ϕ.

The other case we shall consider is that of a volume transmission grating in which the refractive index does not vary, but the absorption constant varies with an amplitude $\Delta\alpha$ about its mean value α. In this case, the diffraction efficiency is given by

$$\eta = \exp\left(\frac{-2\alpha d}{\cos\theta_B}\right) \sinh^2\left(\frac{\Delta\alpha d}{2\cos\theta_B}\right). \tag{2.23}$$

The maximum diffraction efficiency is obtained when

$$\Delta\alpha = \alpha = \frac{(\ln 3)}{d\cos\theta_B} \tag{2.24}$$

and has the value $\eta_{max} = 0.037$.

The distinction between thin gratings and volume gratings is commonly made [2.17] on the basis of a parameter Q defined by the relation

$$Q = 2\pi\lambda d/n\Lambda^2. \tag{2.25}$$

Small values of Q ($Q < 1$) correspond to thin gratings, while large values of Q ($Q > 1$) correspond to volume gratings. However, more detailed studies have shown that the transition between the two regimes is not completely defined by (2.25) and that, as the modulation amplitude increases, an intermediate regime appears and widens [2.18, 19].

2.5.3 Holograms of Diffusely Reflecting Objects

It should be noted that the values of diffraction efficiency obtained with a hologram of a diffusely reflecting object are always much lower than those for a grating, because it is not possible to maintain optimum modulation over the entire area, due to nonuniformity of the object wave. The maximum diffraction efficiencies of transmission phase holograms recorded with a diffuse object beam have been calculated [2.20], on the assumption that the amplitude of the object wave has similar statistics to a speckle pattern, and are presented in Table 2.1.

2.5.4 Multiply Exposed Holograms

With a thick recording medium, it is possible to record two or more holograms in the same medium and read them out separately. In order to do this, the Bragg angles should be sufficiently far apart that the maximum of the angular selectivity curve for one hologram coincides with the first minimum for the other. However, with N amplitude transmission holograms, the diffraction efficiency of each hologram drops to $1/N^2$ of that for a single hologram, since the available dynamic range is divided equally between the N holograms [2.21].

On the other hand, with volume phase holograms whose Bragg angles are far enough apart for coupling between the gratings to be negligible, each hologram diffracts independently of the others [2.22]. However, if the recording medium is nearing saturation, the consequent reduction in modulation can result in a decrease in the diffraction efficiencies of the individual holograms.

Table 2.1. Maximum theoretical diffraction efficiencies for transmission phase holograms

Type of hologram	Thin		Volume	
Object beam	Collimated	Diffuse	Collimated	Diffuse
η_{max}	0.33	0.22	1.00	0.64

Table 2.2. Characteristics of some lasers used for holographic interferometry

Laser	Output	Wavelength [nm]	Power
He–Ne	cw	633	2–50 mW
Ar$^+$	cw	514	~1 W
		488	~1 W
Kr$^+$	cw	647	500 mW
Ruby	Pulsed	694	1–10 J
Diode	cw		~5 mW
Dye	cw	Tunable	~200 mW

2.6 Light Sources and Optical Systems

In order to obtain maximum fringe visibility while recording a hologram, it is essential to use coherent illumination. Lasers are therefore employed almost universally as light sources in holography. The characteristics of some of the lasers used for holographic interferometry are listed in Table 2.2.

2.6.1 Coherence Requirements

Operation of the laser on a single line can be obtained, where necessary, by means of a wavelength selector prism. Spatial coherence is then ensured if the laser oscillates in the lowest order transverse mode (the TEM$_{00}$ mode). However, most lasers will then oscillate (Fig. 2.10) in a number of longitudinal modes lying within the gain profile of the active medium, at which the gain is adequate to overcome the cavity losses. These modes correspond to the resonant frequencies of the laser cavity and are separated by a frequency interval

$$\Delta v = c/2L, \tag{2.26}$$

where c is the speed of light, and L is the length of the laser cavity. If we assume that the output power is divided equally between N longitudinal modes, the effective coherence length of the output is

$$\Delta l = 2L/N. \tag{2.27}$$

Equation (2.27) reveals that the existence of more than one longitudinal mode in the output reduces the coherence length severely. Even if the mean optical paths of the object and reference beams are equalized carefully, severe restrictions are placed on the maximum depth of the object.

Depending on their power, most commercial He–Ne lasers oscillate in two to five longitudinal modes, and the coherence length of the output is limited to

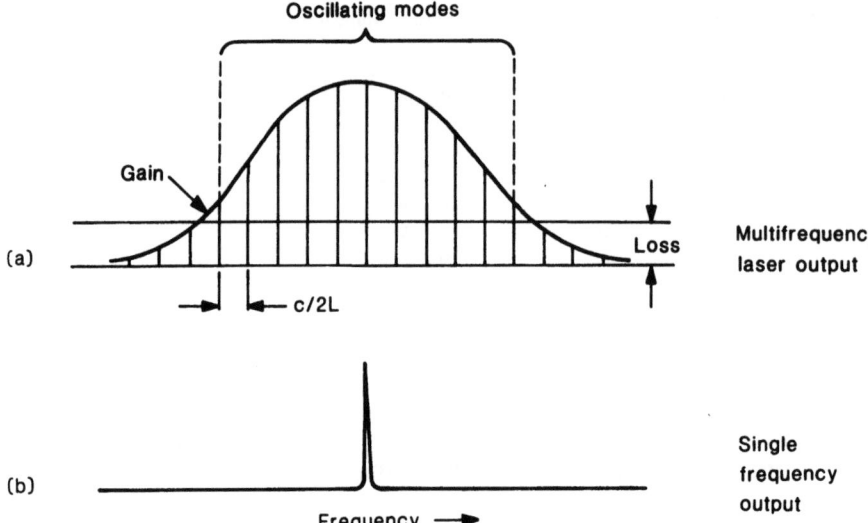

Fig. 2.10. Laser modes: (a) without, and (b) with an intra-cavity etalon

a few centimetres. With high-power Ar^+ and Kr^+ lasers, it is possible to obtain operation in a single longitudinal mode and coherence lengths in excess of a metre by using an intracavity etalon. This etalon is tuned to obtain maximum power output by mounting it in an oven whose temperature can be controlled.

2.6.2 Laser Beam Expansion

Since the beam from a laser typically has a diameter of a couple of millimetres, low-power microscope objectives are commonly used to expand it to illuminate the object as well as the hologram. However, due to the high coherence of laser light, the expanded beam usually exhibits random diffraction patterns (spatial noise) produced by defects and dust on the optical surfaces in the path of the beam. Spatial noise can be eliminated by placing a pinhole at the focus of the microscope objective, as shown in Fig. 2.11. If the laser is oscillating in the TEM_{00} mode, the beam has a Gaussian intensity profile given by the relation

$$I(r) = I(0) \exp(-2r^2/w^2), \tag{2.28}$$

where r is the radial distance from the center of the beam, and w is the distance at which the intensity drops to $(1/e^2)$ of that at the center of the beam. If the aperture of the microscope objective is greater than $2w$, the diameter of the focal spot is

$$d = 2\lambda f/\pi w, \tag{2.29}$$

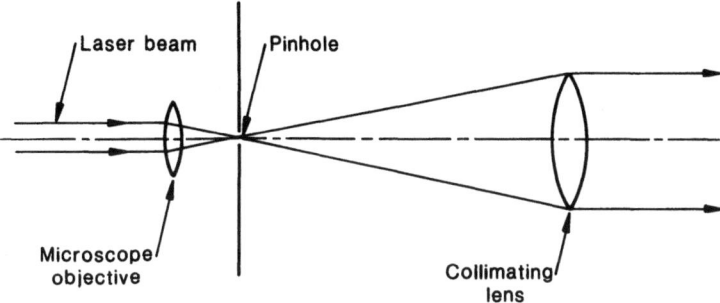

Fig. 2.11. Optical system used to expand and spatially filter a laser beam

where f is the focal length of the microscope objective. With a pinhole whose diameter is less than d, randomly diffracted light is blocked, and the transmitted beam has a smooth profile.

2.6.3 Beam Polarization

The visibility of the interference fringes forming the hologram is a maximum when the electric vectors in the object and reference beams are parallel. This condition is satisfied if the two beams are linearly polarized with their electric vectors normal to the plane containing the beams. If, on the other hand, they are polarized with their electric vectors in the plane containing the beams, the visibility of the hologram fringes can drop to zero when the beams intersect at right angles.

It should be noted that in the case of an object with a rough surface, a substantial fraction of the reflected light can be depolarized. The resulting decrease in the visibility of the interference fringes can be minimized, where necessary, by using a sheet polarizer in front of the hologram to eliminate the cross-polarized component.

2.6.4 Optical Systems for Holography

A typical optical system for recording transmission holograms of a diffusely reflecting object is displayed schematically in Fig. 2.12. Since any change in the phase difference between the two beams while recording the hologram results in a movement of the interference fringes and reduced modulation in the hologram, all the optical components as well as the object and the recording material must be mounted on a stable surface. Most laboratories now use a rigid optical table supported on air bags to isolate it from floor vibrations. In addition, the working area is enclosed to minimize the effects of acoustic waves, air currents and temperature changes.

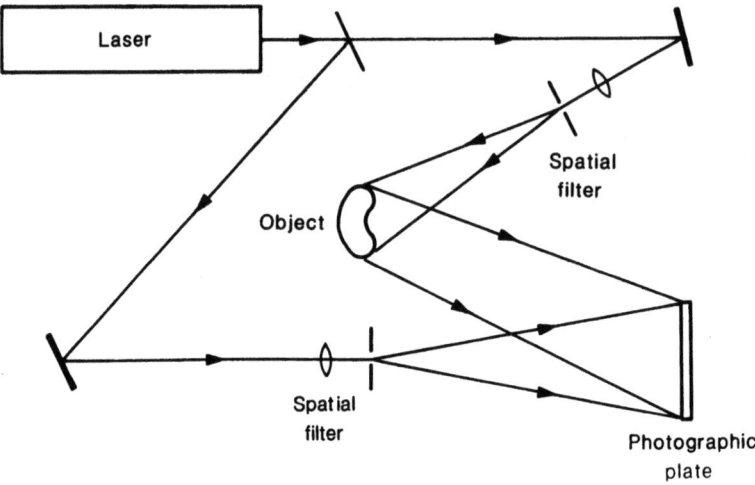

Fig. 2.12. Typical optical arrangement used to record a hologram

2.6.5 Holography with Pulsed Lasers

Very short light pulses (< 20 ns) can be obtained with a pulsed laser, if a Pockels cell is employed as a Q-switch in the laser cavity. As a result, problems of vibration and air currents are largely eliminated. Because their output wavelength is well matched to the peak sensitivity of available photographic materials, and their output energy is fairly large, pulsed ruby lasers are used widely to record holograms in a workshop environment [2.23].

2.6.6 Laser Safety

Since the beam from a laser is focused by the lens of the eye to a very small spot on the retina, direct exposure to low-power lasers can cause eye damage. With pulsed lasers, even stray reflections can be dangerous. It is essential to take due precautions and, where required, to use appropriate eye protection [2.24].

2.7 The Recording Medium

The response of recording materials used for amplitude holograms can be characterized on a macroscopic scale (Fig. 2.13) by plotting the resultant amplitude transmittance against the exposure. Similarly, the response of recording materials utilized for phase holograms can be described, as shown in Fig. 2.14, by a curve showing the effective phase shift as a function of the exposure.

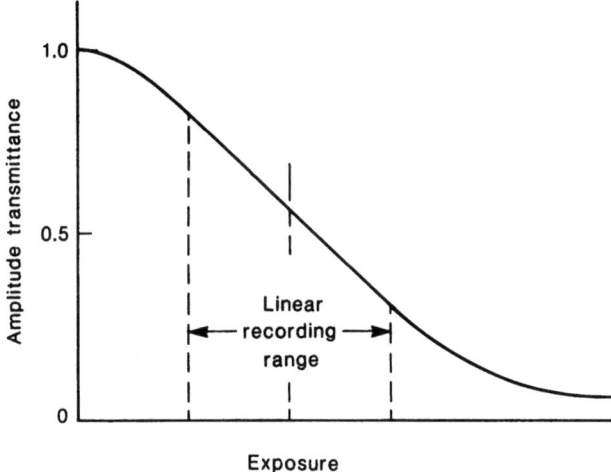

Fig. 2.13. Typical amplitude transmission vs. exposure curve for a recording material

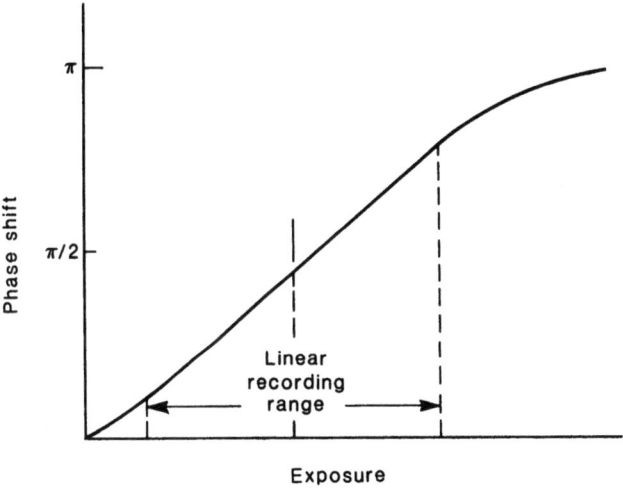

Fig. 2.14. Typical phase-shift vs. exposure curve for a recording material

However, these curves are not adequate to describe the response of the recording medium on a microscopic scale. This is because the actual intensity distribution to which the material is exposed always differs from that incident on it, due to scattering and absorption. In addition, the response of the material to different spatial frequencies may be affected by the type of processing. Accordingly, it is necessary to specify the response of the material as the spatial frequency s is varied, relative to that at low spatial frequencies ($s \to 0$), by a parameter $M(s)$ termed the modulation transfer function.

2.7.1 Effects of Nonlinearity

For simplicity, we have assumed so far that the amplitude transmittance of the hologram film is a linear function of the intensity described by (2.04). However, this assumption is not always valid. The amplitude transmittance of the recording material can then be represented by a polynomial [2.25]

$$\begin{aligned} t &= t_0 + \beta_1 TI + \beta_2 T^2 I^2 + \cdots \\ &= t_0 + \beta_1 T(rr^* + oo^* + r^*o + ro^*) \\ &\quad + \beta_2 T^2 (rr^* + oo^* + r^*o + ro^*)^2 + \cdots \end{aligned} \qquad (2.30)$$

If the hologram is illuminated once again with a plane wave of unit amplitude, the complex amplitude of the wave transmitted by the hologram can be written in the form

$$u = \text{linear terms} + \beta_2 T^2 [(oo^*)^2 + o^2 + o^{*2} + 2o^2 o^* + 2oo^{*2}] + \cdots \qquad (2.31)$$

Nonlinearity leads, therefore, to the production of additional spurious terms. An examination of (2.31) reveals that the term involving $(oo^*)^2$ results in a doubling of the width of the halo surrounding the directly transmitted beam, while the terms involving o^2 and o^{*2} correspond to higher-order diffracted images, and the terms involving $2o^2 o^*$ and $2oo^{*2}$ are intermodulation terms, giving rise to false images.

The effects of nonlinearity are particularly noticeable with phase holograms. Even if we assume that the phase shift produced by the recording medium is proportional to the exposure, the complex amplitude transmittance is given by

$$t = \exp(-i\varphi) = 1 - i\varphi - (\tfrac{1}{2})\varphi^2 + (\tfrac{1}{6})i\varphi^3 \ldots \qquad (2.32)$$

If the phase modulation is increased to obtain higher diffraction efficiency, the effects of the higher-order terms cannot be neglected.

A major advantage of volume holograms is that the effects of nonlinearity are reduced significantly by the angular selectivity of the hologram. If the angle between the beams in the recording setup is large enough that the diffracted beams corresponding to different orders do not overlap, a simple analysis [2.26] shows that the signal-to-noise ratio should improve by a factor approximately equal to $(\psi/\Delta\theta)$, where $2\Delta\theta$ is the width of the passband of the angular selectivity function and ψ is the angle subtended by the object at the hologram.

2.8 Recording Materials

Several recording materials have been used for holography. Table 2.3 lists the principal characteristics of those that have been found most useful for holographic interferometry [2.27].

Table 2.3. Recording materials for holographic interferometry

Material	Exposure [J/m^2]	Resolution [mm^{-1}]	Processing	Type of hologram	η_{max} (gratings)
Photographic emulsions	5×10^{-3} to 5×10^{-1}	1000–10000	Normal Bleach	Amplitude Phase	0.05 0.60
Photothermo plastics	10^{-1}	500–1200 (bandpass)	Charge and heat	Thin phase	0.30
Photorefractive (BSO)	10	>10000	None	Volume phase	0.20

2.8.1 Photographic Emulsions

High-resolution photographic plates and films are the most widely employed recording materials for holographic interferometry because of their relatively high sensitivity. Conventional processing produces an amplitude hologram and results in a reduction in the thickness of the emulsion layer of about 15%, due to the removal of the unexposed silver-halide grains in the fixing bath. This reduction in thickness can cause a rotation of the fringe planes as well as a reduction in their spacing, so that the reference beam is no longer incident on the hologram at the Bragg angle. To minimize the effects of emulsion shrinkage, the object and reference beams should be incident at equal but opposite angles on the hologram, so that the fringe planes are normal to the surface of the photographic emulsion.

Processing can be speeded up and carried out in situ, in a liquid gate, with a monobath in which development and fixing take place simultaneously [2.28]. Higher diffraction efficiencies can be obtained by using a bleach bath to convert the developed silver into a transparent silver salt, yielding a volume phase hologram [2.29].

2.8.2 Photothermoplastics

A hologram can also be recorded in a multilayer structure consisting, as shown in Fig. 2.15, of a glass or Mylar substrate coated with a thin, transparent, conducting layer of indium oxide, a photoconductor, and a thermoplastic [2.30]. The film is initially sensitized in darkness by applying a uniform electric charge to the top surface. On exposure and recharging, a spatially varying electrostatic field is created. The thermoplastic is then heated briefly, so that it becomes soft enough to be deformed by this field, and cooled to fix the variations in thickness.

Photothermoplastics have a reasonably high sensitivity and yield a thin phase hologram with good diffraction efficiency. They have the advantage that

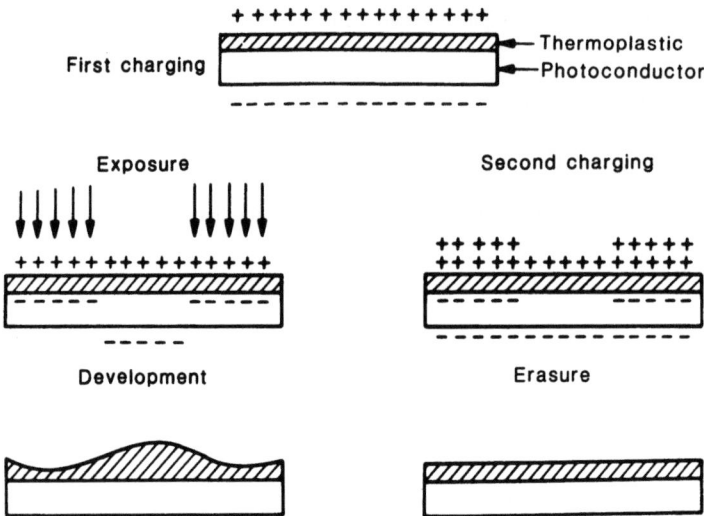

Fig. 2.15. Record-erase cycle for a photothermoplastic [2.30]

they can be processed rapidly in situ; in addition, with a glass substrate, the hologram can be erased by heating the substrate, and the material reused.

2.8.3 Photorefractive Crystals

When a photorefractive crystal is exposed to a spatially varying light pattern, electrons are liberated in the illuminated areas. These electrons migrate to adjacent dark regions and are trapped there. The spatially varying electric field produced by this space-charge pattern modulates the refractive index through the electro-optic effect, producing the equivalent of a phase grating. The space-charge pattern can be erased by uniformly illuminating the crystal, after which another recording can be made.

The photorefractive crystal most commonly used for holographic interferometry has been BSO ($Bi_{12}SiO_{20}$). The best results are obtained with the

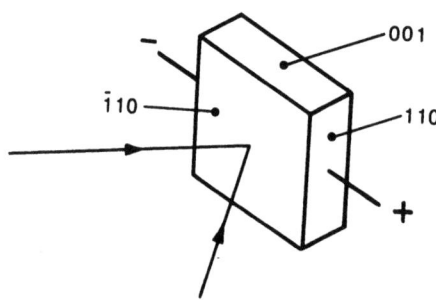

Fig. 2.16. Hologram recording configuration for BSO

recording configuration shown in Fig. 2.16, in which an electric field is applied at right angles to the hologram fringes [2.31]. A diffraction efficiency of 0.1 can be obtained with a field of 500 V/mm. Since readout is destructive, the reconstructed image is best recorded and stored for viewing. Several interesting possibilities have been opened up by such photorefractive materials [2.32–34].

2.9 Holographic Interferometry

Equations (2.07–10) demonstrate that if a hologram is replaced in its original position in the same setup used to record it, and illuminated with the original reference wave, it reconstructs the original object wave. If, then, the shape of the object changes slightly, the directly transmitted object wave will interfere with the reconstructed object wave to produce a fringe pattern that maps the changes in the shape of the object [2.35].

2.9.1 Real-Time Holographic Interferometry

If the change in shape of the object is small, only the phase of the object wave is modified, and the complex amplitude of the wave from the deformed object can be written as

$$o'(x, y) = o(x, y)\exp[-i\Delta\varphi(x, y)], \tag{2.33}$$

where $o(x, y)$ is the complex amplitude of the original object wave, and $\Delta\varphi(x, y)$ is the phase change arising from the deformation. Accordingly, from (2.06), the complex amplitude of the directly transmitted object wave is

$$u_1(x, y) = (t_0 + \beta T r^2)\, o'(x, y), \tag{2.34}$$

while the complex amplitude of the reconstructed object wave is

$$u_2(x, y) = \beta T r^2 o(x, y). \tag{2.35}$$

The intensity in the resultant interference pattern is, therefore,

$$I(x, y) = |u_1(x, y) + u_2(x, y)|^2$$
$$= |o(x, y)|^2 [A^2 + (t_0 - A)^2 - 2A(t_0 - A)\cos\Delta\varphi(x, y)], \tag{2.36}$$

where $A = -\beta T r^2$. Since β is negative, dark fringes are obtained when $\Delta\varphi(x, y) = m\pi$, where m is an integer.

Interference fringes obtained by this technique can be used to study changes in the shape of the object in real time. The problem of replacing the hologram in its original position can be eliminated by in situ processing of the hologram plate, or by using a photothermoplastic or a photorefractive crystal as the recording material.

2.9.2 Double-Exposure Holographic Interferometry

It is also possible to record two holograms on the same photographic plate: one of the object in its original state, and the other of the deformed object. The resultant complex amplitude, due to the superposition of the two reconstructed images, is then, apart from a constant of proportionality,

$$u(x, y) = o(x, y) + o'(x, y) = o(x, y)\{1 + \exp[-i\Delta\varphi(x, y)]\}, \tag{2.37}$$

and the intensity in the image is

$$I(x, y) = |o(x, y)|^2 [1 + \cos \Delta\varphi(x, y)]. \tag{2.38}$$

In this case, bright fringes are obtained when $\Delta\varphi(x, y) = 2m\pi$.

This technique has the advantage that repositioning of the hologram is not critical, since the two interfering waves are always reconstructed in exact register. In addition, the visibility of the fringes is always good, since the two waves have the same polarization and the same amplitude. With a pulsed laser, double-exposure holographic interferometry can be used to study transient phenomena.

Unwanted object movements are often a problem in double-exposure holographic interferometry with a cw laser. Some types of unwanted object motion can be eliminated by reflecting the reference beam from a mirror attached to the object [2.36]. Alternatively, the hologram plate can be attached to the object, and a doubly-exposed reflection hologram can be recorded with the object illuminated through the hologram plate [2.37].

Another disadvantage is that information on intermediate states of the object is not available. This problem can be overcome by making a series of exposures at successive stages of loading, using a set of masks with apertures that overlap in a predetermined order [2.38]. The reconstructed images then yield interference patterns corresponding to any two stages of loading. Another way to overcome this problem is the sandwich hologram [2.39]. In this technique, pairs of photographic plates are exposed in the same plate-holder with their emulsion-coated surfaces facing the object. One pair is exposed with the unstressed object, and successive pairs are exposed at different stages of loading. The back plate from the first pair can then be put together with the front plate from any other pair to produce an interference pattern showing the surface deformation at this stage. In addition, ambiguities can be resolved by tilting the sandwich; this is equivalent to tilting the object between the two exposures.

2.9.3 Phase Difference in the Interference Pattern

Several methods have been employed to evaluate the surface displacements from a holographic interferogram [2.40]. However, the most widely used techniques are based on the fact that, with an object having a rough surface,

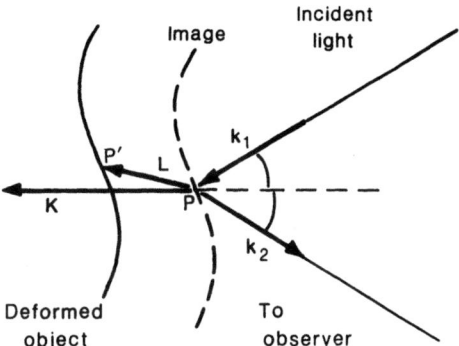

Fig. 2.17. Evaluation of the phase difference produced by a local displacement of the object

the phase varies in a random manner across the object wavefront. As a result, only waves from corresponding points on the object wavefront and the reconstructed wavefront contribute effectively to the interference pattern, and the intensity at any point in it is determined by the phase difference between the waves from these two points. To evaluate this phase difference, we consider a point P on the surface which, as shown in Fig. 2.17, has undergone a vector displacement L to P'. If the displacement of P is small compared to the distances to the source S and the point of observation 0, the phase difference introduced is

$$\Delta\varphi = L \cdot (k_1 - k_2) = L \cdot K, \tag{2.39}$$

where k_1 and k_2 are the propagation vectors of the incident and scattered beams, and $K = k_1 - k_2$ is known as the sensitivity vector [2.41–43].

2.9.4 The Holodiagram

The holodiagram is a useful aid to interpretation of the interference fringes [2.44]. As shown in Fig. 2.18, the holodiagram consists of a set of ellipses whose foci, O and O', correspond to the beam splitter and the viewing point on the photographic plate, respectively, in the recording system (Fig. 2.12). For any object point P, the ellipse on which it lies is the locus for which the distance OPO' is a constant. A displacement of P from one ellipse to the next corresponds to a change in this distance of one wavelength and a shift of one fringe in the interference pattern. The required displacement of P is obviously a minimum when its motion is along the normal to the ellipse, which corresponds to the sensitivity vector K.

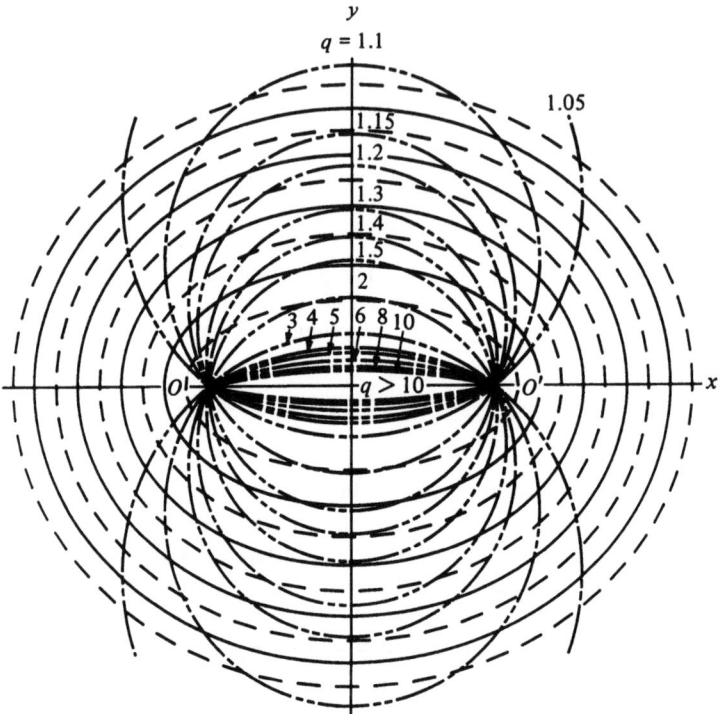

Fig. 2.18. The holodiagram. The ellipses are loci of constant path length; the circles are loci of constant K [2.44]

The circles drawn through O and O′ are curves of constant K. They correspond to the specified values of the parameter $q = 1/\cos\psi$, where the angle OPO′ $= 2\psi$. These curves can be used to optimize a hologram recording system for a particular type of surface displacement.

2.9.5 Localization of the Interference Fringes

With an object having a rough surface, the visibility of the interference fringes is a maximum for a particular position of the plane of observation, known as the plane of localization.

As mentioned earlier, because of the random phase variations across the object wave front, only waves from corresponding points on the two interfering wave fronts contribute effectively to the interference fringes. For a specific viewing direction, the phase difference $\Delta\varphi$ between the waves from two such points, P and P′ (Fig. 2.17), is given by (2.39). This phase difference will therefore vary over the range of viewing directions defined by the aperture of the viewing lens, resulting in a loss of contrast of the fringes. However, it is possible to find a plane in which the variation in the value of $\Delta\varphi$ is a minimum over this range of viewing directions; this is the plane of localization of the fringes [2.45].

The position of the plane of localization depends on the type of displacement [2.46, 47]. Two cases are of particular interest. One is pure translation of the object, which produces fringes localized at infinity; the other is rotation of the object about an axis contained in its surface, which results in fringes localized at the surface.

Observations of fringe localization can be utilized to obtain information on surface displacements and strains [2.48–53]. Such techniques have advantages for some applications, but they are not commonly employed where quantitative measurements are required.

References

2.1 D. Gabor: Nature, **161**, 777–778 (1948)
2.2 E.N. Leith, J. Upatnieks: J. Opt. Soc. Am. **52**, 1123–1130 (1962)
2.3 E.N. Leith, J. Upatnieks: J. Opt. Soc. Am. **53**, 1377–1381 (1963)
2.4 E.N. Leith, J. Upatnieks: J. Opt. Soc. Am. **54**, 1295–1301 (1964)
2.5 R.E. Brooks, L.O. Heflinger, R.F. Wuerker: Appl. Phys. Lett. **7**, 248–249 (1965)
2.6 J.M. Burch, Prodn. Engineer: **44**, 431–442 (1965)
2.7 R.J. Collier, E.T. Doherty, K.S. Pennington: Appl. Phys. Lett. **7**, 223–225 (1965)
2.8 K.A. Haines, B.P. Hildebrand: Phys. Lett. **19**, 10–11 (1965)
2.9 R.L. Powell, K.A. Stetson: J. Opt. Soc. Am. **55**, 1593–1598 (1965)
2.10 L. Rosen: Appl. Phys. Lett. **9**, 337–339 (1966)
2.11 R.W. Meier: J. Opt. Soc. Am. **55**, 987–992 (1965)
2.12 J.W. Goodman: Statistical properties of laser speckle patterns, in *Laser Speckle and Related Phenomena*, ed. by J.C. Dainty, 2nd edn., Topics. Appl. Phys., Vol. 9 (Springer, Berlin, Heidelberg 1984) pp. 9–75
2.13 T.S. McKechnie: Speckle reduction, in *Laser Speckle and Related Phenomena*, ed. by J.C. Dainty, 2nd edn., Topics. Appl. Phys., Vol. 9 (Springer, Berlin, Heidelberg 1984) pp. 123–170
2.14 J.W. Goodman: J. Opt. Soc. Am. **57**, 493–502 (1967)
2.15 H. Kogelnik: Bell Syst. Tech. J. **48**, 2909–2947 (1969)
2.16 L. Solymar, D.J. Cooke: *Volume Holography and Volume Gratings* (Academic, New York 1981) pp. 164–253
2.17 W.R Klein, B.D. Cook: IEEE Trans. SU-14, 123–134 (1967)
2.18 M. Moharam, T.K. Gaylord, R. Magnusson: Opt. Commun. **32**, 14–18 (1980)
2.19 M. Moharam, T.K. Gaylord, R. Magnusson: Opt. Commun. **32**, 19–23 (1980)
2.20 J. Upatnieks, C. Leonard: J. Opt. Soc. Am. **60**, 297–305 (1970)
2.21 R.J. Collier, C.B. Burckhardt, L.H. Lin: *Optical Holography* (Academic, New York 1971) pp. 520–521
2.22 S.K. Case, J. Opt. Soc. Am. **65**, 724–729 (1975)
2.23 W. Koechner: Solid state lasers, in *Handbook of Optical Holography*, ed. by H.J. Caulfield (Academic, New York 1979) pp. 257–267
W. Koechner: *Solid-State Laser Engineering*, 3rd edn., Springer Ser. Opt. Sci., Vol. 1 (Springer, Berlin, Heidelberg 1992)
2.24 D. Sliney, M. Wolbarsht: *Safety with Lasers and Other Optical Sources: a Comprehensive Handbook* (Plenum, New York 1980)
2.25 O. Bryngdahl, A. Lohmann: J. Opt. Soc. Am. **58**, 1325–1334 (1968)

2.26 P. Hariharan: Opt. Acta, **26**, 211–215 (1979)
2.27 H.I. Bjelkhagen: *Silver-Halide Recording Materials for Holography*, Springer Ser. Opt. Sci., Vol. 66 (Springer, Berlin, Heidelberg 1993)
2.28 P. Hariharan, C.S. Ramanathan, G.S. Kaushik: Appl. Opt. **12**, 611–612 (1973)
2.29 P. Hariharan: J. Phot. Sci. **38**, 76–81 (1990)
2.30 L.H. Lin and H.L. Beauchamp: Appl. Opt. **9**, 2088–2092 (1970)
2.31 J.P. Huignard, F. Micheron: Appl. Phys. Lett. **29**, 591–593 (1976)
2.32 J.P. Huignard: "Phase conjugation, real-time holography and degenerate four-wave mixing", in *Current Trends in Optics*, ed. by F.T. Arecchi and F.R. Aussenegg (Taylor & Francis, London 1981) pp. 150–160
2.33 P. Günter, J.-P. Huignard (eds.): *Photorefractive Materials and Their Applications I and II*, Topics Appl. Phys., Vols. 61 and 62 (Springer, Berlin, Heidelberg 1988 and 1989)
2.34 M.P. Petrov, S.I. Stepanov, A.V. Khomenko: *Photorefractive Crystals in Coherent Optical Systems*, Springer Ser. Opt. Sci., Vol. 59 (Springer, Berlin, Heidelberg 1991)
2.35 Yu. I. Ostrovsky, M.M. Butusov, G.V. Ostrovskaya: *Interferometry by Holography*, Springer Ser. Opt. Sci., Vol. 20 (Springer, Berlin, Heidelberg 1980)
2.36 F.M. Mottier: Appl. Phys. Lett. **15**, 44–45 (1969)
2.37 P.M. Boone: Opt. Acta **22**, 579–589 (1975)
2.38 P. Hariharan, Z.S. Hegedus: Opt. Commun. **9**, 152–155 (1973)
2.39 N. Abramson: Appl. Opt. **13**, 2019–2025 (1974)
2.40 J.D. Briers, Opt. Quant. Electron. **8**, 469–501 (1976)
2.41 E.G. Aleksandrov, A.M. Bonch-Bruevich: Sov. Phys: Tech. Phys. **12**, 258–265 (1967)
2.42 A.E. Ennos: J. Phys. E: Sci. Instrum. **1**, 731–743 (1968)
2.43 J.E. Sollid: Appl. Opt. **8**, 1587–1595 (1969)
2.44 N. Abramson: Appl. Opt. **8**, 1235–1240 (1969)
2.45 S. Walles: Arkiv for Fysik **40**, 299–403 (1969)
2.46 N.E. Molin and K.A. Stetson, Optik **31**, 157–177 (1970)
2.47 N.E. Molin and K.A. Stetson, Optik **31**, 281–291 (1970)
2.48 M. Dubas, W. Schumann, Opt. Acta **21**, 547–562 (1974)
2.49 M. Dubas, W. Schumann, Opt. Acta **22**, 807–819 (1975)
2.50 K.A. Stetson, J. Opt. Soc. Am. **66**, 627 (1976)
2.51 W. Schumann, M. Dubas: *Holographic Interferometry*, Springer Ser. Opt. Sci., Vol. 16 (Springer, Berlin, Heidelberg 1979)
2.52 W. Schumann, J.-P. Zürcher, D. Cuche: *Holography and Deformation Analysis*, Springer Ser. Opt. Sci., Vol. 46 (Springer, Berlin, Heidelberg 1985)
2.53 Yu. I. Ostrovsky, V.P. Shchepinov, V.V. Yakovlev: *Holographic Interferometry in Experimental Mechanics*, Springer Ser. Opt. Sci., Vol. 60 (Springer, Berlin, Heidelberg 1991)

3. Quantitative Determination of Displacements and Strains from Holograms

R.J. Pryputniewicz

Center for Holographic Studies and Laser Technology, Department of Mechanical Engineering, Worcester Polytechnic Institute, Worcester, MA 01609-2280, USA

In this chapter, procedures for quantitative determination of displacements and strains directly from holograms are presented. This presentation begins with the definition of projection matrices and discussion of their application in holographic analysis. Then, system geometries used for recording and reconstruction of holograms are defined in terms of the illumination, observation, and sensitivity vectors. This is followed by a discussion of determination of displacements for the cases when the fringe order is known, when the fringe order is unknown, and when multiple holograms are used. Next, the fringe-vector method, relating the holographically determined strains and rotations to the object surface via the projection matrices based on the object's local surface normals, is outlined. Finally, recent advances in electronic acquisition, storage, processing, and display of optical interference information is discussed and their implementation in quantitative measurements of displacements, due to static and dynamic loading conditions, is illustrated with representative examples.

3.1 Projection Matrices: Definition and Properties

Quantitative interpretation of holograms is based on one's ability to delineate various parameters that characterize recording and reconstruction processes in hologram interferometry [3.1]. These parameters, being vectorial in nature, can be most clearly defined by matrix transformations. These transformations map fields of three-dimensional vectors from one space into corresponding fields of vectors in another space by means of projection matrices [3.1–5].

The matrix transformation that is of primary interest in quantitative interpretation of holograms is that which transforms a vector into its shadow on a surface. This transformation may fall into either of two categories: if the direction from which the shadow is cast is parallel to the surface normal, the operation is called a *normal projection*; if it is not, then it is called an *oblique projection*.

3.1.1 Normal Projection

Referring to Fig. 3.1, where A is a vector, $\hat{b}$ is a unit vector normal to a surface, and A_b is the projection of A onto a surface normal to $\hat{b}$, it is seen that all that is

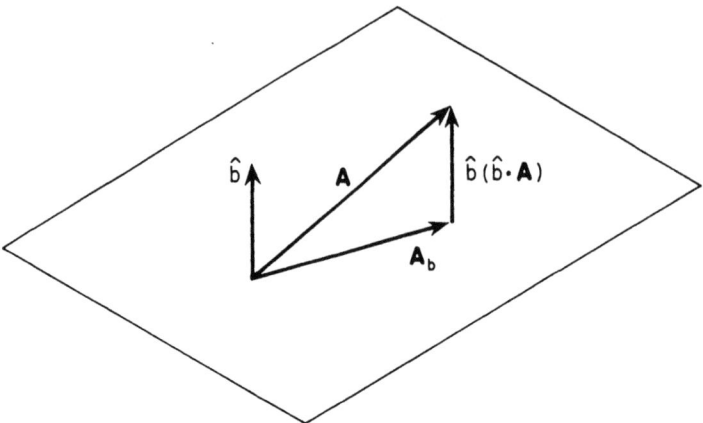

Fig. 3.1. Normal projection onto a surface

required to obtain A_b is to subtract from A its component in the direction of $\hat{b}$. The magnitude of that component is given by a scalar product $\hat{b} \cdot A$, and its direction is that of $\hat{b}$, so that the result is

$$A_b = A - \hat{b}(\hat{b} \cdot A). \tag{3.1}$$

We may recognize the right-hand side of (3.1) as the expansion of the triple vector product, that is,

$$A_b = -\hat{b} \times (\hat{b} \times A). \tag{3.2}$$

Because the vector product of two vectors yields a third vector, vector multiplication may be described as a transformation, and it may be represented by a matrix. This is achieved by arranging the components of the vector $\hat{b}$ as an antisymmetric matrix $\tilde{b}$, which can be written as

$$\tilde{b} = \begin{bmatrix} 0 & -b_z & b_y \\ b_z & 0 & -b_x \\ -b_y & b_x & 0 \end{bmatrix}. \tag{3.3}$$

Using the matrix defined by (3.3), we may rewrite (3.2) as

$$A_b = -\tilde{b}\tilde{b}A = \tilde{P}_b A, \tag{3.4}$$

where

$$\tilde{P}_b = -\tilde{b}\tilde{b} \tag{3.5}$$

is the matrix transformation that projects the vector A onto the surface which is normal to $\hat{b}$ to form the shadow A_b.

In an alternate way, we can rewrite (3.1) in a matrix form such that

$$A_b = \tilde{I}A - \begin{bmatrix} b_x & 0 & 0 \\ b_y & 0 & 0 \\ b_z & 0 & 0 \end{bmatrix} \begin{bmatrix} b_x & b_y & b_z \\ 0 & 0 & 0 \\ 0 & 0 & 0 \end{bmatrix} \begin{bmatrix} A_x & 0 & 0 \\ A_y & 0 & 0 \\ A_z & 0 & 0 \end{bmatrix}, \qquad (3.6)$$

where A_x, A_y, and A_z are the components of A, b_x, b_y, and b_z are the components of $\hat{b}$, and $\tilde{I}$ is the identity matrix. The product of the first two matrices in the second term on the right-hand side of (3.6) yields a 3×3 matrix whose elements are all nine possible products of the three components of $\hat{b}$. This operation defines the third type of product between two vectors, in addition to the well known *scalar product* and *vector product*. In this text, this operation will be called a *matric product* of two vectors, with the word *matric* meaning *of or pertaining to a matrix*, and it will be represented by an encircled cross, that is, ⊗. Thus, using the definition of the matric product, we may rewrite (3.6) as

$$A_b = (\tilde{I} - \hat{b} \otimes \hat{b})A = \tilde{P}_b A, \qquad (3.7)$$

from which it follows that

$$\tilde{P}_b = \tilde{I} - \hat{b} \otimes \hat{b}. \qquad (3.8)$$

Equation (3.8) defines the matrix transformation which describes the normal projection along the unit vector $\hat{b}$ onto a plane whose normal is parallel to the direction of projection. It should be noted that based on (3.5 and 8)

$$\tilde{P}_b = -\tilde{bb} = \tilde{I} - \hat{b} \otimes \hat{b}. \qquad (3.9)$$

3.1.2 Oblique Projection

Now, let us define an oblique projection of vector A along direction $\hat{b}$ onto a surface that is normal to $\hat{c}$, to produce the resultant vector A_{bc}, as shown in Fig. 3.2. In this case, we must subtract from A such a component in the direction of $\hat{b}$ that the resultant has no component in the direction of $\hat{c}$, that is,

$$A_{bc} = A - \frac{\hat{b}(\hat{c} \cdot A)}{\hat{b} \cdot \hat{c}}, \qquad (3.10)$$

where A_{bc} is the projection of A from the direction of $\hat{b}$ onto a plane perpendicular to $\hat{c}$.

Equation (3.10) can be expressed in terms of the vector triple product as

$$A_{bc} = -\frac{1}{\hat{b} \cdot \hat{c}} \hat{c} \times (\hat{b} \times A). \qquad (3.11)$$

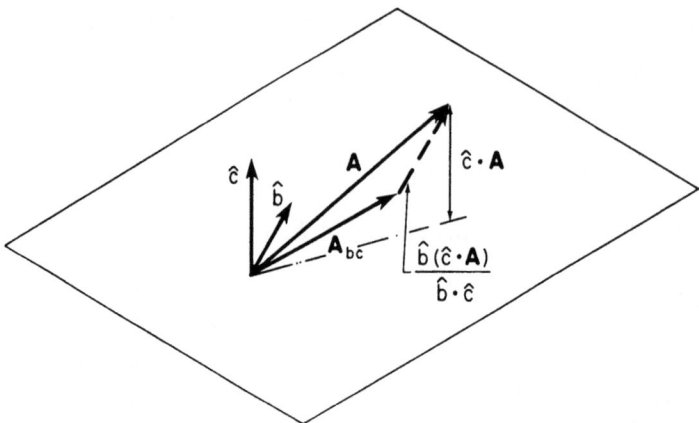

Fig. 3.2. Oblique projection onto a surface

Using (3.3) and defining $\hat{c}$ as an antisymmetric matrix $\tilde{c}$, expressed as

$$\tilde{c} = \begin{bmatrix} 0 & -c_z & c_y \\ c_x & 0 & -c_x \\ -c_y & c_x & 0 \end{bmatrix}, \qquad (3.12)$$

we can rewrite (3.11) to obtain

$$A_{bc} = -\frac{1}{\hat{b}\cdot\hat{c}} \tilde{c}\tilde{b} A = \tilde{P}_{bc} A, \qquad (3.13)$$

where

$$\tilde{P}_{bc} = -\frac{1}{\hat{b}\cdot\hat{c}} \tilde{c}\tilde{b} \qquad (3.14)$$

is the transformation matrix that projects a vector along the direction of $\hat{b}$ onto a plane normal to $\hat{c}$, to form the shadow A_{bc}. Furthermore, we can easily express (3.10) in the form analogous to (3.7), that is,

$$A_{bc} = \tilde{I} A - \frac{1}{\hat{b}\cdot\hat{c}} \hat{b} \otimes \hat{c} A = \left(\tilde{I} - \frac{1}{\hat{b}\cdot\hat{c}} \hat{b} \otimes \hat{c} \right) A = \tilde{P}_{bc} A, \qquad (3.15)$$

from which it clearly follows that

$$\tilde{P}_{bc} = \tilde{I} - \frac{1}{\hat{b}\cdot\hat{c}} \hat{b} \otimes \hat{c}. \qquad (3.16)$$

Finally, comparing (3.14 and 16), we obtain

$$\tilde{P}_{bc} = -\frac{1}{\hat{b}\cdot\hat{c}} \tilde{c}\tilde{b} = \tilde{I} - \frac{1}{\hat{b}\cdot\hat{c}} \hat{b} \otimes \hat{c}. \qquad (3.17)$$

Equation (3.17) represents the transformation matrix defining the oblique projection that projects a vector along the direction of $\hat{b}$ onto the plane normal to $\hat{c}$, to form the shadow A_{bc}. It should be noted that the order of subscripts b and c in specification of the transformation matrix $\tilde{P}_{bc}$, given by (3.17), is that of vectors $\hat{b}$ and $\hat{c}$ in the matric product $\hat{b} \otimes \hat{c}$. It should also be noted that in the special case when $\hat{b} = \hat{c}$, (3.17) reduces to (3.9).

It must be noted that the order of multiplication is not interchangeable for the matric products because $\hat{c} \otimes \hat{b}$ yields a matrix not equal to $\hat{b} \otimes \hat{c}$, but equal instead to its transpose. From this it may be deduced that the transpose of an oblique projection matrix simply exchanges the role of the two unit vectors. Therefore,

$$\tilde{P}_{bc}^{T} = \tilde{P}_{cb}, \tag{3.18}$$

where $\tilde{P}_{cb}$ is the transformation matrix that projects a vector along the direction of $\hat{c}$ onto a plane normal to $\hat{b}$.

In addition, all projection matrices are singular and do not possess inverses. There is one property, however, that they do have and which is extensively used in holographic analysis. This property is based on the fact that if a vector is projected from a given direction onto a plane, and then if this is projected from the same direction onto the second plane, the final result is the same as having projected the original vector onto the second plane directly. Thus, for example, if a vector is first projected from the direction $\hat{b}$ onto a plane normal to $\hat{c}$, and then if this result is projected again from the same direction $\hat{b}$ onto a plane normal to $\hat{b}$, the result is

$$\tilde{P}_b \tilde{P}_{bc} = \tilde{P}_b, \tag{3.19}$$

where it should be noted that the second projection premultiplies the first projection.

In another case, when $\tilde{P}_{bc}$ is the transformation matrix defining the first projection and $\tilde{P}_{ba}$ is the matrix characterizing the second projection, then

$$\tilde{P}_{ba} \tilde{P}_{bc} = \tilde{P}_{ba}. \tag{3.20}$$

It is important to keep in mind that in these sequences the direction along which the projection is made does not change. However, it should be observed that the sequence of two normal projections, first onto one surface, $\tilde{P}_a$, and then onto another, $\tilde{P}_b$, that is,

$$\tilde{P}_b \tilde{P}_a \neq \tilde{P}_{ba}, \tag{3.21}$$

is not the same as the oblique projection $\tilde{P}_{ba}$.

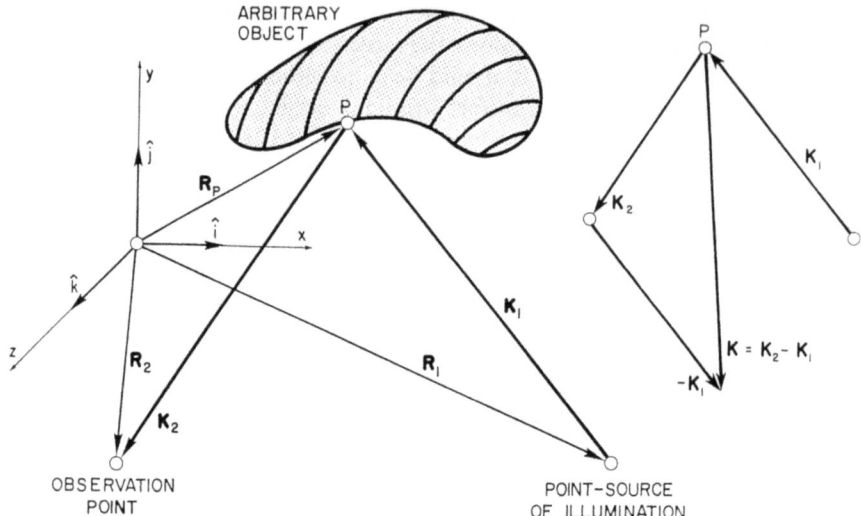

Fig. 3.3. Illumination and observation geometry in hologram interferometry

3.2 Illumination, Observation, and Sensitivity Vectors

Quantitative interpretation of holograms depends on knowledge of the illumination and observation directions used during recording and reconstruction of the holograms [3.6, 7]. These directions are defined by illumination and observation vectors, K_1 and K_2, respectively, as shown in Fig. 3.3.

The vectors K_1 and K_2 are defined as propagation vectors of light from the point source of illumination to the object and from the object to the "observer" (photosensitive medium, or the sensing element of a CCD camera), respectively. Therefore, K_1 and K_2 can be readily described in terms of position vectors, R_1, R_P, and R_2, which are defined with respect to the origin of the Cartesian x–y–z coordinate system. These position vectors, defining the locations of the point source of illumination, a point on the object, and a point of observation, are

$$R_1 = x_1\hat{i} + y_1\hat{j} + z_1\hat{k}, \tag{3.22}$$

$$R_P = x_P\hat{i} + y_P\hat{j} + z_P\hat{k}, \tag{3.23}$$

and

$$R_2 = x_2\hat{i} + y_2\hat{j} + z_2\hat{k}, \tag{3.24}$$

respectively, where $\hat{i}, \hat{j},$ and $\hat{k}$ represent unit vectors, which are parallel to the axis of the coordinate system.

Using (3.22–24) and the geometry shown in Fig. 3.3, the vectors K_1 and K_2 can be expressed as

$$K_1 = K_{1_x}\hat{i} + K_{1_y}\hat{j} + K_{1_z}\hat{k} = k\frac{R_P - R_1}{|R_P - R_1|}$$

$$= k\frac{(x_P - x_1)\hat{i} + (y_P - y_1)\hat{j} + (z_P - z_1)\hat{k}}{[(x_P - x_1)^2 + (y_P - y_1)^2 + (z_P - z_1)^2]^{1/2}}$$

$$= k\hat{K}_1 = k(|\hat{K}_{1_x}|\hat{i} + |\hat{K}_{1_y}|\hat{j} + |\hat{K}_{1_z}|\hat{k}) \qquad (3.25)$$

and

$$K_2 = K_{2_x}\hat{i} + K_{2_y}\hat{j} + K_{2_z}\hat{j} = k\frac{R_2 - R_P}{|R_2 - R_P|}$$

$$= k\frac{(x_2 - x_P)\hat{i} + (y_2 - y_P)\hat{j} + (z_2 - z_P)\hat{k}}{[(x_2 - x_P)^2 + (y_2 - y_P)^2 + (z_2 - z_P)^2]^{1/2}}$$

$$= k\hat{K}_2 = k(|\hat{K}_{2_x}|\hat{i} + |\hat{K}_{2_y}|\hat{j} + |\hat{K}_{2_z}|\hat{k}). \qquad (3.26)$$

In (3.25 and 26), $\hat{K}_1$ and $\hat{K}_2$ are the unit illumination and observation vectors, respectively, and $|\hat{K}_{1_x}|, |\hat{K}_{1_y}|, \ldots, |\hat{K}_{2_z}|$ indicate magnitudes of their Cartesian components, while k is the magnitude of the K_1 and K_2 vectors defined as

$$|K_1| = |K_2| = k = \frac{2\pi}{\lambda}, \qquad (3.27)$$

with λ being the wavelength of the laser light.

Finally, definition of the sensitivity vector K, as a difference between the observation and illumination vectors, yields

$$K = K_x\hat{i} + K_y\hat{j} + K_z\hat{k}$$

$$= K_2 - K_1 = (K_{2_x} - K_{1_x})\hat{i} + (K_{2_y} - K_{1_y})\hat{j} + (K_{2_z} - K_{1_z})\hat{k}$$

$$= k(|\hat{K}_{2_x} - \hat{K}_{1_x}|\hat{i} + |\hat{K}_{2_y} - \hat{K}_{1_y}|\hat{j} + |\hat{K}_{2_z} - \hat{K}_{1_z}|\hat{k}), \qquad (3.28)$$

where K_1 and K_2 are as defined in (3.25) and (3.26), respectively.

3.3 Determination of Displacements

Hologram interferometry is used to measure displacements of objects subjected to static and/or dynamic loads. Depending on the loading method used and the methods for recording/reconstruction of holograms and readout of the interferometric information, fringe patterns produced during these reconstructions can be classified either as cosinusoidal fringes or Bessel (J_0) fringes. Typical appearances of these two types of holographic fringes are shown in Fig. 3.4. The cosinusoidal fringes are of equal brightness across the image, regardless of the

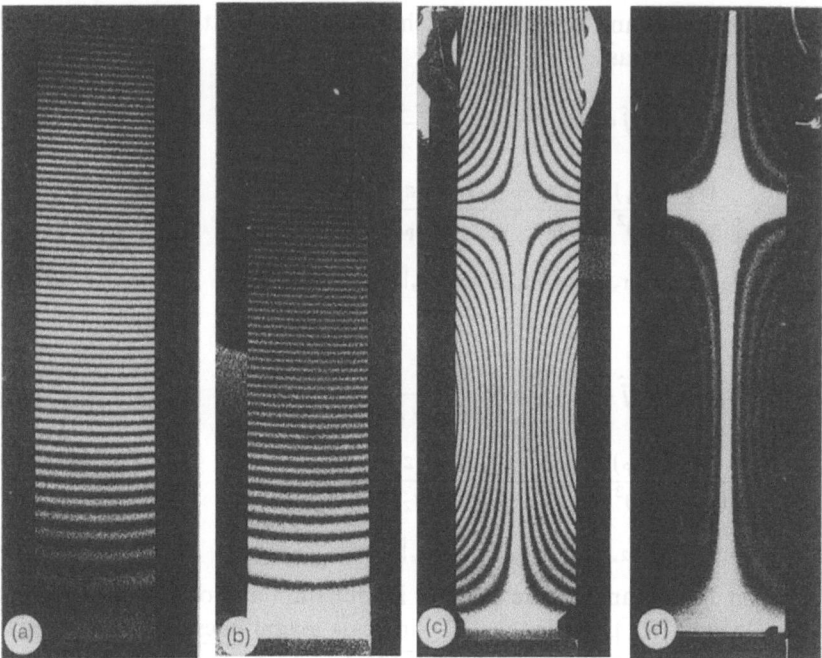

Fig. 3.4a–d. Typical appearances of the two types of holographic fringes: (a) and (c) show cosinusoidal fringes – note that the zero-order fringes are as bright as the higher-order cosinusoidal fringes, while (b) and (d) show J_0 fringes – note that the zero-order fringes are much brighter than the higher order J_0 fringes

fringe order, and are normally associated with a response of the object to the static load. The Bessel fringes are of unequal brightness, which decreases as the fringe order increases, across the image. These fringes are normally associated with a response of the object to dynamic loading, especially periodic excitation causing objects to resonate. Interpretation of the cosinusoidal fringes is a subject of this section, while interpretation of the Bessel fringes is discussed in Chap. 8.

Holographic numerical analysis depends on knowledge of the illumination and the observation directions used while recording and/or reconstructing the holograms. These directions are described by the illumination and observation vectors, K_1 and K_2, respectively, as discussed in Sect. 3.2, and define propagation of light from the point source of illumination via the object to the observation point.

As the object displaces/deforms, while recording a hologram, the phase difference between light beams arriving at a specific observation point, from the object and its displaced/deformed copy, is encoded in form of fringe patterns. These fringes are described by a *fringe-locus function*, Ω, constant values of which define fringe loci on the object's surface. The fringe-locus function relates directly to the fringe orders, n, via

$$\Omega = 2\pi n. \tag{3.29}$$

The fringe-locus function can also be related to the scalar product of the sensitivity vector $\boldsymbol{K}$ with the displacement vector $\boldsymbol{L}$, that is,

$$\boldsymbol{K} \cdot \boldsymbol{L} = \Omega. \tag{3.30}$$

The sensitivity vector $\boldsymbol{K}$, appearing in (3.30), can be varied by changing either the direction of illumination, $\boldsymbol{K}_1$, or the direction of observation, $\boldsymbol{K}_2$, or both $\boldsymbol{K}_1$ and $\boldsymbol{K}_2$ simultaneously. Because of different approaches employed to vary $\boldsymbol{K}$ and because of different approaches used in determination of Ω, the problem of extracting displacement vectors, $\boldsymbol{L}$, directly from cosinusoidal fringes has been solved in a number of ways [3.8–33]. In this section, we will discuss some of these ways for determination of the object displacements directly from the fringe patterns observed during reconstruction of holograms. We will begin with an interpretation of a single hologram when the fringe order is known, then, we will follow with the interpretation of a single hologram when the fringe order is unknown. Finally, we will present a method, based on the projection matrices, for determination of displacements from multiple holograms.

3.3.1 Determination of Displacements when Fringe Order is Known

Determination of displacements when the fringe order is known represents a special case, in holographic numerical analysis, when the zero-order fringe is identifiable within the holographically reconstructed image. Then, orders are unambiguously assigned to fringes, with respect to the zero-order fringe (Fig. 3.4). The most popular of the existing techniques, used to assign the fringe-order numbers, is based on multiple observations of images reconstructed from a single hologram, i.e., by varying $\boldsymbol{K}_2$.

Multiple observations of a holographically reconstructed image must be made from various, noncoplanar directions. This can be accomplished by noting that the fringes and the surface of the object can nearly always be viewed in focus by simply using a sufficiently small observing aperture, i.e., one having a sufficiently large f/number.

The three-dimensional displacement vector, $\boldsymbol{L}$, appearing in (3.30), is defined in terms of its Cartesian components as

$$\boldsymbol{L} = L_x \hat{\boldsymbol{i}} + L_y \hat{\boldsymbol{j}} + L_z \hat{\boldsymbol{k}}. \tag{3.31}$$

Therefore, in order to completely determine its unknown components L_x, L_y, and L_z, an "ideal" system of three equations of the type of (3.30) must be solved simultaneously. Such equations can be generated using the observation

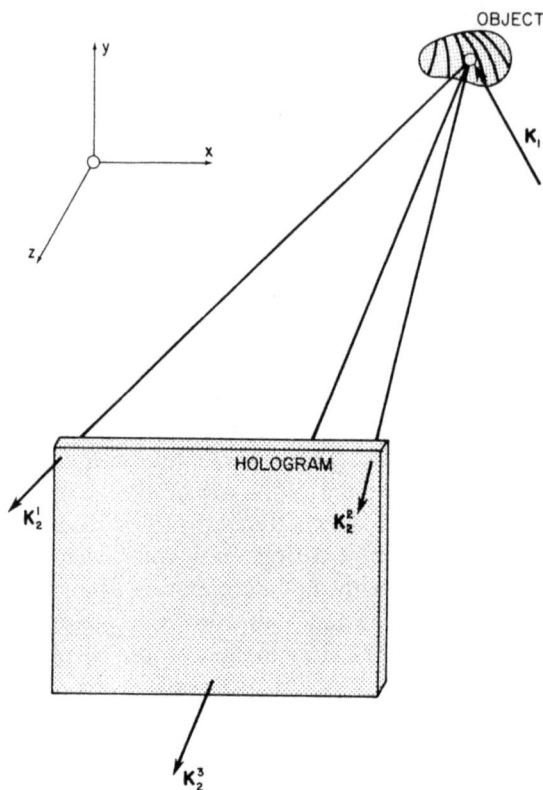

Fig. 3.5. Multiple observations of a holographically reconstructed image from three different directions

geometry shown in Fig. 3.5. Therefore, writing one equation of the type of (3.30), for each of the three observations depicted in Fig. 3.5, we obtain

$$\boldsymbol{K}^1 \cdot \boldsymbol{L} = \Omega^1, \tag{3.32}$$

$$\boldsymbol{K}^2 \cdot \boldsymbol{L} = \Omega^2, \tag{3.33}$$

and

$$\boldsymbol{K}^3 \cdot \boldsymbol{L} = \Omega^3, \tag{3.34}$$

where the superscripts identify quantities corresponding to the specific directions of observation and where, according to (3.28), the sensitivity vectors are

$$\begin{aligned}\boldsymbol{K}^1 &= K_x^1 \hat{\boldsymbol{i}} + K_y^1 \hat{\boldsymbol{j}} + K_z^1 \hat{\boldsymbol{k}} \\ &= \boldsymbol{K}_2^1 - \boldsymbol{K}_1 = (K_{2x}^1 - K_{1x})\hat{\boldsymbol{i}} + (K_{2y}^1 - K_{1y})\hat{\boldsymbol{j}} + (K_{2z}^1 - K_{1z})\hat{\boldsymbol{k}}, \end{aligned} \tag{3.35}$$

$$\begin{aligned}\boldsymbol{K}^2 &= K_x^2 \hat{\boldsymbol{i}} + K_y^2 \hat{\boldsymbol{j}} + K_z^2 \hat{\boldsymbol{k}} \\ &= \boldsymbol{K}_2^2 - \boldsymbol{K}_1 = (K_{2x}^2 - K_{1x})\hat{\boldsymbol{i}} + (K_{2y}^2 - K_{1y})\hat{\boldsymbol{j}} + (K_{2z}^2 - K_{1z})\hat{\boldsymbol{k}}, \end{aligned} \tag{3.36}$$

and

$$\mathbf{K}^3 = K_x^3 \hat{i} + K_y^3 \hat{j} + K_z^3 \hat{k}$$
$$= \mathbf{K}_2^3 - \mathbf{K}_1 = (K_{2x}^3 - K_{1x})\hat{i} + (K_{2y}^3 - K_{1y})\hat{j} + (K_{2z}^3 - K_{1z})\hat{k}. \quad (3.37)$$

Therefore, based on (3.35–37), we have

$$K_x^1 = K_{2x}^1 - K_{1x},$$
$$K_y^1 = K_{2y}^1 - K_{1y},$$
$$K_z^1 = K_{2z}^1 - K_{1z},$$
$$K_x^2 = K_{2x}^2 - K_{1x}, \quad (3.38)$$
$$\vdots$$
$$K_z^3 = K_{2z}^3 - K_{1z}.$$

Expanding (3.32–34) we obtain

$$K_x^1 L_x + K_y^1 L_y + K_z^1 L_z = \Omega^1, \quad (3.39)$$
$$K_x^2 L_x + K_y^2 L_y + K_z^2 L_z = \Omega^2, \quad (3.40)$$

and

$$K_x^3 L_x + K_y^3 L_y + K_z^3 L_z = \Omega^3. \quad (3.41)$$

The set of (3.39–41) can be rewritten as the matrix equation

$$\begin{bmatrix} K_x^1 & K_y^1 & K_z^1 \\ K_x^2 & K_y^2 & K_z^2 \\ K_x^3 & K_y^3 & K_z^3 \end{bmatrix} \begin{pmatrix} L_x \\ L_y \\ L_z \end{pmatrix} = \begin{pmatrix} \Omega^1 \\ \Omega^2 \\ \Omega^3 \end{pmatrix}. \quad (3.42)$$

Providing that $\mathbf{K}^1$, $\mathbf{K}^2$, and $\mathbf{K}^3$ are independent, the 3×3 matrix of the sensitivity vectors is non-singular and the solution to (3.42) becomes

$$\begin{pmatrix} L_x \\ L_y \\ L_z \end{pmatrix} = \begin{bmatrix} K_x^1 & K_y^1 & K_z^1 \\ K_x^2 & K_y^2 & K_z^2 \\ K_x^3 & K_y^3 & K_z^3 \end{bmatrix}^{-1} \begin{pmatrix} \Omega^1 \\ \Omega^2 \\ \Omega^3 \end{pmatrix}, \quad (3.43)$$

where the superscript -1 indicates the inverse of the matrix of the sensitivity vectors; the elements of this matrix are defined by (3.38).

Equation (3.43) gives solution for the three components of $\mathbf{L}$ when three noncoplanar sensitivity vectors are used to obtain the corresponding fringe-locus functions. This is the ideal case. In general, however, the matrix of the sensitivity vectors used in (3.42) is ill-conditioned because of the limited size of the hologram. As a result of this, substantial errors result in L. In order to overcome this drawback, more than three observations of the reconstructed image are made. This generates an overdetermined set of simultaneous equations, with one equation for each observation, that is,

$$K^1 \cdot L = \Omega^1,$$
$$K^2 \cdot L = \Omega^2, \qquad (3.44)$$
$$\vdots$$
$$K^r \cdot L = \Omega^r,$$

where r is the total number of observations, e.g., five, as shown in Fig. 3.6. To facilitate solution of this overdetermined set of equations, (3.44) can be rewritten by means of the index notation as

$$K^m \cdot L = \Omega^m, \quad m = 1, 2, \ldots, r. \qquad (3.45)$$

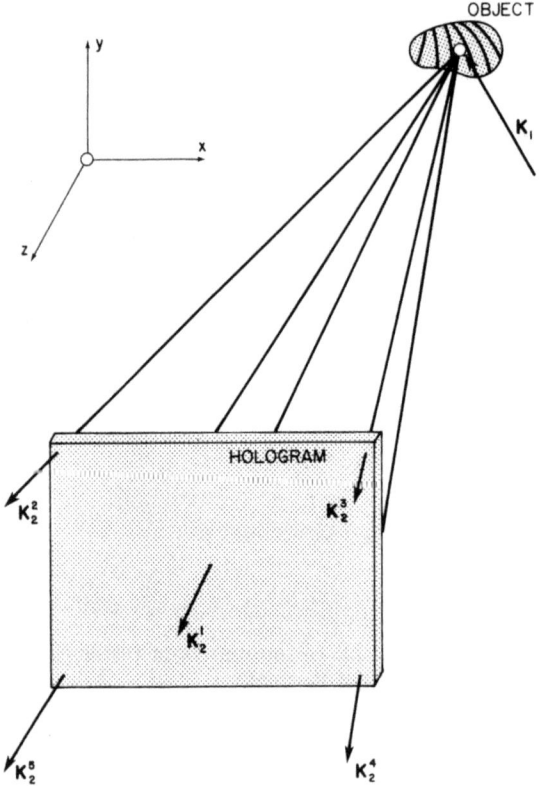

Fig. 3.6. Multiple observations of a holographically reconstructed image from five different directions

3.3 Determination of Displacements   45

Obviously, no matter how careful and how many observations are made, there will always be some error E inherent in the measurements. Therefore, (3.45) can be expressed as

$$E^m = \mathbf{K}^m \cdot \mathbf{L} - \Omega^m, \quad m = 1, 2, \ldots, r. \tag{3.46}$$

The goal of the analysis is to adjust the coefficients in (3.46) in such a way as to minimize the sum of all the errors squared, that is,

$$\sum_{m=1}^{r} (E^m)^2 = \sum_{m=1}^{r} (\mathbf{K}^m \cdot \mathbf{L} - \Omega^m)^2, \quad m = 1, 2, \ldots, r. \tag{3.47}$$

In order for the

$$\sum_{m=1}^{r} (E^m)^2$$

to be minimum, the partial derivatives of (3.47) with respect to the components L_x, L_y, and L_z of the displacement vector $\mathbf{L}$ must be zero, that is,

$$\frac{\partial}{\partial L_i}\left[\sum_{m=1}^{r} (E^m)^2\right] = \frac{\partial}{\partial L_i}\left[\sum_{m=1}^{r} (\mathbf{K}^m \cdot \mathbf{L} - \Omega^m)^2\right] = 0,$$

$$m = 1, 2, \ldots, r, \quad i = x, y, z, \tag{3.48}$$

which results in the following system of three simultaneous equations:

$$\sum_{m=1}^{r} (K_x^m L_x + K_y^m L_y + K_z^m L_z)K_x^m = \sum_{m=1}^{r} K_x^m \Omega^m,$$

$$\sum_{m=1}^{r} (K_x^m L_x + K_y^m L_y + K_z^m L_z)K_y^m = \sum_{m=1}^{r} K_y^m \Omega^m, \tag{3.49}$$

$$\sum_{m=1}^{r} (K_x^m L_x + K_y^m L_y + K_z^m L_z)K_z^m = \sum_{m=1}^{r} K_z^m \Omega^m.$$

Equations (3.49) can be rewritten in the matrix form as

$$\begin{bmatrix} \sum_{m=1}^{r} K_x^m K_x^m & \sum_{m=1}^{r} K_x^m K_y^m & \sum_{m=1}^{r} K_x^m K_z^m \\ \sum_{m=1}^{r} K_y^m K_x^m & \sum_{m=1}^{r} K_y^m K_y^m & \sum_{m=1}^{r} K_y^m K_z^m \\ \sum_{m=1}^{r} K_z^m K_x^m & \sum_{m=1}^{r} K_z^m K_y^m & \sum_{m=1}^{r} K_z^m K_z^m \end{bmatrix} \begin{pmatrix} L_x \\ L_y \\ L_z \end{pmatrix} = \begin{pmatrix} \sum_{m=1}^{r} K_x^m \Omega^m \\ \sum_{m=1}^{r} K_y^m \Omega^m \\ \sum_{m=1}^{r} K_z^m \Omega^m \end{pmatrix},$$

(3.50)

and can be readily solved to obtain

$$\begin{pmatrix} L_x \\ L_y \\ L_z \end{pmatrix} = \begin{bmatrix} \sum_{m=1}^{r} K_x^m K_x^m & \sum_{m=1}^{r} K_x^m K_y^m & \sum_{m=1}^{r} K_x^m K_z^m \\ \sum_{m=1}^{r} K_y^m K_x^m & \sum_{m=1}^{r} K_y^m K_y^m & \sum_{m=1}^{r} K_y^m K_z^m \\ \sum_{m=1}^{r} K_z^m K_x^m & \sum_{m=1}^{r} K_z^m K_y^m & \sum_{m=1}^{r} K_z^m K_z^m \end{bmatrix}^{-1} \begin{pmatrix} \sum_{m=1}^{r} K_x^m \Omega^m \\ \sum_{m=1}^{r} K_y^m \Omega^m \\ \sum_{m=1}^{r} K_z^m \Omega^m \end{pmatrix}. \tag{3.51}$$

The result shown in (3.51) can be obtained in a much more direct way if we write (3.45) in a condensed matrix form as

$$\tilde{K} L = \bar{\Omega}, \tag{3.52}$$

where $\tilde{K}$ is a rectangular $r \times 3$ matrix of the sensitivity vectors and $\bar{\Omega}$ is a column $r \times 1$ matrix of the fringe-locus functions.

Defining the error vector E in much the same way as it was done in (3.46) we can rewrite (3.52) as

$$E = \tilde{K} L - \bar{\Omega} \tag{3.53}$$

and write the equivalent of (3.47) as

$$E^2 = E \cdot E = E^T E = (\tilde{K} L - \bar{\Omega})^T (\tilde{K} L - \bar{\Omega}). \tag{3.54}$$

Noting that the transpose of a product of two matrices is equal to the product of the transposes of the two matrices with the order of multiplication reversed, (3.54) can be rewritten to obtain

$$E^2 = (L^T \tilde{K}^T - \bar{\Omega}^T)(\tilde{K} L - \bar{\Omega}) = L^T \tilde{K}^T \tilde{K} L - \bar{\Omega}^T \tilde{K} L - L^T \tilde{K}^T \bar{\Omega} + \bar{\Omega}^T \bar{\Omega}. \tag{3.55}$$

In order to minimize E^2, the partial derivatives of (3.55) with respect to the components of L must be zero, that is,

$$\frac{\partial(E^2)}{\partial L_i} = 0, \quad i = x, y, z. \tag{3.56}$$

Therefore, applying (3.56) to (3.55) we obtain

$$\tilde{K}^T \tilde{K} L + L^T \tilde{K}^T \tilde{K} - \bar{\Omega}^T \tilde{K} - \tilde{K}^T \bar{\Omega} = 0, \tag{3.57}$$

wherefrom, after rearranging, we get

$$(\tilde{K}^T \tilde{K} L - \tilde{K}^T \bar{\Omega}) + (L^T \tilde{K}^T \tilde{K} - \bar{\Omega}^T \tilde{K}) = 0, \tag{3.58}$$

or

$$\tilde{K}^T (\tilde{K} L - \bar{\Omega}) + (\tilde{K} L - \bar{\Omega})^T \tilde{K} = 0. \tag{3.59}$$

Examining (3.59), it becomes apparent that both terms on the left-hand side of the equality sign are identical to each other. Therefore, in order for (3.59) to be satisfied, both of these terms must be zero. Hence, we obtain

$$\tilde{K}^T(\tilde{K}L - \bar{\Omega}) = \tilde{K}^T\tilde{K}L - \tilde{K}^T\bar{\Omega} = 0, \tag{3.60}$$

or

$$\tilde{K}^T\tilde{K}L = \tilde{K}^T\bar{\Omega}. \tag{3.61}$$

It should be noted, at this time, that (3.61) is nothing more than (3.52) with both sides premultiplied by the transpose of the rectangular matrix $\tilde{K}$. This procedure decreases the rank of the rectangular matrix $\tilde{K}$ to a square 3×3 matrix $\tilde{K}^T\tilde{K}$ and yields the solution for the displacement vector L which has the least-squares-error, that is,

$$L = [\tilde{K}^T\tilde{K}]^{-1}(\tilde{K}^T\bar{\Omega}). \tag{3.62}$$

It is interesting to note that the square matrix $\tilde{K}^T\tilde{K}$ of (3.62) may be expressed as the sum of all r matric products of the sensitivity vectors K^m, that is,

$$\tilde{K}^T\tilde{K} = \sum_{m=1}^{r} K^m \otimes K^m, \quad m = 1, 2, \ldots, r. \tag{3.63}$$

Therefore, combining (3.62 and 63) we obtain

$$L = \left[\sum_{m=1}^{r} K^m \otimes K^m\right]^{-1} (\tilde{K}^T\bar{\Omega}), \quad m = 1, 2, \ldots, r. \tag{3.64}$$

Comparing (3.51 and 64) it becomes obvious that both equations are identical. However, (3.64) is much more compact than (3.51) and, therefore, greatly facilitates holographic numerical analysis.

3.3.2 Determination of Displacements when Fringe Order is Unknown

In holographic analysis of objects for which the entire surface has moved and/or deformed it is often impossible to identify the zero-order fringe and, therefore, procedures of Sect. 3.3.1 do not apply. Instead, we may only determine the fringe orders to within an additive constant. With this in mind, we will set up a system of equations of the type of (3.52) and solve for the displacement vector. To reduce measurement errors, we will use an overdetermined set of equations and solve for L which has the least-squares-error.

To set up the system of equations, we must determine the fringe-locus functions to within an additive constant. Let us call this constant Ω_0 and let us bear in mind that this constant comes from the lack of knowledge of the absolute fringe order. Then, for the first "observation" of the holographically reconstructed image, as shown in Fig. 3.7, we can write an equation relating Ω_0 to the

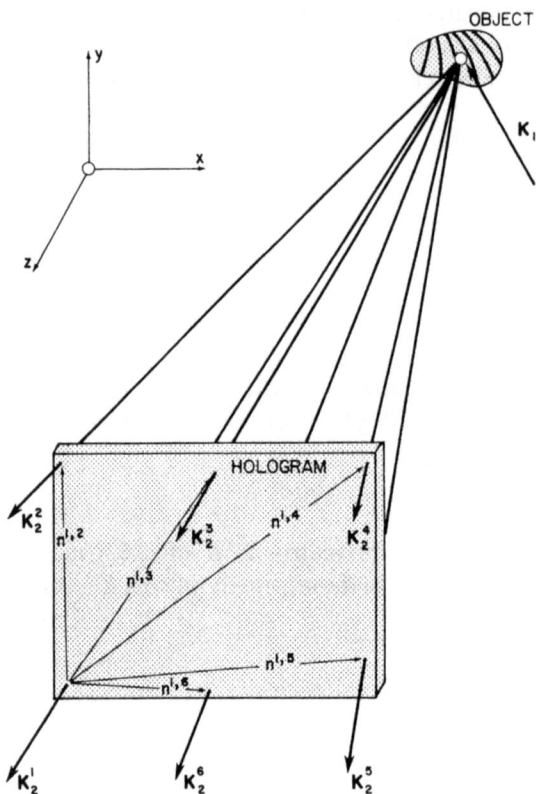

Fig. 3.7. Multiple observations of a holographically reconstructed image with the first direction of observation chosen arbitrarily

scalar product of L with K^1, corresponding to the observation along the direction specified by the vector K_2^1, as

$$K^1 \cdot L = \Omega_0, \qquad (3.65)$$

where the sensitivity vector is as defined by (3.35). Next, "counting" the fringes that pass across the point of interest on the object as the view is being changed from the observation along the direction of K_2^1 to the observation along the direction of K_2^2, while continuously observing the point of interest on the object, we determine the (observed) fringe order, more commonly known as the fringe shift, $n^{1,2}$, which relates to the change, $\Delta\Omega^{1,2}$, in the fringe-locus function via

$$\Delta\Omega^{1,2} = 2\pi n^{1,2}, \qquad (3.66)$$

thus giving

$$K^2 \cdot L = \Omega_0 + \Delta\Omega^{1,2}. \qquad (3.67)$$

Following the procedure used to obtain (3.67), we can determine the fringe shifts $n^{1,m}$ for other changes in the direction of observation from that along the direction of K_2^1 to those along the directions of K_2^m, one at a time. The sensitivity

3.3 Determination of Displacements

vectors corresponding to these observations can be described by the system of equations, similar to (3.35–37), that is,

$$\begin{aligned}\boldsymbol{K}^m &= K_x^m \hat{\boldsymbol{i}} + K_y^m \hat{\boldsymbol{j}} + K_z^m \hat{\boldsymbol{k}} \\ &= \boldsymbol{K}_2^m - \boldsymbol{K}_1 = (K_{2x}^m - K_{1x})\hat{\boldsymbol{i}} + (K_{2y}^m - K_{1y})\hat{\boldsymbol{j}} + (K_{2z}^m - K_{1z})\hat{\boldsymbol{k}}, \\ m &= 1, 2, \ldots, r, \end{aligned} \qquad (3.68)$$

where r is the total number of observations.

Based on (3.68), the components of the sensitivity vectors can be defined as

$$\begin{aligned} K_x^m &= K_{2x}^m - K_{1x}, \quad m = 1, 2, \ldots, r, \\ K_y^m &= K_{2y}^m - K_{1y}, \quad m = 1, 2, \ldots, r, \\ K_z^m &= K_{2z}^m - K_{1z}, \quad m = 1, 2, \ldots, r. \end{aligned} \qquad (3.69)$$

The changes in the fringe-locus function, corresponding to the sensitivity vectors defined by (3.68), can be expressed as

$$\Delta \Omega^{1,m} = 2\pi n^{1,m}, \quad m = 1, 2, \ldots, r, \qquad (3.70)$$

where it should be realized that when $m = 1$, $\Delta \Omega^{1,1} = 0$.

Equations relating the sensitivity vectors of (3.68) and the changes in the fringe-locus functions of (3.70) to the displacement vector $\boldsymbol{L}$ are

$$\boldsymbol{K}^m \cdot \boldsymbol{L} = \Omega_0 + \Delta \Omega^{1,m}, \quad m = 1, 2, \ldots, r. \qquad (3.71)$$

It should be noted that in the approach described by (3.68–71), the number of observations must be equal to or be greater than four (i.e. $r \geq 4$) because in addition to the three unknown components of the displacement vector $\boldsymbol{L}$ we must also account for Ω_0 which is an unknown. Therefore, from (3.71) we have

$$\boldsymbol{K}^m \cdot \boldsymbol{L} - \Omega_0 = \Delta \Omega^{1,2}, \quad m = 1, 2, \ldots, r. \qquad (3.72)$$

Since $\boldsymbol{L}$ and Ω_0 are common to all r equations, (3.72) can be rewritten in a matrix form as

$$[\tilde{\boldsymbol{K}}, -1](\boldsymbol{L}, \Omega_0) = \Delta \bar{\boldsymbol{\Omega}}, \qquad (3.73)$$

where $[\tilde{\boldsymbol{K}}, -1]$ is a rectangular $r \times 4$ matrix, $(\boldsymbol{L}, \Omega_0)$ is a column 4×1 matrix, and $\Delta \bar{\boldsymbol{\Omega}}$ is a column $r \times 1$ matrix. Renaming the $[\tilde{\boldsymbol{K}}, -1]$ matrix of the sensitivity vectors as $\tilde{\boldsymbol{G}}$, that is,

$$[\tilde{\boldsymbol{K}}, -1] = \tilde{\boldsymbol{G}}, \qquad (3.74)$$

we can rewrite (3.73) as

$$\vec{\tilde{\boldsymbol{G}}}(\boldsymbol{L}, \Omega_0) = \Delta \bar{\boldsymbol{\Omega}}. \qquad (3.75)$$

Finally, (3.75) can be solved to obtain

$$(\boldsymbol{L}, \Omega_0) = [\tilde{\boldsymbol{G}}^T \tilde{\boldsymbol{G}}]^{-1} (\tilde{\boldsymbol{G}}^T \Delta \bar{\boldsymbol{\Omega}}), \qquad (3.76)$$

50   3. Quantitative Determination of Displacements and Strains from Holograms

or

$$(L, \Omega_0) = \left[\sum_{m=1}^{r} G^m \otimes G^m \right]^{-1} (\tilde{G}^T \Delta \bar{\Omega}), \quad m = 1, 2, \ldots, r. \tag{3.77}$$

It should be noted that the usual methods of holographic displacement analysis eliminate the unknown Ω_0 from the system of equations by subtracting one member of the set from the rest, or by subtracting pairs of equations. The method given in this section is preferred because, in the subtraction process, the effects of some measurement errors can be escalated. However, we should realize that in the case when the first observation is made through a point lying within the center of all points through which observations are made, as shown in Fig. 3.8, it does not matter whether we introduce the additive constant Ω_0 and employ (3.77) in determination of displacements, or simply use the fringe shifts alone. The results will be the same.

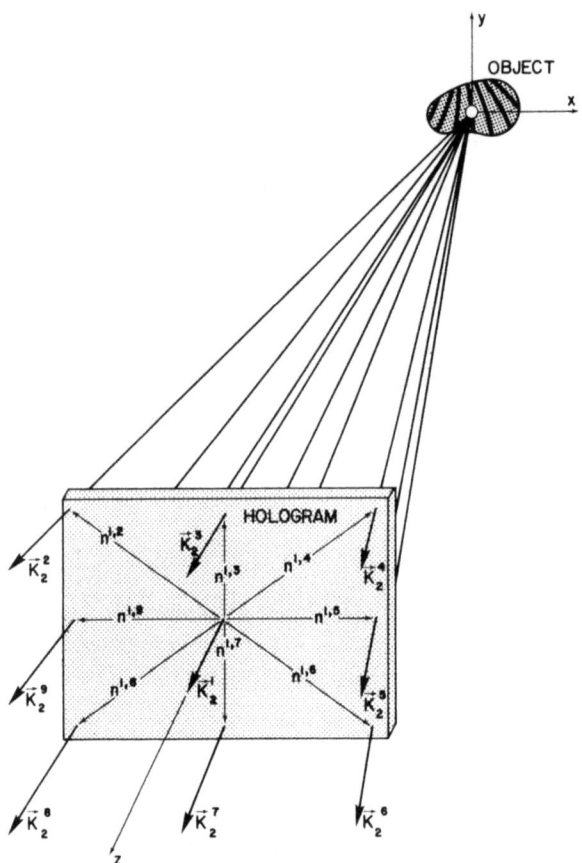

Fig. 3.8. Multiple observations of a holographically reconstructed image with the first direction of observation being in the center of all points through which the observations are made

It is stressed that in the cases when we want to make observations in an arbitrary fashion, with the first observation not being through the central point, only (3.77) can be applied to determine the displacements. Use of other equations may give erratic results.

3.3.3 Determination of Displacements from Multiple Holograms

In certain applications, it is either not practical or, even, not possible to record holograms large enough to allow multiple observations of the reconstructed images. In some other applications, the holograms do not exhibit any parallax at all. This is certainly the case with electro-optic (also referred to as electronic, and/or TV) holography [3.34, 35], where holograms are recorded electronically and exhibit no parallax. Therefore, procedures discussed in Sects. 3.3.1 and 2 cannot be used in these applications. However, even in the cases of holograms that do not exhibit parallax, we can easily determine two components of the object displacement vectors in the directions transverse to the direction from which the hologram was recorded. The vector sum of these two components may be referred to as the *observed displacement vector* and it can be denoted as L_{ob}. This vector is a result of a projection of the actual (unknown) displacement vector L onto a plane normal to the direction of observation, i.e., the plane normal to the direction from which the hologram was recorded. As such L_{ob} provides only a two-dimensional information on the object's displacement. However, if we are able to record more than one hologram, each from a different, non collinear direction, then each of these holograms can be interpreted to determine the corresponding observed displacement vectors. A set of simultaneous equations representing these observed displacement vectors can be solved for the unknown L using projection matrices, as follows.

Let us assume that r observations are made of the holographically reconstructed image. If each of these observations is made along a direction of a unit vector $\hat{K}_2^m$, which, based on (3.26), can be defined as

$$\hat{K}_2^m = \frac{\lambda}{2\pi} K_2^m, \tag{3.78}$$

where K_2^m is the average direction of the mth ($m = 1, 2, \ldots, r$) recording, then the corresponding projection matrices $\tilde{P}_{K_2}^m$ can be expressed as

$$\tilde{P}_{K_2}^m = \tilde{I} - \hat{K}_2^m \otimes \hat{K}_2^m. \tag{3.79}$$

Therefore, for each of the r observations, the observed displacements L_{ob}^m (i.e., the displacement vectors that are completely defined by two components transverse to the direction of observation) can be related to the actual (unknown) displacement vector L (i.e., the vector requiring three components for a complete definition) by means of the corresponding projection matrices $\tilde{P}_{K_2}^m$, that is,

$$L_{ob}^1 = \tilde{P}_{K_2}^1 L,$$
$$L_{ob}^2 = \tilde{P}_{K_2}^2 L,$$
$$\vdots \qquad\qquad\qquad\qquad\qquad\qquad\qquad\qquad\qquad (3.80)$$
$$L_{ob}^r = \tilde{P}_{K_2}^r L.$$

Since displacement L is common to all equations of (3.80), this system of equations can be written in a matrix form as

$$\begin{bmatrix} L_{ob}^1 \\ L_{ob}^2 \\ \vdots \\ L_{ob}^r \end{bmatrix} = \begin{bmatrix} \tilde{P}_{K_2}^1 \\ \tilde{P}_{K_2}^2 \\ \vdots \\ \tilde{P}_{K_2}^r \end{bmatrix} L, \qquad (3.81)$$

and (3.81) can be solved to obtain

$$L = \left[\sum_{m=1}^{r} \tilde{P}_{K_2}^m \right]^{-1} \left(\sum_{m=1}^{r} L_{ob}^m \right). \qquad (3.82)$$

Equation (3.82) shows that the actual object displacement L is equal to the product of the inverse of a 3×3 matrix formed by summation of all r projection matrices $\tilde{P}_{K_2}^m$ with a column matrix of the sum of all r observed displacements L_{ob}^m.

Inspection of (3.82) indicates that even two observations, along two non-collinear directions, provide enough data for the solution for L. In fact, the two observations yield four equations while only three components of the object's displacement may be specified independently. The overdetermined set of four simultaneous equations, resulting in this case, can be solved to obtain

$$L = [\tilde{P}_{K_2}^1 + \tilde{P}_{K_2}^2]^{-1}(L_{ob}^1 + L_{ob}^2), \qquad (3.83)$$

which is just the special case of (3.82) when $r = 2$. Equation (3.83) as well as (3.82) yields the components of L which have the least-squares-error.

The only parameters that are needed to evaluate (3.82) or (3.83) are the unit vectors $\hat{K}_2^m$ and the corresponding observed displacements L_{ob}^m. The unit observation vectors are easily determined from the illumination/observation geometry, as discussed in Sect. 3.2, while the observed displacements relate directly to the fringe patterns seen during reconstruction of the holographic interferograms and are determined from the spatial distributions of the fringe-locus function Ω.

It should be noted, at this time, that (3.82 and 83) apply not only to the cases when the observations are made through two or more holograms (Fig. 3.9) but also when fringe parallax is used in analysis of the holographic images (Fig. 3.10).

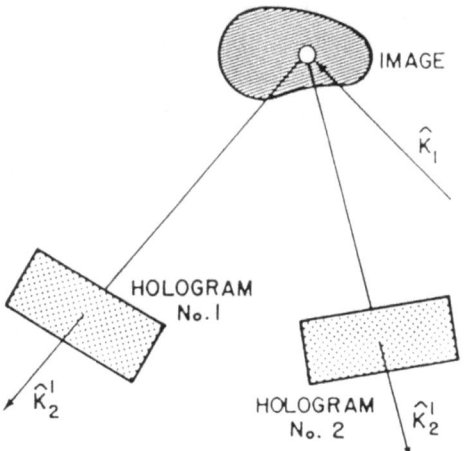

Fig. 3.9. Use of multiple holograms in analysis of a single image

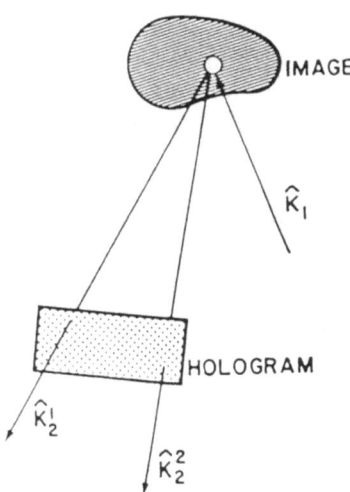

Fig. 3.10. Multiple observations of an image through a single hologram

The displacements obtained using the methods discussed in Sects. 3.3.1–3 can be subject to unwanted rigid-body motions. A procedure for determination of such motions has been worked out and is detailed in [3.36].

3.4 Determination of Strains and Rotations

Accurate determination of object's strain is of great interest in mechanics because it may correlate with the structural failure. As a direct outcome of this need, a number of methods were developed to perform the necessary diagnosis. Some of these methods are based on the principles of hologram interferometry which, with its high sensitivity, has found strong applications in analysis of small strains. This becomes obvious when we realize that object deformations, on the order of a fraction of a wavelength, will cause fringes to be formed during reconstruction of a hologram which recorded this deformation. In this way, any changes in the object's state of stress, which cause changes in its shape and/or deformation, can be measured.

Holographic strain analysis has been introduced in a classical paper by *Ennos* [3.37] in 1968. Since then, a few different applications for measurement of small strains were developed [3.38–49]. For example, *Dändliker* et al. [3.39–41] have developed an optoelectronic fringe interpolation method, *Schumann* et al. [3.42, 43] have advanced a method of fringe localization, while *Stetson* [3.44, 45] presented a method, based on a concept of a fringe-vector. The *fringe-vector method* permits determination of strain when it can be assumed to be homogeneous over a sufficiently large region of a sufficiently three-dimensional object, and is the subject of this section.

The fringe-vector method recognizes that any combination of a homogeneous strain, shear, and rotation of an object, yields fringes on the object's surface, which can be described by a single vector, the *fringe-vector*. The fringe-vector method was generalized by *Pryputniewicz* and *Stetson* [3.46–48] by introducing a procedure which accounts for variations of the sensitivity vector across the object and this, therefore, allows application of the method even in the presence of a perspective variation in the illumination or observation directions.

If an object undergoes a homogeneous deformation and/or rotation while a hologram is being recorded, then, during the reconstruction of the hologram, the object will be seen covered by a pattern of fringes that would appear to be generated along the lines of intersection of the object's surface with a set of parallel, equally spaced planes called *fringe-locus planes*. For example, holographically reconstructed three-dimensional object with flat surfaces will be seen covered by a pattern of parallel, equally spaced fringes (Fig. 3.11) as though it were intersected by a set of the fringe-locus planes. These fringe-locus planes are uniquely defined by a fringe-vector K_f, whose magnitude is inversely proportional to the spacing between these planes and whose direction is normal to them (Fig. 3.12). Therefore, observation of the fringe patterns allows determination of the fringe-vector. Since fringe-vectors are related to the holographic sensitivity vectors by a matrix transformation which describes strains, shears, and rotations of the object, multiple observations, from noncoplanar directions, of the holographically reconstructed image allow determination of the desired strain–rotation matrix.

In Sects. 3.4.1 and 2, the procedures for determination of the fringe-vectors, strains, and rotations, from holographically reconstructed images are discussed.

3.4 Determination of Strains and Rotations   55

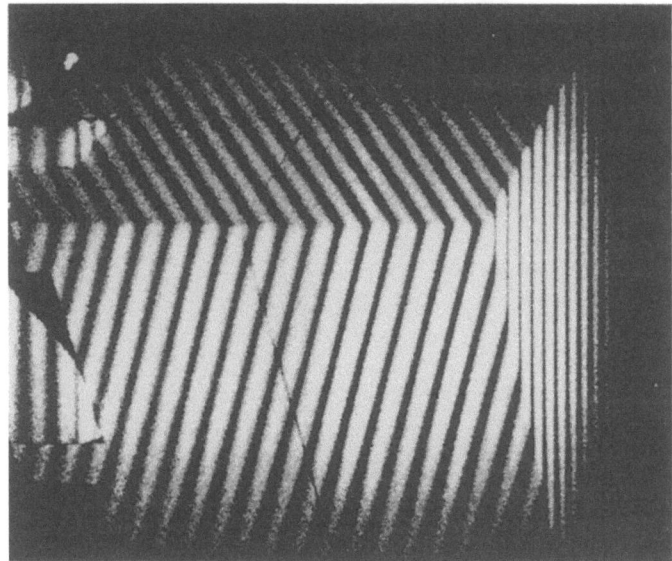

Fig. 3.11. A photograph of a reconstruction from a double-exposure hologram recording a rigid-body rotation of the object

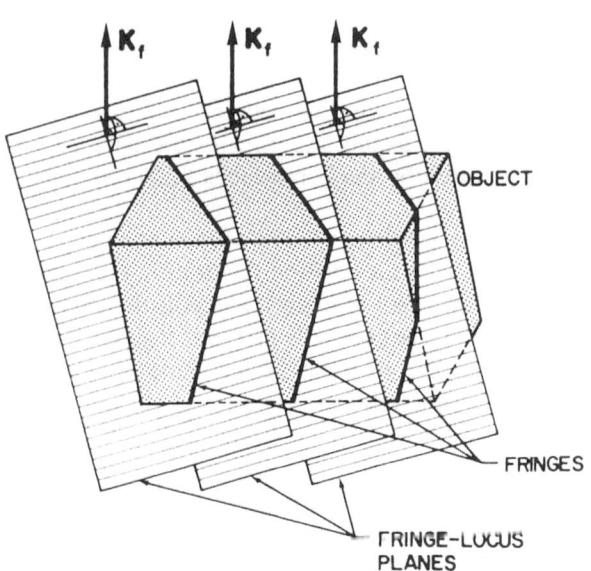

Fig. 3.12. Formation of fringes in hologram interferometry: the object's surface appears as if it were intersected by a set of equally spaced fringe-locus planes, which are represented by K_f

56   3. Quantitative Determination of Displacements and Strains from Holograms

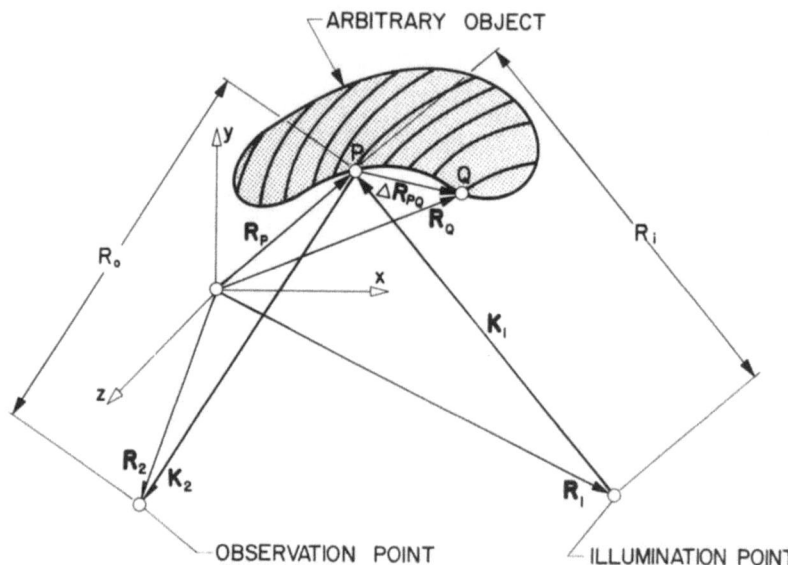

Fig. 3.13. Illumination and observation geometry for interpretation of holograms using the fringe-vector theory

3.4.1 Spatial Dependence of the Fringe-Locus Function

Let us assume that an object is illuminated with spherical wavefronts from a point defined by a position vector R_1 and observed with a spherical perspective from a point given by R_2 (Fig. 3.13).

As a result of some deformation and/or rotation between exposures of, e.g., a double-exposure hologram, the object would be reconstructed covered with fringes. These fringes will be generated by the fringe-locus function which is a scalar function of space.

Let us also assume that the value of this fringe-loucs function at a point P on the object, $\Omega(K, R_P)$, is known, where R_P is the position vector from the origin of the Cartesian coordinate system to the point of interest and K is the sensitivity vector, as defined in Sect. 3.2. Then, the fringe-locus function Ω_Q, at a nearby point Q on the object may be expressed in terms of the value of Ω and its derivatives at P, by means of the Taylor series expansion, that is,

$$\Omega(K, R_Q) = \Omega(K, R_P + \Delta R_{PQ})$$
$$= \Omega(K, R_P) + \Delta R_{PQ} \cdot K_f + \tfrac{1}{2} \Delta R_{PQ} \, \tilde{T}_f \, \Delta R_{PQ} + \cdots , \qquad (3.84)$$

where the third and the higher-order terms were neglected.

The second term on the right-side of (3.84) represents a scalar product of the differential position vector,

$$\Delta R_{PQ} = \Delta x_{PQ} \hat{i} + \Delta y_{PQ} \hat{j} + \Delta z_{PQ} \hat{k} \qquad (3.85)$$

3.4 Determination of Strains and Rotations   57

with the fringe vector, which is defined as the gradient of Ω, that is,

$$K_\mathrm{f} = K_{\mathrm{f}_x}\hat{i} + K_{\mathrm{f}_y}\hat{j} + K_{\mathrm{f}_z}\hat{k}$$

$$= \Omega^{x_\mathrm{P}}\hat{i} + \Omega^{y_\mathrm{P}}\hat{j} + \Omega^{z_\mathrm{P}}\hat{k} = \frac{\partial \Omega}{\partial x_\mathrm{P}}\hat{i} + \frac{\partial \Omega}{\partial y_\mathrm{P}}\hat{j} + \frac{\partial \Omega}{\partial z_\mathrm{P}}\hat{k}$$

$$= \nabla_\mathrm{R}\Omega = \nabla_\mathrm{R}(K \cdot L) = (\nabla_\mathrm{R} \otimes L)K + (\nabla_\mathrm{R} \otimes K)L, \qquad (3.86)$$

where ∇_R is the gradient operator in the real space,

$$\nabla_\mathrm{R} = \hat{i}\frac{\partial}{\partial x_\mathrm{P}} + \hat{j}\frac{\partial}{\partial y_\mathrm{P}} + \hat{k}\frac{\partial}{\partial z_\mathrm{P}}, \qquad (3.87)$$

and where $\otimes$ denotes the matric product.

It should be noted that (3.86) takes into account the first-order (i.e., linear) variations in Ω. The second-order variations in Ω are accounted for by the last term on the right-side of (3.84), where the *fringe-tensor* $\tilde{T}_\mathrm{f}$ represents the linear variations in K_f. Therefore, expansion of the first two terms on the right side of (3.84) leads to formulations allowing holographic numerical analysis of rigid-body motions and homogeneous deformations, while all three terms let us interpret inhomogeneous strains:

$$\Omega(\bar{K}, \bar{R}_\mathrm{Q}) = \Omega(\bar{K}, \bar{R}_\mathrm{P}) + \Delta\bar{R}_\mathrm{PQ} \cdot \bar{K}_\mathrm{f} + \tfrac{1}{2}\Delta\bar{R}_\mathrm{PQ}\,\tilde{T}_\mathrm{f}\,\Delta\bar{R}_\mathrm{PQ}. \qquad (3.88)$$

3.4.2 Rigid-Body Rotations and Homogeneous Strains

The second term on the right side of (3.84) describes a change in Ω between points P and Q. On the surface of the object, where fringes are visible, Ω is defined as

$$\Omega = K \cdot L = (K_x\hat{i} + K_y\hat{j} + K_z\hat{k}) \cdot (L_x\hat{i} + L_y\hat{j} + L_z\hat{k})$$

$$= K_xL_x + K_yL_y + K_zL_z. \qquad (3.89)$$

Differentiating (3.89), as required by (3.86), we obtain eighteen (18) terms which can be described as a sum of the products of two vectors and two matrices, that is,

$$K_\mathrm{f} = K\tilde{f} + L\tilde{g}. \qquad (3.90)$$

The two matrices on the right side of (3.90) are the sensitivity vector K and the displacement vector L, as discussed in Sects. 3.2 and 3, respectively, while the two matrices are the matrix $\tilde{g}$ accounting for the perspective variations in K and the transformation matrix $\tilde{f}$.

The matrix $\tilde{g}$ is defined as

$$\tilde{g} = \tilde{g}_2 - \tilde{g}_1 = \frac{k}{R_o}(\tilde{I} - \hat{K}_2 \otimes \hat{K}_2) - \frac{k}{R_i}(\tilde{I} - \hat{K}_1 \otimes \hat{K}_1)$$

$$= \frac{k}{R_o} \tilde{P}_{K_2} - \frac{k}{R_i} \tilde{P}_{K_1}. \tag{3.91}$$

In (3.91) R_o and R_i are radii of curvature of illumination and observation perspectives, respectively (Fig. 3.13) while $\tilde{P}_{K_1}$ and $\tilde{P}_{K_2}$ represent the projection matrices based on the unit vectors defining average illumination and observation directions, respectively.

What is of interest in (3.90) is the transformation matrix $\tilde{f}$, which can be decomposed into a matrix of strains and shears, $\tilde{\varepsilon}$, and a matrix of rotations, $\tilde{\theta}$, that is,

$$\tilde{f} = \tilde{\varepsilon} + \tilde{\theta}. \tag{3.92}$$

In order to solve (3.90) for the transformation matrix $\tilde{f}$, multiple observations of the holographically produced image must be made. For each observation, the sensitivity vector K and the fringe vector K_f must be determined that best fit data for the entire region examined. Also, multiple views are used to obtain displacement L at a point of interest on the object. For each view, the matrix $\tilde{g}$ is computed and multiplied by L to obtain perspective correction to K_f. From multiple views, a set of equations of the type of (3.90), with the matrix $\tilde{f}$ common to all, is generated and solved to obtain

$$\tilde{f} = [\tilde{K}^T \tilde{K}]^{-1} [\tilde{K}^T \tilde{K}_{fc}], \tag{3.93}$$

where

$$\tilde{K}_{fc} = \tilde{K}_f - L\tilde{g} \tag{3.94}$$

is the matrix formed by the fringe-vectors corrected for perspective. Decomposition of the matrix $\tilde{f}$, computed from (3.93), into a symmetric part $\tilde{\varepsilon}$ and the antisymmetric part $\tilde{\theta}$, that is,

$$\tilde{\varepsilon} = \tfrac{1}{2}[\tilde{f} + \tilde{f}^T] \tag{3.95}$$

and

$$\tilde{\theta} = \tfrac{1}{2}[\tilde{f} - \tilde{f}^T], \tag{3.96}$$

gives strains and shears, and rotations, respectively.

When the object deformations are not homogeneous over the entire body under study, they may, nonetheless, be approximately so *over small* regions of its surface, and projection matrices are very helpful in formulating the solution to this problem. In this case, the surface strain–rotation matrix $\tilde{f}$ is

$$\tilde{f}_s = \tilde{f} \tilde{P}_n, \tag{3.97}$$

where $\tilde{P}_n$ is the projection matrix defined as

$$\tilde{P}_n = \tilde{I} - \hat{n} \otimes \hat{n}, \tag{3.98}$$

with $\hat{n}$ being the local surface normal.

It should be noted that derivatives of the observed displacement are not generally equal to surface strains and rotations of the object. They become approximately equal to the extent that the viewing direction can be made parallel to the surface normal.

Among numerous applications of holographic methods, their use in the field of computational analysis is gaining popularity [3.50–53].

3.5 Electro-Optic Holography

Measurements of displacements of objects undergoing static and dynamic loads have been solved in a number of ways. One of these ways, based on recent advances in the phase step hologram interferometry, speckle metrology, and computer technology, allows direct electronic recording of holograms and transmission of holographic interferograms by television systems for real-time display of interference fringes [3.30, 34, 35, 54–60]. It is known as the Electro-Optic Holography (EOH) method.

3.5.1 Fundamentals of EOH

In the EOH method, also known as electronic holography, or TV holography, the interfering beams are combined by a speckle interferometer, which produces speckles large enough to be resolved by the TV camera. The output of the TV camera is fed to a system that computes and stores the magnitude and phase, relative to the reference beam, of each picture element in the image of the illuminated object.

The EOH method allows automated processing of fringes of statically and dynamically loaded objects [3.30, 34, 35, 61]. In this method, measurements of irradiances produced by mutual interference of the object and the reference fields are made electronically by a CCD camera (Fig. 3.14). Processing of this interferometric information and display of the computational results are carried out concomitantly with measurements of irradiation. The EOH method does not depend on recording of holograms in conventional media, but rather relies on electronic acquisition, processing, and display of optical interference information.

In the following subsections, principles of the EOH method are outlined and its implementation to static and dynamic measurements is presented.

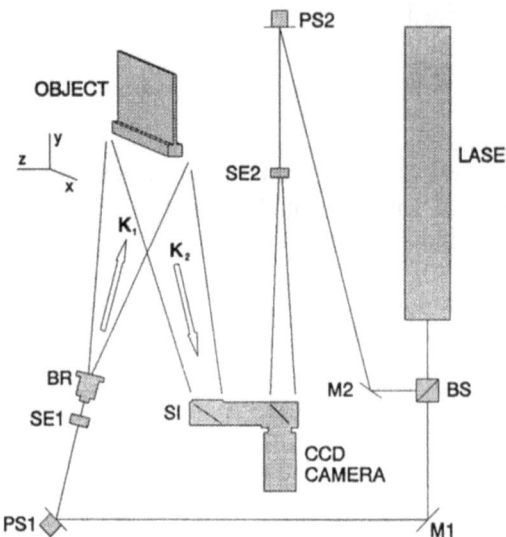

Fig. 3.14. Optical configuration of the EOH system: BS is the beamsplitter, M1 and M2 are the mirrors, PS1 and PS2 are the phase steppers, SE1 and SE2 are the spatial filter beam expander assemblies, BR is the object beam rotator, and SI is the speckle interferometer

3.5.2 Electronic Processing of Holograms

The EOH system is capable of performing either static or dynamic measurements [3.35]. Static measurements are implemented using double-exposure hologram interferometry method, while dynamic measurements are implemented by time-average method.

3.5.2.1 Static Measurements

Static measurements are characterized by recording "single-exposure" holograms of an object at two different states of stress. As a result of interference between a set of two "single-exposure" holograms, fringes form, if there are any optical path differences between the corresponding points on the object as recorded in the two holograms.

In EOH, this process is carried out by recording sequential frames of images of the object, corresponding to the two states of stress. Typically, four sequential frames are recorded, with a finite phase step – imposed on the reference beam – between each frame, for every single-exposure image of the object. In the following discussion, in order to simplify derivation of equations describing the EOH process for static measurements, the object will be initially unstressed.

The image of an unstressed (i.e., unloaded) object can be described by the irradiance distribution for the nth sequential frame, $I_n(x, y)$, at the detector array of a CCD camera in the EOH system setup, as

$$I_n(x, y) = I_o(x, y) + I_r(x, y) + 2A_o(x, y)A_r(x, y)\cos\left[\Delta\varphi(x, y) + \Delta\theta_n\right], \quad (3.99)$$

while the corresponding image of the stressed (i.e., loaded) object can be described by the irradiance distribution, $I'_n(x, y)$, as

$$I'_n(x, y) = I'_o(x, y) + I_r(x, y)$$
$$+ 2A'_o(x, y)A_r(x, y)\cos[\Delta\varphi(x, y) + \Omega(x, y) + \Delta\theta_n]. \quad (3.100)$$

In (3.99 and 100), x and y identify coordinates of the detectors in the array, I_o and I_r denote irradiances of the object and reference fields, whose amplitudes are A_o and A_r, respectively, $\Delta\varphi$ is the phase difference between the two fields, $\Delta\theta_n$ is the finite phase step imposed on the reference beam between sequential frames recording individual images, and Ω is the fringe-locus function, constant values of which define fringe loci on the object's surface.

Since I and I' are measured at known coordinates x and y, (3.99 and 100) contain four unknowns, that is, irradiances (which are squares of the amplitudes) of the two fields, the phase differences between these fields, and the fringe-locus function. The goal of the analysis is to determine Ω because it relates directly to displacements and deformations of the object.

In EOH, $\Delta\varphi$ is eliminated by recording sequentially four TV frames with an introduction of a 90° phase step between each frame. That is, $\Delta\theta_n$, appearing in (3.99 and 100), takes on the values of 0°, 90°, 180°, and 270°. This process can be represented by two sets of four simultaneous equations corresponding to (3.99 and 100), respectively, that, is,

$$I_1 = I_o + I_r + 2A_oA_r\cos\Delta\varphi, \quad (3.101)$$

$$I_2 = I_o + I_r + 2A_oA_r\sin\Delta\varphi, \quad (3.102)$$

$$I_3 = I_o + I_r - 2A_oA_r\cos\Delta\varphi, \quad (3.103)$$

$$I_4 = I_o + I_r - 2A_oA_r\sin\Delta\varphi, \quad (3.104)$$

and

$$I'_1 = I'_o + I_r + 2A'_oA_r\cos(\Delta\varphi + \Omega), \quad (3.105)$$

$$I'_2 = I'_o + I_r + 2A'_oA_r\sin(\Delta\varphi + \Omega), \quad (3.106)$$

$$I'_3 = I'_o + I_r - 2A'_oA_r\cos(\Delta\varphi + \Omega), \quad (3.107)$$

$$I'_4 = I'_o + I_r - 2A'_oA_r\sin(\Delta\varphi + \Omega), \quad (3.108)$$

where the arguments (x, y) were omitted for simplification. It should be noted that systems of equations similar to (3.101–104) and (3.105–108) could be obtained utilizing any value of the phase step, however, use of the 90° phase step results in the simplest computations.

Subtracting (3.101 and 103) and (3.102 and 104) gives, for the unstressed object, the following set of two equations:

$$(I_1 - I_3) = 4A_oA_r\cos\Delta\varphi \quad (3.109)$$

and

$$(I_2 - I_4) = 4A_o A_r \sin \Delta\varphi. \tag{3.110}$$

Following the above procedure and subtracting (3.105 and 107) and (3.106 and 3.108), a set of two equations is obtained for the stressed object, that is,

$$(I'_1 - I'_3) = 4A'_o A_r \cos(\Delta\varphi + \Omega) \tag{3.111}$$

and

$$(I'_2 - I'_4) = 4A'_o A_r \sin(\Delta\varphi + \Omega). \tag{3.112}$$

Addition of (3.109 and 111) yields

$$(I_1 - I_3) + (I'_1 - I'_3) = 4A_o A_r \cos \Delta\varphi + 4A'_o A_r \cos(\Delta\varphi + \Omega). \tag{3.113}$$

Because object displacements and deformations are small, it can be assumed that $A'_o \approx A_o$. Therefore, (3.113) becomes

$$(I_1 - I_3) + (I'_1 - I'_3) = 4A_o A_r [\cos \Delta\varphi + \cos(\Delta\varphi + \Omega)]. \tag{3.114}$$

Recognizing that $\cos(\Delta\varphi + \Omega) = \cos \Delta\varphi \cos \Omega - \sin \Delta\varphi \sin \Omega$, (3.114) can be rewritten as

$$D_1 = (I_1 - I_3) + (I'_1 - I'_3)$$
$$= 4A_o A_r [(1 + \cos \Omega) \cos \Delta\varphi - \sin \Delta\varphi \sin \Omega]. \tag{3.115}$$

In a similar way, addition of (3.110 and 112) simplifies to

$$D_2 = (I_2 - I_4) + (I'_2 - I'_4)$$
$$= 4A_o A_r [(1 + \cos \Omega) \sin \Delta\varphi + \cos \Delta\varphi \sin \Omega]. \tag{3.116}$$

Finally, addition of the squares of (3.115 and 116) yields

$$D_1^2 + D_2^2 = \{4A_o A_r [(1 + \cos \Omega) \cos \Delta\varphi - \sin \Delta\varphi \sin \Omega]\}^2$$
$$+ \{4A_o A_r [(1 + \cos \Omega) \sin \Delta\varphi + \cos \Delta\varphi \sin \Omega]\}^2,$$

which reduces to

$$D_1^2 + D_2^2 = 16 A_o^2 A_r^2 [(1 + \cos \Omega)^2 + \sin^2 \Omega],$$

wherefrom

$$D_1^2 + D_2^2 = 32 A_o^2 A_r^2 (1 + \cos \Omega). \tag{3.117}$$

Furthermore, recognizing that $(1 + \cos \Omega) = 2 \cos^2(\Omega/2)$, (3.117) can be reduced to

$$\sqrt{D_1^2 + D_2^2} = 8 A_o A_r \cos\left(\frac{\Omega}{2}\right), \tag{3.118}$$

which represents the static viewing image displayed by EOH. In (3.118) Ω is the fringe-locus function corresponding to the object's static displacements and/or

3.5 Electro-Optic Holography

deformations. The fringe-locus function can be determined by processing the sequential EOH images as described below.

In order to obtain data from the EOH images, we will again employ (3.109–113) and follow the procedure used to derive (3.117). The result of this procedure is

$$D_3 = (I_1 - I_3) - (I'_1 - I'_3)$$
$$= 4A_o A_r [(1 - \cos \Omega) \cos \Delta\varphi + \sin \Delta\varphi \sin \Omega], \tag{3.119}$$

$$D_4 = (I_2 - I_4) - (I'_2 - I'_4)$$
$$= 4A_o A_r [(1 - \cos \Omega) \sin \Delta\varphi - \cos \Delta\varphi \sin \Omega], \tag{3.120}$$

and

$$D_3^2 + D_4^2 = 32 A_o^2 A_r^2 (1 - \cos \Omega). \tag{3.121}$$

Subtracting (3.121) from (3.117) we obtain

$$D = (D_1^2 + D_2^2) - (D_3^2 + D_4^2) = 32 A_o^2 A_r^2 (1 + \cos \Omega) - 32 A_o^2 A_r^2 (1 - \cos \Omega)$$

or

$$D = 64 A_o^2 A_r^2 \cos \Omega \tag{3.122}$$

Starting with (3.109–112), we can also determine

$$N_1 = (I_1 - I_3) + (I'_2 - I'_4)$$
$$= 4A_o A_r [(1 + \sin \Omega) \cos \Delta\varphi + \sin \Delta\varphi \cos \Omega], \tag{3.123}$$

$$N_2 = (I_2 - I_4) - (I'_1 - I'_3)$$
$$= 4A_o A_r [(1 + \sin \Omega) \sin \Delta\varphi - \cos \Delta\varphi \cos \Omega], \tag{3.124}$$

$$N_3 = (I_1 - I_3) - (I'_2 - I'_4)$$
$$= 4A_o A_r [(1 - \sin \Omega) \cos \Delta\varphi - \sin \Delta\varphi \cos \Omega], \tag{3.125}$$

$$N_4 = (I_2 - I_4) + (I'_1 - I'_3)$$
$$= 4A_o A_r [(1 - \sin \Omega) \sin \Delta\varphi + \cos \Delta\varphi \cos \Omega], \tag{3.126}$$

$$N_1^2 + N_2^2 = 32 A_o^2 A_r^2 (1 + \sin \Omega), \tag{3.127}$$

$$N_3^2 + N_4^2 = 32 A_o^2 A_r^2 (1 - \sin \Omega), \tag{3.128}$$

and

$$N = (N_1^2 + N_2^2) - (N_3^2 + N_4^2) = 64 A_o^2 A_r^2 \sin \Omega. \tag{3.129}$$

Finally, dividing (3.129) by (3.122), we obtain

$$\frac{N}{D} = \frac{64 A_o^2 A_r^2 \sin \Omega}{64 A_o^2 A_r^2 \cos \Omega},$$

from which it follows that

$$\Omega = \tan^{-1}\left(\frac{N}{D}\right). \tag{3.130}$$

It should be noted that Ω, computed from (3.130), is a spatial function that depends on coordinates x and y. Therefore, its values are determined for every coordinate pair (x, y) in the object space. Once the values of Ω are determined, they can be used to compute object displacements using procedures discussed in Sect. 3.3.

3.5.2.2 Dynamic Measurements

Application of EOH to dynamic measurements is made based on the time-average hologram interferometry. To facilitate this presentation, the time-average recording of a sinusoidally vibrating object will be considered. For this case, the irradiance distribution for the nth sequential frame, I_{t_n}, can be represented by a relationship similar to those shown in (3.99 and 100), that is,

$$I_{t_n} = I_{t_o} + I_r + 2A_{t_o}A_r \cos(\Delta\varphi_t + \Delta\theta_n)M(\Omega_t). \tag{3.131}$$

In (3.131) the arguments (x, y) were omitted for simplification, subscript t indicates time varying parameters, M is the characteristic fringe function [3.62] that modulates the interference of the two fields due to the object's motion, Ω_t is fringe-locus function defining fringe loci on the surface of a vibrating object, and other parameters are as defined for (3.99 and 100).

The EOH system is capable of operating in either a viewing mode, utilized for visual examination of the interference patterns, or a data mode, employed for quantitative analysis of vibrating objects. The viewing mode produces an 8-bit image, while the data mode produces a 16-bit image. These images are produced by look-up tables resident in the EOH system. If the viewing mode is selected, the resident look-up table for the viewing mode is loaded into the operating system and, following procedure of Sect. 3.5.2, it can be shown that the viewing image can be represented as

$$\sqrt{(I_1 - I_3)^2 + (I_2 - I_4)^2} = 4A_o A_r |M(\Omega_t)|. \tag{3.132}$$

The image described by (3.132) is displayed live on a TV monitor, and it can be stored in processor memory at any time. This storage can be of two types, as selected by menu options. If the image is to be recalled in the future for visual observation, then an 8-bit image is stored and occupies approximately one-quarter megabyte of memory – this is image storage.

If the image is to be processed quantitatively, then the look-up table for the data mode is loaded into the operating system and produces a data image which can be represented by

$$(I_1 - I_3)^2 + (I_2 - I_4)^2 = 16 I_o I_r M^2(\Omega_t). \tag{3.133}$$

3.5 Electro-Optic Holography

The result represented by (3.133) is stored as the 16-bit data image and occupies one-half megabyte of memory – this is data storage.

Equations (3.132 and 133) indicate that the viewing and the data images are proportional to the characteristic function and to the square of the characteristic function, respectively. The characteristic function is determined by the object's temporal motion, and for sinusoidal vibrations, assuming that the vibration period is much shorter than the TV framing time,

$$M[\Omega_t(x,y)] = J_0[\Omega_t(x,y)], \tag{3.134}$$

where J_0 is the zero-order Bessel function of the first kind. Therefore, (3.132 and 133) become

$$\sqrt{(I_1 - I_3)^2 + (I_2 - I_4)^2} = 4A_o A_r |J_0(\Omega_t)| \tag{3.135}$$

and

$$(I_1 - I_3)^2 + (I_2 - I_4)^2 = 16 I_o I_r J_0^2(\Omega_t), \tag{3.136}$$

respectively. Equation (3.135) results in a viewed image that is modulated by a system of fringes described by the zero-order Bessel function of the first kind, while (3.136) shows that the data image is modulated by the square of this function. Thus, centers of the dark fringes are located at those points on the object's surface where $J_0(\Omega_t)$ equals zero, as shown in Fig. 3.15. This figure indicates that the zero-order fringe is much brighter than the other J_0 fringes. Since the zero-order fringes represent the stationary points on the vibrating object they allow easy identification of nodes. The brightness of other fringes

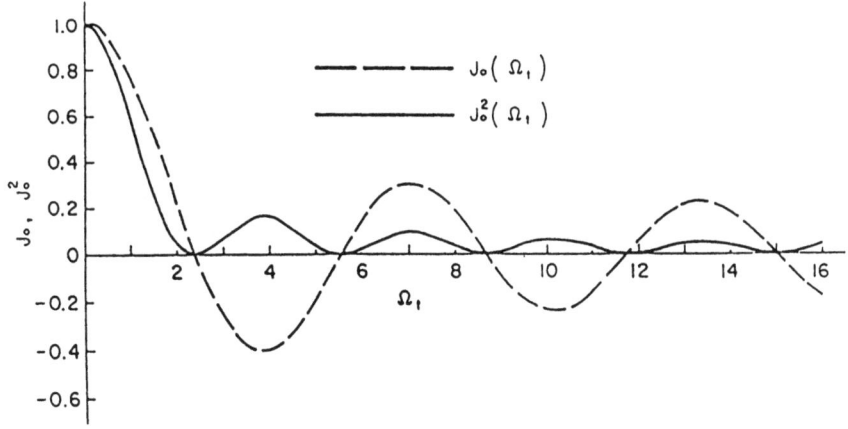

Fig. 3.15. The zero-order Bessel function of the first kind and its square, defining locations of the centers of dark fringes seen during reconstruction of the time-average holograms of the vibrating objects

decreases with increasing fringe order and can be directly related to the mode shapes. It should be noted that higher order zeros are nearly equally spaced giving the J_0 function an almost periodic nature that is utilized in quantitative interpretation of images recorded by the EOH system as discussed in Sect. 3.5.2.2.1.

In the EOH system, the data provided by the CCD camera are processed to produce results shown by (3.136) for every pixel in the image frame at the rate of 30 frames per second. Each frame contains 512×480 8-bit numbers so that each image consists of 245 760 points. For visual examination of the vibration modes, time-average hologram images corresponding to (3.135) are displayed on the TV monitor. These images are generated concomitantly by the pipeline processor of the EOH system. To produce data suitable for quantitative analysis of time-average holograms, 16-bit images represented by (3.136) are stored. These data are stored in two 8-bit bytes per pixel and produce a frozen image which can be displayed on the TV monitor one byte at a time, that is, either as a high-byte image or a low-byte image.

3.5.2.2.1 Determination of the Fringe-Locus Function for the Vibrating Object. To interpret electronically recorded time-average holograms quantitatively, the argument of the J_0^2 function, appearing in (3.136), must be determined. One method to determine this argument, suitable for the time-average images recorded by EOH, uses the fact that it is possible to shift J_0 fringes in a manner similar to that in which phase modulation shifts consinusoidal fringes in conventional double-exposure hologram interferometry [3.59]. In time-average holography, this is done by modulating the phase of either the object or the reference beams sinusoidally at the same frequency and phase as the object vibration. Such a process can be represented mathematically by addition of a phasor bias, B, to the argument of the Bessel function, resulting in the characteristic function

$$M[\Omega_t(x, y), B] = J_0[\Omega_t(x, y) - B]. \tag{3.137}$$

For purposes of analysis, the object must be made to vibrate in only one vibration mode at a time so that the motions of its various parts are either in or out of phase with one another. If the phase of the sinusoidal beam modulation is adjusted to coincide with that of the object vibration, the phasor bias becomes a simple additive term within the argument of the Bessel function, that is,

$$M[\Omega_t(x, y), B] = J_0[\Omega_t(x, y) - B]. \tag{3.138}$$

Therefore, (3.136) becomes

$$[I_1(x, y) - I_3(x, y)]^2 + [I_2(x, y) - I_4(x, y)]^2$$
$$= 16I_o(x, y)I_r(x, y)J_0^2[\Omega_t(x, y) - B]. \tag{3.139}$$

For comparison, the general equation representing the irradiance I_{h_t} of an image reconstructed from a time-average hologram is

$$I_{h_t}(x, y) = I_{a_t}(x, y) + I_{m_t}(x, y) J_0^2[\Omega_t(x, y) - B] \tag{3.140}$$

where I_{a_t} represents local average background irradiance from scattered light and I_{m_t} is the local maximum irradiance. Therefore, (3.139) is the special case of (3.140) with

$$I_{h_t}(x, y) = [I_1(x, y) - I_3(x, y)]^2 + [I_2(x, y) - I_4(x, y)]^2, \tag{3.141}$$

$$I_{a_t}(x, y) = 0, \tag{3.142}$$

and

$$I_{m_t}(x, y) = 16 I_o(x, y) I_r(x, y). \tag{3.143}$$

The output of the processor in the data mode, I_{h_t}, is stored in the host computer for different values of B, while I_{a_t}, I_{m_t}, and Ω_t constitute three unknowns, and the goal of the analysis is to determine Ω_t. Unfortunately, the Bessel function is not separable in terms of Ω_t and B, so a straightforward solution is not possible. However, the nearly periodic nature of the J_0 function allows an approximate solution for the fringe-locus function. This approximate solution recognizes that (3.139) is similar to the general equation for the irradiance distribution, I_h, for an image reconstructed from a conventional double-exposure hologram with cosinusoidal fringes, that is,

$$I_h(x, y) = I_a(x, y) + I_m(x, y) \cos^2[\Omega(x, y) - B], \tag{3.144}$$

where J_0^2 in (3.140) has been replaced by $\cos^2$ and Ω_t has been replaced by Ω.

Examination of (3.144) shows that it, just like (3.140), also has three unknowns: I_a, I_m, and Ω. However, the $\cos^2[\Omega(x, y) - B]$ term, appearing in (3.144), unlike the $J_0^2[\Omega_t(x, y) - B]$ term of (3.140), is separable in its component arguments. To facilitate solution for Ω, (3.144) is rewritten as

$$I_h(x, y) = I'_a(x, y) + I'_m(x, y) \cos[2\Omega(x, y) - 2B], \tag{3.145}$$

where

$$I'_a(x, y) = I_a(x, y) + \frac{I_m(x, y)}{2} \tag{3.146}$$

and

$$I'_m(x, y) = \frac{I_m(x, y)}{2}. \tag{3.147}$$

With three values of B, three simultaneous equations of the type of (3.145) can be solved uniquely for Ω. The three simultaneous equations are

$$I_{h_1}(x, y) = I'_a(x, y) + I'_m(x, y) \cos[2\Omega(x, y)], \tag{3.148}$$

$$I_{h_2}(x, y) = I'_a(x, y) + I'_m(x, y) \cos[2\Omega(x, y) - 2B], \tag{3.149}$$

$$I_{h_3}(x, y) = I'_a(x, y) + I'_m(x, y) \cos[2\Omega(x, y) + 2B], \tag{3.150}$$

corresponding to the zero-, positive-, and negative-shifts, respectively. Solution of (3.148–150) yields

$$\Omega(x, y) = \frac{1}{2}\tan^{-1}\left(\frac{1 - \cos(2B)}{\sin(2B)} \times \frac{I_{h_3}(x, y) - I_{h_2}(x, y)}{2I_{h_1}(x, y) - I_{h_2}(x, y) - I_{h_3}(x, y)}\right). \tag{3.151}$$

If the three irradiance distributions $I_{h_{t_1}}(x, y)$, $I_{h_{t_2}}(x, y)$, and $I_{h_{t_3}}(x, y)$, corresponding to three time-average holograms, are substituted into (3.151) the result is $\Omega_{t_{\text{approx}}}(x, y)$. This value of $\Omega_{t_{\text{approx}}}$ differs from the correct argument, Ω_t, of the J_0 function, because of inequality between the J_0^2 and $\cos^2$ functions, and should be expressed as

$$\Omega_{t_{\text{approx}}}(x, y) = \Omega_t(x, y) + \varepsilon(x, y), \tag{3.152}$$

where ε is the error representing this difference.

Equation (3.151) yields values of $\Omega_{t_{\text{approx}}}$ modulo 180°. By adding or subtracting 180°, depending on the sign of the numerator in (3.151), whenever the denominator is negative, $\Omega_{t_{\text{approx}}}$ can be obtained modulo 360°. The image can be searched by the computer to locate discontinuities to define areas where the missing multiples of the 360° should be added to unwrap function $\Omega_{t_{\text{approx}}}$. By further identifying pixels within the zero-order fringe, an overall level shift can be applied to make those pixels have values between $\pm 180°$.

Errors ε can be computed for any value of Ω_t for specific values of B to create a look-up table. This look-up table is used to correct the values computed from (3.151) that have been unwrapped and level shifted. In this way, vibratory deformations can be obtained from time-average hologram reconstructions with little more mathematical computation than is required for static deformations. Once the correct values of Ω_t are determined, they can be used in any one of the equations for quantitative interpretation of time-average holograms [3.63–65] (Sect. 3.3 and Chap. 8).

3.5.3. Representative Applications of EOH

Representative results obtained using EOH are displayed in Figs. 3.16 to 19. All interferograms shown in these figures were produced electronically and were recorded by video printing the displays on the TV monitor. All wire frame displacement plots were based on the irradiances measured from the sets of three holograms corresponding to the zero, the positive, and the negative bias modulations added to the object beam at the object's vibration frequency.

It should be noted that because of the plate's aspect ratio only $(250 \times 400 =)$ 100 000 pixels out of the total available of $(512 \times 480 =)$ 245 760 pixels were used in the interpretation of the images produced by the EOH. Furthermore, in order to produce visually acceptable plots, wire frame grid of 5 mm × 5 mm for representation of displacements was selected. In this way, displacements at only

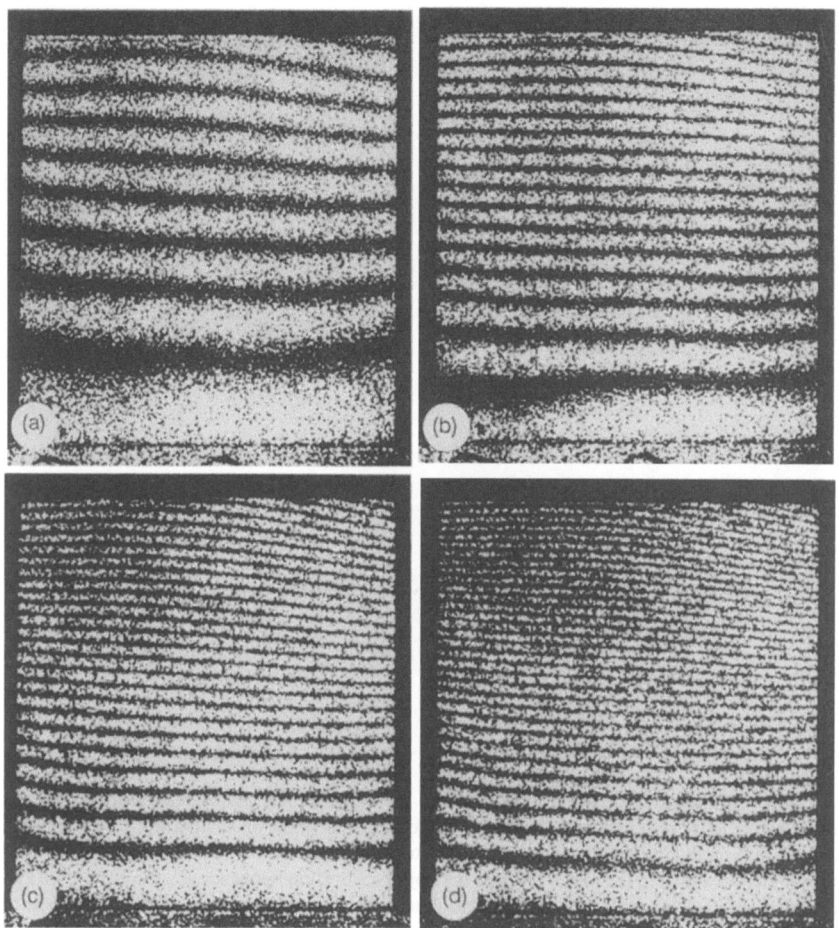

Fig. 3.16a–d. EOH images of the double-exposure holograms of the cantilever plate: the static load increases from (**a**) to (**d**). The image in (**d**) shows 33 fringes along the plate

a very small number of points ($24 \times 28 = 672$), out of the total of 100 000 points considered on the object, were displayed.

Figure 3.16 depicts the cosinusoidal fringe patterns of the statically loaded cantilever plate, under four different loads. The electronic hologram corresponding to the maximum load used in this study shows 33 fringes along the plate.

Figure 3.17 exhibits the cantilever plate vibrating in its second torsional mode at 966 Hz. More specifically, Fig. 3.17a shows the time-average hologram produced by the EOH system without the bias modulation of the object beam, whereas Figs. 3.17b and c display the holograms with the equal positive and negative bias modulations added to the object beam at the object's vibration frequency, respectively. Note the symmetric location of the zero-order fringe in

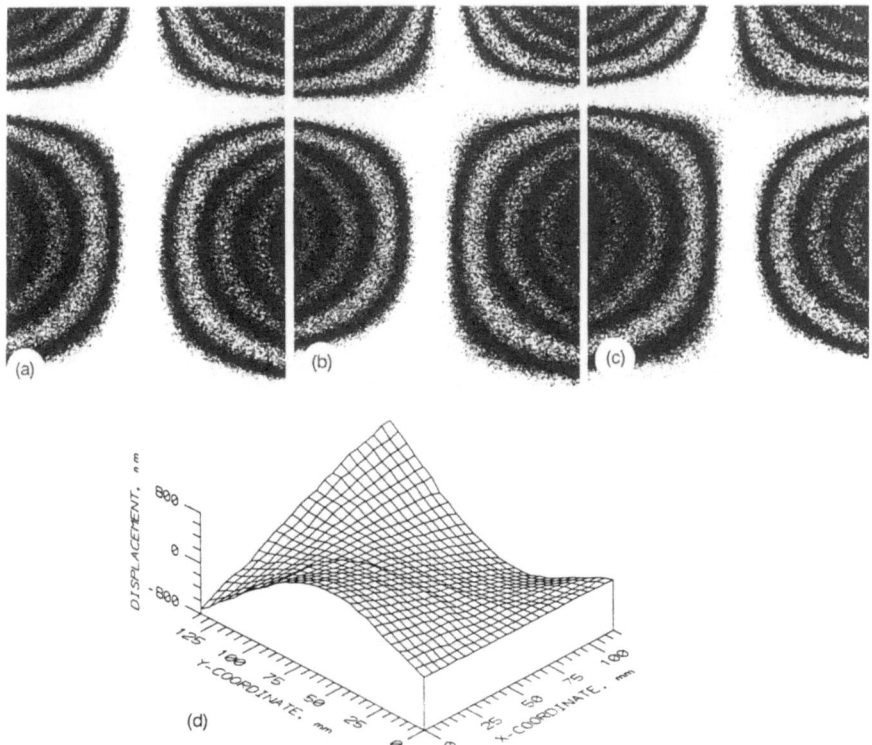

Fig. 3.17a–d. Cantilever plate vibrating at 966 Hz: (a) image of the plate and the fringe pattern produced by the EOH system during the time-average recording of the plate vibrating at its resonance without the bias modulation of the object beam, (b) the same plate and the vibration as in (a) but with a bias modulation added to the object beam at the object's vibration frequency, (c) the same plate and the vibration as in (a) but with the bias modulation added to the object beam at 180° with respect to (b), (d) wire frame representation of displacements computed from the images shown in (a)–(c)

Fig. 3.17a and the offsets in the symmetry, due to the bias modulation, in Figs. 3.17b and c.

Based on the irradiance values from the corresponding points in Figs. 3.17a–c, displacements were computed as a function of x and y coordinates on the vibrating plate. These displacements are shown in Fig. 3.17d and correlate well with the image displayed.

Figures 3.18 and 19 present a comparison between the wire-frame displacements obtained by the EOH and those computed by the finite element method (FEM). The comparison of the EOH and the FEM results shows good correlation.

3.6 Conclusions

This chapter outlines procedures for a quantitative interpretation of holographically recorded images. Implementation of these procedures is illustrated by

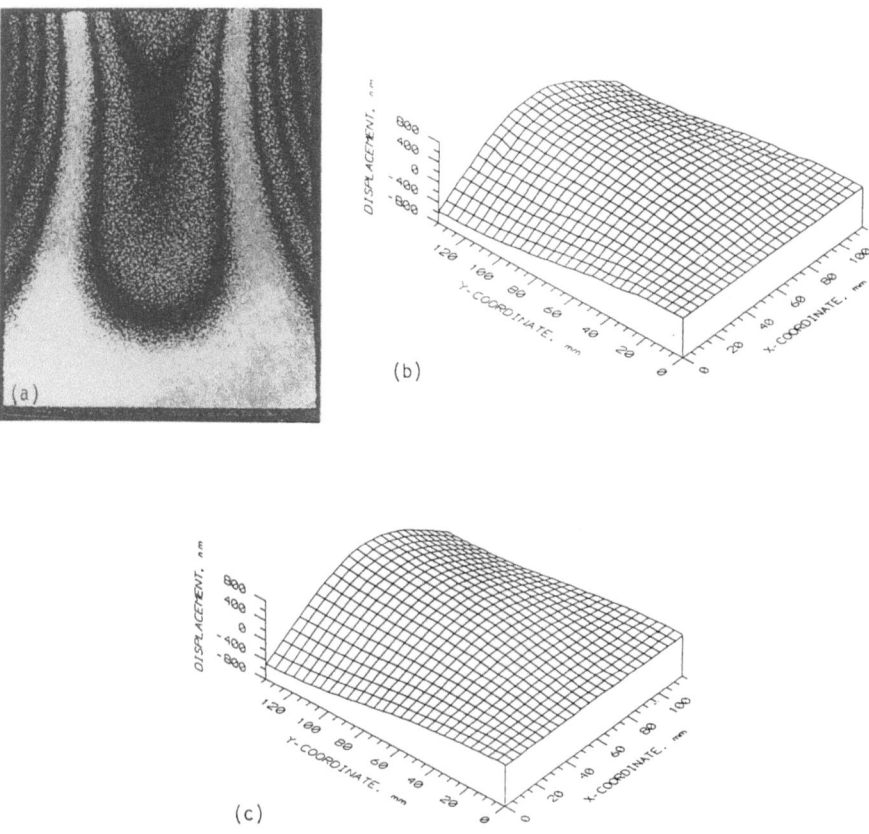

Fig. 3.18a–c. Comparison of the EOH and the FEM results for the cantilever plate vibrating at 1082 Hz: (**a**) the EOH image, (**b**) the EOH displacements, (**c**) the FEM displacements

means of the Electro-Optic Holography (EOH) system allowing electronic recording, storage, processing, and display of holographic interferograms in real-time. Using this system, the displacements can be extracted from the electronically recorded hologram by a method analogous to optical fringe shifting.

Using the electro-optic holography the results are obtained in a truly automated manner. The interferometric information is recorded at the rate of 30 frames per second, it is processed in a pipeline fashion, and produces results which have very high spatial density – currently up to 512×480 points per frame. These results correlate well with the holographically produced fringe patterns, and they also correlate well with the FEM predictions of the object's load–displacement characteristics.

Currently, work is underway to merge, or unify, within the host computer, the displacements determined by the EOH system with the computational

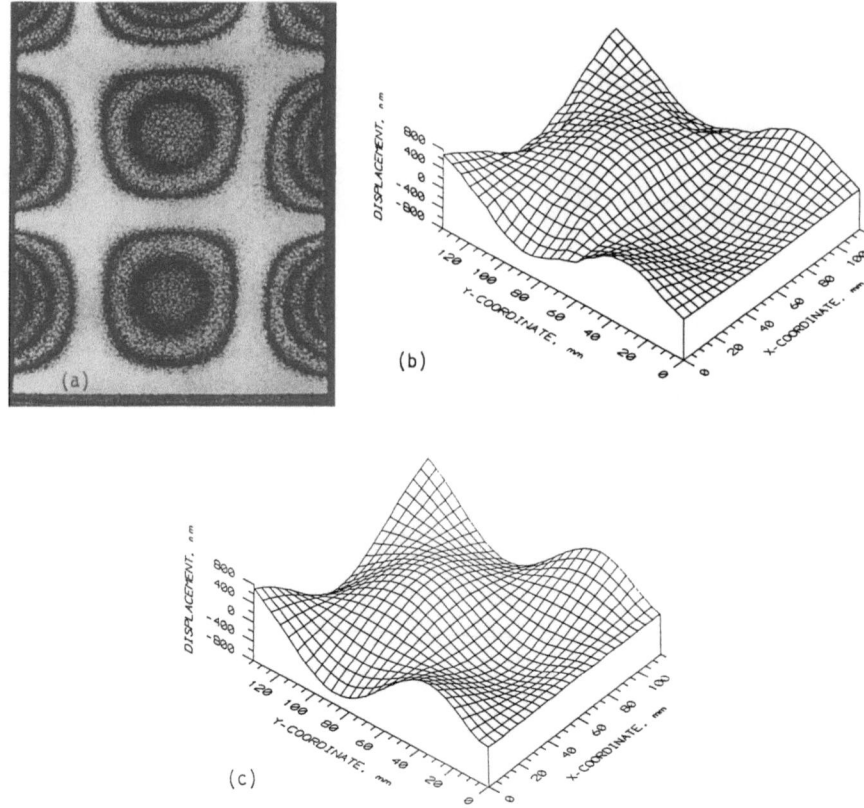

Fig. 3.19a–c. Comparison of the EOH and the FEM results for the cantilever plate vibrating at 3148 Hz: (a) the EOH image, (b) the EOH displacements, (c) the FEM displacements

procedures of FEM. This feat will result in a hybrid system which will allow automated, quantitative analysis of structural deformations.

References

3.1 R.J. Pryputniewicz, K.A. Stetson: Fundamentals and applications of laser speckle and hologram interferometry. Center for Holographic Studies and Laser Technology, Department of Mechanical Engineering, Worcester Polytechnic Institute, Worcester, MA (1980)
3.2 W. Schumann: *Exp. Mech.* **13**, 225–231 (1973)
3.3 R.J. Pryputniewicz: *Proc. SPIE*, **243**, 158–164 (1980)
3.4 K.A. Stetson: *J. Opt. Soc. Am.*, **71**, 1248–1257 (1981)
3.5 W. Schumann, J.-P. Zürcher, D. Cuche: *Holography and Deformation Analysis*, Springer Ser. Opt. Sci., Vol. 46 (Springer, Berlin, Heidelberg 1985)
3.6 R.J. Pryputniewicz: *J. Opt. Soc. Am.*, **67**, 1351–1353 (1977)

3.7 R.J. Pryputniewicz, K.A. Stetson: *Appl. Opt.* **19**, 2201–2205 (1980)
3.8 E.B. Aleksandrv, A.M. Bonch-Bruevich: *Sov. Phys. Tech. Phys.*, **12**, 258–265 (1967)
3.9 J.E. Sollid: *Appl. Opt.* **8**, 1587–1595 (1969)
3.10 K. Shibayama, H. Uchiyama: *Appl. Opt.*, **10**, 2150–2154 (1971)
3.11 E. Archbold, A.E. Ennos: *Optica Acta*, **19**, 253–271 (1972)
3.12 S.K. Dhir, J.P. Sikora: *Exp. Mech.* **12**, 323–327 (1972)
3.13 P.M. Boone, L.C. De Backer: *Optik* **37**, 61–81 (1973)
3.14 K.A. Stetson: *J. Opt. Soc. Am.* **64**, 1–10 (1974)
3.15 V. Fossati Bellani, A. Sona: *Appl. Opt.* **13**, 1337–1341 (1974)
3.16 L. Ek, K. Biedermann: *Appl. Opt.* **16**, 2535–2542 (1977)
3.17 R.J. Pryputniewicz, W.W. Bowley: *Appl. Opt.* **17**, 1748–1756 (1978)
3.18 C.M. Vest: *Holographic Interferometry* (Wiley, New York 1979)
3.19 P. Hariharan, B.F. Oreb, N. Brown: *Appl. Opt.* **22**, 876–880 (1983)
3.20 W. Jüptner, T. Kreis, H. Kreitlow: *Proc. SPIE*, **398**, 22–29 (1983)
3.21 D.W. Robinson: *Appl. Opt.* **22**, 2169–2176 (1983)
3.22 J.B. Schemm, C.M. Vest: *Appl. Opt.* **22**, 2850–2853 (1983)
3.23 P. Hariharan: *Optical Holography: Principles, Techniques, and Applications* (Cambridge Univ. Press, Cambridge 1984)
3.24 R.J. Pryputniewicz: *Proc. SPIE* **673**, 250–257 (1986)
3.25 T. Kreis: *J. Opt. Soc. Am.* A**3**, 847–855 (1986)
3.26 W. Osten, R. Höfling, J. Saedler: *Proc. SPIE* **863**, 105–113 (1987)
3.27 D.R. Matthys, T.D. Dudderar, J.A. Gilbert: *Exp. Mech.* **28**, 86–91 (1988)
3.28 R. Jones, C. Wykes: *Holographic and Speckle Interferometry* (Cambridge Univ. Press, Cambridge 1989)
3.29 W. Osten, R.J. Pryputniewicz, G.T. Reid, H. Rottenkolber (eds): *Automatic Processing of Fringe Patterns* (Akademie, Berlin 1989)
3.30 R.J. Pryputniewicz: *SPIE Institutes for Advanced Optical Technologies*, IS**8**, 215–246 (1990)
3.31 G.M. Brown: *Proc. SPIE* **1553**, 204–208 (1991)
3.32 T. Kreis, W. Jüptner: *Proc. SPIE* **1553**, 263–273 (1991)
3.33 Y.I. Ostrovsky, V.P. Shchepinov, V.V. Yakovlev: *Holographic Interferometry in Experimental Mechanics*, Springer Ser. Opt. Sci., Vol. 60 (Springer, Berlin, Heidelberg 1991)
3.34 R.J. Pryputniewicz, K.A. Stetson: *Proc. SPIE* **1162**, 456–467 (1989)
3.35 R.J. Pryputniewicz: *Proc. SPIE* **1554B**, 790–798 (1991)
3.36 R.J. Pryputniewicz: *Appl. Opt.* **18**, 1442–1444 (1979)
3.37 A.E. Ennos: *J. Phys. E* **1**, 731–746 (1968)
3.38 A.E. Ennos: Strain measurement, in *Holographic Nondestructive Testing*, ed. by R.K. Erf (Academic, New York 1974) pp. 275–287
3.39 R. Dändliker, B. Ineichen, F.M. Mottier: *Opt. Commun.* **9**, 412–416 (1973)
3.40 R. Dändliker, B. Ineichen: *Proc. SPIE* **99**, 90–98 (1976)
3.41 R. Thalmann, R. Dändliker: *Appl. Opt.* **26**, 1964–1971 (1987)
3.42 W. Schumann, M. Dubas: *Optica Acta* **22**, 807 (1975)
3.43 W. Schumann, M. Dubas: *Holographic Interferometry*, Springer Ser. Opt. Sci., Vol. 16 (Springer, Berlin, Heidelberg 1979)
3.44 K.A. Stetson: *Appl. Opt.* **14**, 272–273 (1975)
3.45 K.A. Stetson: *Appl. Opt.* **14**, 2256–2259 (1975)
3.46 R.J. Pryputniewicz, K.A. Stetson: *Appl. Opt.* **5**, 725–728 (1976)
3.47 R.J. Pryputniewicz: *Appl. Opt.* **17**, 3613–3618 (1978)
3.48 K.A. Stetson: *Exp. Mech.* **7**, 273–275 (1981)
3.49 S. Toyooka, N. Nishida, J. Takezaki: *Opt. Engr.* **28**, 55–60 (1989)
3.50 S. Jüptner, T. Bischof: *Proc. 10th Int'l Invitational UFEM Symposium*, ed. by R.J. Pryputniewicz, (Worcester Polytechnic Institute, Worcester, MA 1991) pp. 237–247
3.51 P. Hariharan: *Proc. 10th Int'l Invitational UFEM Symposium*, ed. by R.J. Pryputniewicz (Worcester Polytechnic Institute, Worcester, MA 1991) pp. 471–490

3.52 G.M. Brown, J. Zhang: *Proc. 10th Int'l. Invitational UFEM Symposium*, ed. by R.J. Pryputniewicz (Worcester Polytechnic Institute, Worcester, MA 1991) pp. 493–504
3.53 R.J. Pryputniewicz, D.G. Grabbe: *Proc. 10th Int'l. Invitational UFEM Symposium*, ed. by R.J. Pryputniewicz (Worcester Polytechnic Institute, Worcester, MA 1991) pp. 505–532
3.54 J.E. Berrang: *Bell System Techn. J.* **49**, 879–887 (1970)
3.55 A. Makovski, S.D. Ramsey, L.F. Schaefer: Appl. Opt. **10**, 2711–2727 (1971)
3.56 K.A. Stetson, W.R. Brohinsky: *Appl. Opt.* **24**, 3631–3637 (1985)
3.57 S. Nakadate: *Appl. Opt.* **25**, 4162–4167 (1986)
3.58 K.A. Stetson, W.R. Brohinsky: *Proc. SPIE* **746**, 44–51 (1987)
3.59 K.A. Stetson, W.E. Brohinsky: *J. Opt. Soc. Am. A* **5**, 1472–1476 (1988)
3.60 K. Creath: in *Progress in Optics* **26**, 349–393 (North-Holland, Amsterdam 1988)
3.61 R.J. Pryputniewicz: *Proc. Spring Conf. on Exp. Mech.*, SEM, Bethel, CT (1991) pp. 912–919
3.62 K.A. Stetson: *J. Opt. Soc. Am.* **60**, 1378–1384 (1970)
3.63 R.J. Pryputniewicz: *Opt. Engr.* **24**, 843–848 (1985)
3.64 R.J. Pryputniewicz: *Proc. SPIE* **599**, 54–62 (1985)
3.65 R.J. Pryputniewicz: in *Optical Metrology*, ed. by O.D.D. Soares, NATO ASI Series (Martinus Nijhoff, Dodrecht 1987) pp. 296–316

4. Two-Reference-Beam Holographic Interferometry

R. Dändliker

Institute of Microtechnology, University of Neuchâtel, CH-2000 Neuchâtel, Switzerland

Since its invention in 1965, holographic interferometry has found many applications, mainly in deformation and vibration analysis of solid objects, but also in the investigation of transparent objects with refractive-index changes and in holographic contouring. In numerous works, all kind of methods for the observation and interpretation of the interference fringe pattern have been elaborated. However, in spite of many propositions for quantitative evaluation of holographic interferograms, the major applications of holographic interferometry in industry are still based on qualitative interpretation of the interferograms. The historical reasons are: insufficient accuracy and density of data obtained from interference fringe counting, lack of automated fringe evaluation methods which avoid time consuming manual evaluation, and furthermore quantitative methods often required complicated optical arrangements to simplify the calculation and interpretation.

The introduction of electronic fringe interpolation techniques had an important impact on interferometry, because it offers high accuracy and automated data acquisition. However, the application of electronic interference phase measurement to double-exposure holographic interferometry requires independent access to the phase of the two interfering reconstructions. This is only possible when the two exposures are recorded with two different reference beams.

In this chapter we will present the special aspects of two-reference-beam holographic interferometry, assuming that the reader is familiar with the standard techniques of holographic interferometry discussed by Hariharan (Chap. 2) and Pryputniewicz (Chap. 3). In Sect. 4.1, the basics of the electronic interference phase measurement and the recording of double exposure holograms with two reference beams will be introduced. Section 4.2 deals with the statistical aspects of interferometry with speckle fields, as obtained from diffusely scattering objects. In view of the high-accuracy fringe evaluation, Sect. 4.3 is devoted to the different sources of error due to two-reference-beam holography and speckle effects. Section 4.4 describes some practical examples of what can be accomplished with two-reference-beam holographic interferometry.

4.1 Electronic Interference Phase Measurement

The introduction of electronic fringe-interpolation techniques had an important impact on interferometry, because it offers high accuracy and automated data acquisition [4.1]. These methods are based on the principle of shifting the relative phase between the interfering wave fields, either linearly in time by introducing a frequency offset (heterodyne) or stepwise (quasi-heterodyne or phase-shifting). These techniques can readily be applied to real-time holographic interferometry. However, the application of electronic interference phase measurement in double-exposure holographic interferometry requires a setup with two reference beams [4.2] to have independently access to the relative phase between the two interfering reconstructions, which allows shifting the fringes in the interferogram. In the following, we will first review the basic concepts of electronic interference phase measurements by phase-shifting and heterodyne detection and then introduce the principle of two-reference-beam holographic interferometry.

4.1.1 Phase-Shifting and Heterodyne Detection

Two different approaches for quasi-heterodyne (phase shifting) phase measurement are known. In the *phase-step* method, the local intensity of the interference pattern

$$I(x) = a(x)\{1 + m(x)\cos[\phi(x) + \psi_n]\} \tag{4.1}$$

is sampled at fixed-phase steps ψ_n. At least three intensity measurements $I_n(x)$ have to be carried out to determine all the three unknowns, i.e. the local mean intensity $a(x)$, the fringe contrast $m(x)$ and the interference phase $\phi(x)$. The *integrating-bucket* method is intended primarily for use with charge-coupled devices, where the optical power is integrated by the detector. The relative phase is varying linearly in time and the sampled intensity is integrated from $(\psi_n - \Delta\psi/2)$ to $(\psi_n + \Delta\psi/2)$, which yields

$$\bar{I}_n(x) = a(x)\{1 + \text{sinc}(\Delta\psi/2)\, m(x)\cos[\phi(x) + \psi_n]\}. \tag{4.2}$$

Thus, the only effect of integrating the intensity compared to the phase-step method is a reduction of the fringe modulation by the factor $\text{sinc}(\Delta\psi/2) = \sin(\Delta\psi/2)/(\Delta\psi/2)$. For data processing, both methods can therefore be handled in an identical fashion [4.3].

To calculate the unknowns, we assume that $N \geq 3$ interferograms $I_n(x)$ have been sampled, where the relative phases ψ_n are equidistantly distributed over one 2π period, i.e. $\psi_n = n2\pi/N$. For each picture element x only three accumulating registers

$$\Sigma_c(x) = \sum_{n=1}^{N} I_n(x)\cos(\psi_n), \quad \Sigma_s(x) = \sum_{n=1}^{N} I_n(x)\sin(\psi_n),$$

$$\Sigma(x) = \sum_{n=1}^{N} I_n(x) \tag{4.3}$$

have to be stored. The best solution in the sense of least squares is found to be [4.4]

$$a(x) = \frac{1}{N}\Sigma(x), \quad m(x) = \frac{1}{\Sigma}\sqrt{\Sigma_c^2 + \Sigma_s^2}, \quad \tan\phi(x) = -\frac{\Sigma_s(x)}{\Sigma_c(x)}. \tag{4.4}$$

For the particular cases of 3 and 4 phase steps, the equations for the phase $\phi(x)$ become very simple, namely for $N = 3$

$$\tan\phi(x) = \sqrt{3}\,\frac{I_2 - I_1}{2I_3 - I_2 - I_1} \tag{4.5}$$

and for $N = 4$

$$\tan\phi(x) = \frac{I_3 - I_1}{I_2 - I_4}. \tag{4.6}$$

Obviously, quasi-heterodyne methods allow only to determine the interference phase modulo 2π. The complete phase is evaluated using the continuity of the phase function, assuming that $\phi(x)$ changes less than π between two adjacent points. Other phase-shifting techniques and algorithms for the phase evaluation have been developed and are described in the literature [4.5] (see also Chap. 5). The accuracy is typically $\delta\phi \approx 3°$ or 1/100 of a fringe, but it might be limited by additional sources of statistical and systematic errors, depending on the particular application.

The mean intensity $a(x)$ and the fringe contrast $m(x)$ determined from (4.4) allow to control the quality of the interferogram evaluation. In particular, they can be employed to distinguish between true interference fringes and other structures in the image, such as shadows, contours, holes, etc. Considerable fluctuations of the fringe contrast versus position may also indicate inaccurate reference-phase shifts and intensity measurements, or extraneous fringe patterns in the image.

In heterodyne interferometry, the two interfering wave fields are reconstructed with different optical frequencies ω_1 and ω_2 [4.6], i.e., the optical frequency of one of the two reference beams is shifted (e.g., with the help of acousto-optical modulators). The local intensity of the interference pattern is then varying sinusoidally at the beat frequency $\Delta\omega = \omega_2 - \omega_2$ and (4.1) becomes

$$I(x) = a(x)\{1 + m(x)\cos[\Delta\omega t + \phi(x) + \psi]\}, \tag{4.7}$$

where ψ is an additional constant phase. The interference phase $\phi(x)$ is transformed into the phase of the beat frequency signal. As the beat frequency $\Delta\omega/2\pi$ is chosen low enough (< 100 MHz) to be resolved by the opto-electronic detector employed, the interference phase can be measured with high accuracy independently of $a(x)$ and $m(x)$ using an electronic phasemeter. By this way, both the interpolation problem and the sign ambiguity of classical interferometry are solved. This method requires special equipment, such as acousto-optical modulators and phasemeters. The image is scanned mechanically by photodetectors to measure the interference phase $\phi(x)$ locally. Therefore the speed is relatively low (≈ 1 s per point) but the accuracy ($\delta\phi \approx 0.3°$ or 1/1000 of a fringe) and the spatial resolution ($>10^6$ resolvable points) are extremely high [4.7].

4.1.2 Double-Exposure with Two Reference Beams

The application of electronic interference phase measurement in double-exposure holographic interferometry requires a setup with two reference beams to have independently access to the relative phase between the two interfering reconstructions. The optical arrangements for two-reference-beam holographic interferometry are sketched in Fig. 4.1a for well separated reference sources and in Fig. 4.1b for reference sources close together. Except for the two references, this setup is the same as for classical holographic interferometry. In both arrangements, the object is illuminated by a point source, and an imaging system permits to observe the interferogram on a screen.

First and second object fields O_1 and O_2 are recorded on the same hologram in the same setup, but using two different reference waves R_1 and R_2, respectively. The two object fields are stored and accessible independently by the corresponding reference. Interferometry takes place during reconstruction with both

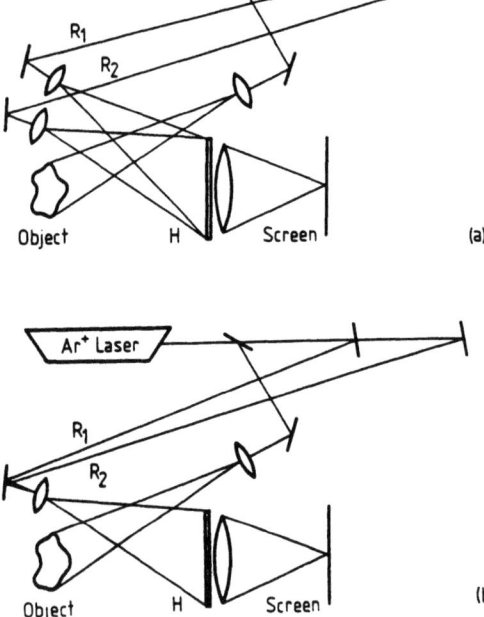

Fig. 4.1a,b. Arrangement for two-reference-beam holographic interferometry
(a) Well separated reference sources
(b) References source close together

references simultaneously. Relative phase and fringe position can be controlled during reconstruction, which can now be used for accurate fringe interpolation by electronic interference phase measurement. This technique can be applied to double-exposure, pulsed and stroboscopic holographic interferometry. Assuming that the reconstructing reference waves R'_1 and R'_2 are given by

$$R'_1 = R_1 e^{i\phi_1} \quad \text{and} \quad R'_2 = R_2 e^{i\phi_2}, \tag{4.8}$$

where R_1 and R_2 are the recording references, and ϕ_1 and ϕ_2 are the phases of the reconstructing references, and that the two object fields differ only by an additional phase $\phi(x)$, i.e. $O_2(x) = O_1(x) e^{i\phi(x)}$, one gets for the intensity of the superposition of the reconstructions

$$I(x) = |O_1(x) e^{i\phi_1} + O_2(x) e^{i\phi_2}|^2 = 2|O_1(x)|^2 \{1 + \cos[\phi(x) + \Delta\phi]\}, \tag{4.9}$$

where $\Delta\phi = \phi_2 - \phi_1$ is the phase difference of two reconstructing reference waves. This phase difference $\Delta\phi$ can be controlled during the reconstruction.

Two-reference-beam holography requires special attention to the multiplicity of the reconstructed images and the influence of misalignment of the hologram with respect to the reference beams [4.2, 7]. Illuminating the hologram with both reference beams R_1 and R_2 yields not only two, but four primary reconstructions, namely the two desired self-reconstructions ($R_1 R_1^* O_1$ and $R_2 R_2^* O_2$), which give rise to the interference pattern, and the two undesired cross-reconstructions ($R_2 R_1^* O_1$ and $R_1 R_2^* O_2$). The direction of propagation of the various reconstructed waves depends on the geometry of the optical setup. The primary reconstructions are shown for both cases in Figs. 4.2a and b.

To avoid disturbing overlapping of the different reconstructions (Fig. 4.2a), the two reference sources must be chosen on the same side of the object with a mutual separation larger than the angular size of the object in the corresponding direction (Fig. 4.1a). However, the consequence of a large separation of the reference sources is high sensitivity to hologram misalignment and to changes of the wavelength between hologram recording and reconstruction, as described hereafter. Therefore, reference sources close together (Fig. 4.1b) would be preferred if, under certain conditions, overlapping (Fig. 4.2b) of the cross-reconstructions could be tolerated without loss of accuracy for the interference phase measurements.

The sensitivity of two-reference-beam holographic interferometry to repositioning and wavelength changes occurs because the propagation of the two reconstructed wave fields are differently affected, as discussed in detail in [4.2 and 7]. In holographic interferometry, the interferogram is usually observed in the image of the holographically recorded virtual object, as sketched in Fig. 4.3. The imaging lens is placed close behind the holographic plate. The virtual image of the object is reconstructed by the two reference beams R_1 and R_2, which are assumed to be plane waves given by the wave vectors k_1 and k_2, respectively. For small changes of the hologram position and the wavelength from recording

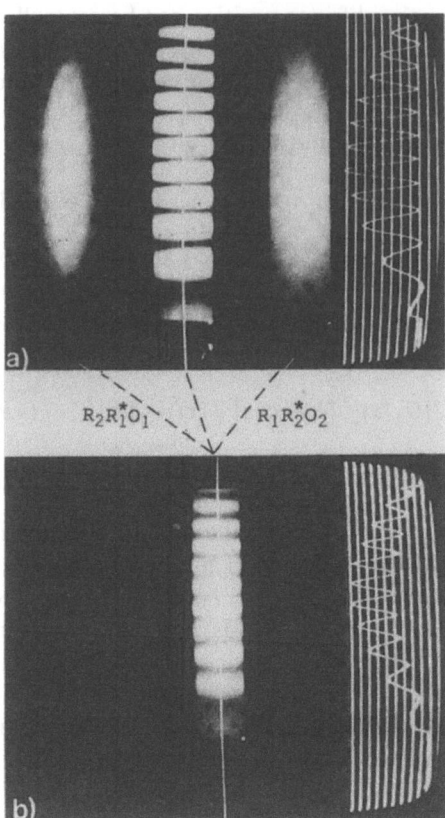

Fig. 4.2a,b. Primary reconstructed images on TV screen. (a) Well separated reference sources yield separated cross-reconstructions $R_2 R_1^* O_1$ and $R_1 R_2^* O_2$. (b) References close together yield overlapping cross-reconstructions

to reconstruction, the resulting additional phase difference $\phi(x_H)$ in the hologram plane between the reconstructed waves corresponding to the objects O_1 and O_2 is given by

$$\phi(x_H) = [(k_1 - k_2) \times w] \cdot x_H + (\Delta\lambda/\lambda)(k_1 - k_2) \cdot x_H = M_H \cdot x_H, \quad (4.10)$$

where $w = (\Delta\alpha, \Delta\beta, \Delta\gamma)$ is the rotation vector for small hologram rotations around the x, y, z axes, respectively, $\Delta\lambda$ is the change of the wavelength, and x_H are the coordinates in the hologram plane. Note that a pure translation of the hologram causes only a constant phase shift and can be ignored. Equation (4.10) shows that $\phi(x_H)$ can be expressed in terms of a fringe vector M_H which describes the density and the direction of the fringes appearing in the hologram plane. Since both contributions in (4.10) depend on the difference vector $\Delta k = k_1 - k_2$ of the two references, the sensitivity to repositioning and to wavelength changes is much smaller for references close together.

Normally, the hologram and the pupil of the imaging lens are not in the same plane. As shown in [Ref. 4.7, pp. 27–28] by applying the Fresnel diffraction

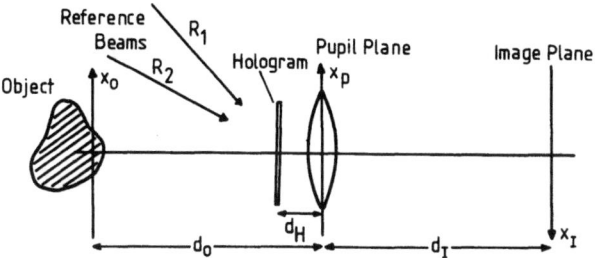

Fig. 4.3. Imaging system for the reconstruction of a holographic interferogram

formula, the phase function is equivalent to a linear phase $\phi_O(x_O)$ in the object plane and $\phi_P(x_P)$ in the pupil plane, given by

$$\phi_O(x_O) = - M_H \cdot x_O \, d_H/d_O, \qquad (4.11a)$$

$$\phi_P(x_P) = - M_H \cdot x_P (d_O - d_H)/d_O, \qquad (4.11b)$$

where d_H is the distance between the hologram and the pupil plane, and d_O is the object distance (Fig. 4.3). The resulting phase $\phi_O(x_O)$ in the object plane will be observed in the interferogram as a linear phase deviation. Since the optical fields in the pupil plane x_P and in the image plane x_I are related by a Fourier transformation, the linear phase deviation $\phi_P(x_P)$ produces a lateral shift between the two object reconstructions in the image plane of

$$u_I = (1/k) M_H (d_O - d_H)(d_I/d_O), \qquad (4.12)$$

where d_I is the image distance and $k = 2\pi/\lambda$. The effect of this lateral shift is quite different for smooth light fields from transparent objects and for speckled light fields from diffusely scattering surfaces. In the first case the effect is similar to a shearing in the interferometer, in the second case the speckle patterns will become decorrelated, which causes a reduction of the fringe contrast (Sect. 4.2.3).

From (4.10) one gets for the magnitude of the fringe vector M_H due to misalignment of the hologram plate, without change of the wavelength and with the hologram plane perpendicular to the z-axis,

$$|M_H| = [(\Delta k_y \, \Delta \gamma - \Delta k_z \, \Delta \beta)^2 + (\Delta k_z \, \Delta \alpha - \Delta k_x \, \Delta \gamma)^2]^{1/2}. \qquad (4.13)$$

The second term of (4.10) reveals that a wavelength change $\Delta\lambda$ introduces a linear phase deviation across the hologram which is proportional to Δk. Thus, reference sources close together reduce also the sensitivity to wavelength changes. Moreover, the effect of a wavelength change between hologram recording and reconstruction can be corrected by adjusting the angle between the two reference beams during reconstruction. In an arrangement with reference sources very close together, such an adjustment can be easily controlled by observing the macroscopic fringe pattern produced on the hologram by the two reference waves. The possibility to admit wavelength changes in two-reference-beam holographic interferometry is very important for pulsed holography. Its feasibility has been demonstrated for double-pulse holograms recorded with a ruby laser ($\lambda = 693$ nm) and reconstructed either with a cw HeNe ($\lambda = 633$ nm) or an argon ($\lambda = 514$ nm) laser [4.8].

It has been shown that for diffusely scattering objects overlapping of the desired self-reconstructions with the cross-reconstructions can be tolerated, as long as they are shifted laterally by more than the average speckle size [4.9] (Sect. 4.4.3). Therefore, the two reference sources can be placed very close together, which reduces drastically the requirements for hologram repositioning and the sensitivity to wavelength changes.

It turns out that an optical arrangement with the references close together makes the use of two-reference-beam holography nearly as simple as classical hologram interferometry. Particularly in an industrial environment, it is the ideal arrangement for double-exposure and double-pulse holography, when high resolution fringe interpolation is required. However, the overlapping with the uncorrelated cross-reconstructions reduces the overall fringe contrast by a factor of two, as seen in Fig. 4.2, and introduces an additional statistical error to the interference phase measurement. This error can be adequately reduced by spatially averaging over a detection area which covers many speckles [4.10] (Sect. 4.3).

4.2 Interferometry with Speckle Fields (Diffusely Scattering Objects)

When a diffusely scattering object is illuminated with coherent light, which is necessary for both interferometry and holography, its image has a granular appearance. It seems to be covered with fine, randomly distributed light and dark speckles. The rough surface can be considered as an ensemble of scattering centers producing light with random phases, which will vary from point to point in proportion to the local surface height. The resulting speckle pattern is characteristic for the microscopic structure of the surface roughness. Therefore, two objects of similar shape but individually different surface roughness cannot be compared interferometrically; nevertheless, two different appearances of the same diffusely scattering object can be compared interferometrically with the help of holography, which allows to store the wave fields for later use. That is the reason why the main importance of holographic interferometry lays in its application to diffusely scattering objects. In this section, the fundamental statistical properties of the speckle pattern in an image and particularly in the holographic reconstruction are discussed. The formation of interference fringes between two interfering speckle fields and the detection of intensity and interference phase are investigated.

4.2 Interferometry with Speckle Fields   83

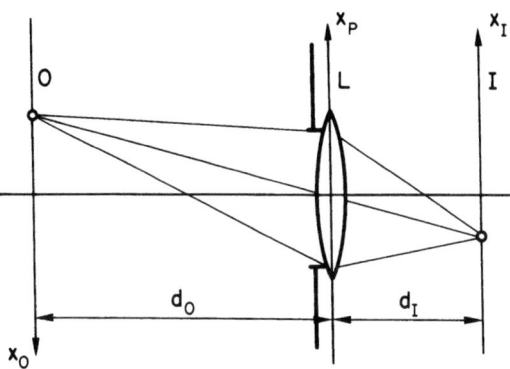

Fig. 4.4. Coherent imaging of the object plane x_O onto the image plane x_I by the lens L

4.2.1 Speckle Statistics for Coherent Imaging

In holographic interferometry, the interferogram is usually observed in the image of the holographically recorded virtual object (Fig. 4.3). The statistical properties of the speckle field in the interferogram will therefore be calculated for an imaging geometry, as sketched in Fig. 4.4. Note, that the hologram in Fig. 4.3 is involved only as an intermediate wave front recording medium, and that the following discussion is generally valid for coherent image formation.

The complex amplitude of the diffused light on the object surface is described by $O(x_O)\rho(x_O)$, where $O(x_O)$ is the amplitude of the object illumination and $\rho(x_O)$ is a stationary and Gaussian distributed variable describing the surface roughness [4.7]. The amplitude in the coherent image of the object reconstruction is then given by the convolution integral [4.11]

$$U(x_I) = \int d^2 x_O \, O(x_O) \rho(x_O) h(x_I - x_O), \tag{4.14}$$

where $h(x_I)$ denotes the impulse response function of the imaging system (for the sake of simplicity, the magnification is assumed to be unity), which is essentially given by the Fourier transform of the aperture function of the imaging lens [4.12].

The interference between speckle fields and the detection of intensity in the coherent image with a detector of finite size involves the calculation of second-order statistics, which relates the statistical properties of different points within the speckle pattern [4.7, 11, 13]. For this purpose, the auto-correlation of the amplitude and the intensity in the image are calculated. For the amplitude, one obtains the twofold integral

$$\langle U(x_I)U^*(x_I')\rangle = \int d^2 x_O \int d^2 x_O' \, O(x_O)O^*(x_O')\langle \rho(x_O)\rho^*(x_O')\rangle$$
$$\times h(x_I - x_O)h^*(x_I' - x_O'), \tag{4.15}$$

where $\langle \cdots \rangle$ denotes the average over an ensemble of rough surfaces. The object amplitude is assumed to be slowly varying compared with the width of $h(x)$ and will therefore be taken out of the integral in all further calculations. For an object surface with sufficiently fine grain, the auto-correlation of $\rho(x)$ can be approximated by a Dirac function, viz.

$$\langle \rho(x)\rho^*(x')\rangle = \delta(x-x'). \tag{4.16}$$

So one gets for (4.15)

$$\langle U(x_\mathrm{I})U^*(x'_\mathrm{I})\rangle = |O(x_\mathrm{O})|^2 \int \mathrm{d}^2 x_\mathrm{O}\, h(x_\mathrm{I}-x_\mathrm{O})h^*(x'_\mathrm{I}-x_\mathrm{O}) = \langle I(x_\mathrm{I})\rangle C_h(X), \tag{4.17}$$

with $X = x'_\mathrm{I} - x_\mathrm{I}$. The auto-correlation $C_h(x)$ of the impulse response function $h(x)$ plays an important role in holographic interferometry, as will be seen in the following. In (4.17), $|O(x_\mathrm{O})|^2$ has been replaced by $\langle I(x_\mathrm{I})\rangle = \langle U(x_\mathrm{I})U^*(x_\mathrm{I})\rangle$, since $C_h(0) = 1$.

The auto-correlation of the intensity is calculated in a similar manner, using the four-fold correlation of the surface roughness

$$\langle \rho(x)\rho^*(x')\rho(x'')\rho^*(x''')\rangle = \delta(x-x')\delta(x''-x''') + \delta(x-x''')\delta(x'-x''). \tag{4.18}$$

Thus one gets

$$\langle I(x_\mathrm{I})I(x'_\mathrm{I})\rangle = \langle I(x_\mathrm{I})\rangle\langle I(x'_\mathrm{I})\rangle + |\langle U(x_\mathrm{I})U^*(x'_\mathrm{I})\rangle|^2 \tag{4.19}$$

and with (4.17)

$$C_I(x_\mathrm{I}, x'_\mathrm{I}) = \langle I(x_\mathrm{I})\rangle^2 [1 + |C_h(X)|^2]. \tag{4.20}$$

For a circular, binary pupil function of the imaging lens, the auto-correlation C_h is given by the well-known Airy function [4.12]

$$C_h(x_\mathrm{I}) = \frac{2J_1(\pi r)}{\pi r} = \mathrm{Airy}(\pi r), \tag{4.21}$$

where $J_1(\pi r)$ is the 1st order Bessel function and $r = |x_\mathrm{I}|D/\lambda d_\mathrm{I}$, with the diameter D of the lens pupil, the wavelength λ and the image distance d_I (Fig. 4.4). In holographic interferometry $C_h(u_\mathrm{I})$ describes the fringe contrast versus the in-plane displacement component u_I, as will be shown in Sect. 4.2.3.

4.2.2 Measuring Intensity with a Detector of Finite Size

The only measurable quantity in the image is the local intensity. It is therefore useful to look for some statistical properties of intensity detection in a speckle field. In holographic interferometry, the detector resolution is normally much larger than the speckle size. Thus all further calculations will be restricted to this case. Following Goodman [4.13], the detected optical power P can be expressed by the integral of the intensity over the detector surface A_d, viz.

$$P = \int_{A_d} d^2 x_I \, I(x_I) = \int d^2 x_I \, D(x_I) I(x_I), \tag{4.22}$$

where $D(x_I)$ is a binary function representing the detector area, so that

$$\int d^2 x_I D(x_I) = A_d. \tag{4.23}$$

The average contrast of the variations (or the reciprocal rms signal-to-noise-ratio) of the detected optical power due to the speckle noise is defined as

$$\frac{\langle \delta P^2 \rangle}{\langle P \rangle^2} = \frac{\langle P^2 \rangle - \langle P \rangle^2}{\langle P \rangle^2}. \tag{4.24}$$

The average detected optical power is

$$\langle P \rangle = \bar{P} = \int d^2 x_I \, D(x_I) \langle I(x_I) \rangle = A_d \langle I \rangle. \tag{4.25}$$

The second moment of P becomes

$$\langle P^2 \rangle = \int d^2 x_I \int d^2 x'_I \, D(x_I) D(x'_I) \langle I(x_I) I(x'_I) \rangle$$
$$= \int d^2 X \, C_d(X) C_I(X), \tag{4.26}$$

where $C_d(X)$ is the auto-correlation of the detector function $D(x_I)$. Assuming the function $D(x_I)$ to be binary and much larger than the speckle size, and introducing (4.20 and 25), one obtains

$$\langle P^2 \rangle = \langle P \rangle^2 \left(1 + \frac{\int d^2 x |C_h(x)|^2}{A_d} \right). \tag{4.27}$$

The integral in (4.27) can be interpreted as the average area of a speckle correlation cell. This leads to the definition of the number of speckles N within the detector surface A_d [4.10, 13], viz.

$$N = \frac{A_d}{\int d^2 x |C_h(x)|^2}. \tag{4.28}$$

Due to the assumption made, this equation is only correct for $N > 1$. With (4.24, 27 and 28) the mean square variation of the detected optical power is now found to be

$$\frac{\langle \delta P^2 \rangle}{\langle P \rangle^2} = \frac{1}{N}, \tag{4.29}$$

where obviously the number of speckles N cannot be smaller than one, i.e. $N \geq 1$. For a binary aperture of the imaging lens one gets

$$\int d^2 x |C_h(x)|^2 = \frac{(\lambda d_I)^2}{A_p}, \tag{4.30}$$

where d_I is the image distance and A_P is the area of the lens aperture. The number of speckles N can now be expressed in terms of the diameter D or the F-number (f/D) of the lens as

$$N = \frac{A_d A_P}{(\lambda d_I)^2} = A_d \pi \left(\frac{D}{2\lambda d_I}\right)^2 = A_d \pi \left(\frac{f}{2\lambda F d_I}\right)^2. \tag{4.31}$$

A reasonable definition for the average speckle size $(\Delta x)_s$ is obtained from the integral in (4.30), if one sets the value of this integral equal to the surface of a small disk with diameter $(\Delta x)_s$. For a circular lens aperture of diameter D one gets

$$(\Delta x)_s = \frac{4}{\pi} \frac{\lambda d_I}{D} = \frac{4}{\pi} \frac{\lambda F d_I}{f}. \tag{4.32}$$

The number N of speckles within the detector area is an important quantity for the intensity and the interference detection in a speckle field. The accuracy of the local mean intensity measurement is limited by the the statistical variation of the detected optical power. From (4.29) one sees that the relative accuracy is given by $\delta P/\bar{P} = 1/\sqrt{N}$. For $\delta P/\bar{P} = 10\%$ one needs therefore $N = 100$, or a detector diameter of $10 \times (\Delta x)_s$, and for $\delta P/\bar{P} = 1\%$ even $N = 10^4$ or $100 \times (\Delta x)_s$. In conclusion, accurate measurement of the local mean intensity in a speckle pattern is only possible by spatially averaging over many speckles, and therefore loosing spatial resolution.

4.2.3 Interference Fringe Formation

In holographic interferometry of diffusely scattering objects, two object states (before and after deformation) are compared interferometrically. According to (4.14), the complex amplitudes $V_1(x_I)$ and $V_2(x_I)$ of the two object reconstructions in the image are given by

$$V_n(x_I) = \int d^2 x_O \, O_n(x_O) \, \rho_n(x_O) \, e^{i[\omega t + \phi_n(x_O)]} h(x_I - x_O), \tag{4.33}$$

where ω is the circular frequency of the optical wave. The deformation of the object between the two states 1 and 2 is described by the displacement vector field $L(x_O)$. The object illumination O_2 can thus be expressed in terms of O_1 by $O_2(x_O) = O_1(x_O + u_I)$, where u_I is the in-plane component of the displacement vector L. The interference phase $\phi(x_O) = \phi_1(x_O) - \phi_2(x_O)$ is related to the object displacement by $\phi = K \cdot L$, where K is the sensitivity vector, which is determined by the geometry of object illumination and observation (Chap. 2).

The two interfering fields V_1 and V_2 yield the intensity $|V_1 + V_2|^2$ in the image plane. The significant quantity for intensity measurements and visual observation is the average intensity

$$\langle |V_1 + V_2|^2 \rangle = \langle I_1 \rangle + \langle I_2 \rangle + 2 \operatorname{Re}\{\langle V_1 V_2^* \rangle\}, \tag{4.34}$$

where Re{··} stands for the real part. The information on the interference phase is given by the complex interference term $V_1 V_2^*$, also called *mutual intensity*

$$I_{12} = V_1(x_I) V_2^*(x_I)$$
$$= \int d^2 x_O \int d^2 x'_O \, O(x_O) O^*(x'_O) e^{i\phi(x_O)} \rho(x_O) \rho^*(x'_O) h(x_I - x_O)$$
$$\times h^*(x_I - x'_O + u_I). \tag{4.35}$$

Assuming again that $O(x_O)$ and also $\phi(x_O)$ are approximately constant within the width of $h(x)$, one obtains for the average mutual intensity

$$\langle I_{12}(x_I) \rangle = \langle I(x_I) \rangle e^{i\phi(x_I)} C_h(u_I), \tag{4.36}$$

and for the total average intensity

$$\langle |V_1 + V_2|^2 \rangle = 2 \langle I(x_I) \rangle [1 + C_h(u_I) \cos \phi(x_I)]$$
$$= a(x_I) \times [1 + m(x_I) \cos \phi(x_I)], \tag{4.37}$$

where $\langle I_1 \rangle = \langle I_2 \rangle = \langle I \rangle$. Thus the average intensity depends sinusoidally on the interference phase $\phi(x_I)$. $C_h(u_I)$, the degree of correlation between the two interfering speckle fields (4.17), determines the visibility (or contrast) $m(x_I)$ of the interference fringes. Figure 4.5 shows experimentally determined fringe contrast values in a holographic interferogram versus in-plane displacement u_I, fitted by the theoretically expected Airy function (4.21). Negative values show up as contrast reversal of the fringes. For in-plane displacements u_I comparable with

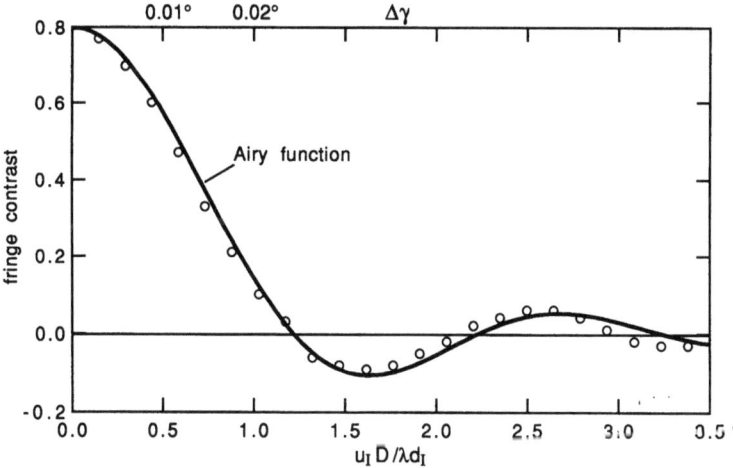

Fig. 4.5. Fringe contrast versus displacement u_I in terms of the diffraction limited resolution $\lambda d_I/D$ for a circular lens aperture of diameter D. The theoretical Airy function is normalized to the maximum measured contrast of 0.8. The displacement is produced by hologram misalignment (rotation $\Delta\gamma$ around the hologram normal axis) in two-reference-beam holography

the speckle size, the fringe visibility $C_h(u_I)$ vanishes (Fig. 4.5). This effect can also be interpreted as de-localization of the interference fringes [4.14].

The quantity which is of interest in interferometry, the interference phase $\phi(x_I)$, cannot be deduced directly from the intensity measured in the interferogram, because the fringe pattern, given by (4.37), is described by two more unknowns, namely the local intensity $a(x_I)$ and the local fringe contrast $m(x_I)$. As described in Sect. 4.1.1, electronic interference phase measurement overcomes this problem.

4.3 Sources of Errors in Interference Phase Measurement

The accuracy of fringe interpolation in holographic interferometry is subject to physical and experimental limitations. In this section, the sources of error of particular importance for interference measurements from double exposure holograms recorded with two reference beams will be discussed. They are divided in systematic errors, caused by misalignment, cross-talk and spurious fringe patterns, and in statistical phase errors that are due to speckle noise. Errors associated to the opto-electronic phase detection, either by phase-shifting or heterodyning, are not considered here (for phase-shifting detection, see Chap. 5).

4.3.1 Systematic Errors from Holographic Interferometry with Two Reference Sources

The use of an optical arrangement with two reference beams can give rise to important systematic deviations in the interference phase. By careful operation, however, these errors can be completely avoided.

4.3.1.1 Misalignment of Hologram and Reconstructing Reference Waves

The sensitivity of two-reference-beam holographic interferometry to repositioning and wavelength changes is discussed in Sect. 4.1.2. The realignment of the hologram is very critical if a setup with two well separated reference sources is to be used. This is required for holographic interferometry of transparent (non-diffusing) objects and for high accuracy measurements (heterodyne detection), when overlapping reconstruction cannot be tolerated. In this case, the same setup should be used for recording and reconstruction, similar to real-time holographic interferometry. The repositioning sensitivity is reduced by orders of magnitude, if an arrangement with reference sources close together can be used, which is the case in most applications with diffusely scattering objects and moderate accuracy (phase-shifting detection).

As seen from (4.11a), the phase error $\phi_0(x_0)$ on the object resulting from misalignment fringes M_H in the hologram can be minimized by placing the imaging lens (Fig. 4.3) as close as possible to the hologram ($x_H/d_0 \ll 1$). The effect of the lateral shift u_1, shown in (4.12), is quite different for smooth light fields from transparent objects and for speckled light fields from diffusely scattering surfaces. In the first case the effect is similar to a shearing in an interferometer, in the second case the speckle patterns will become decorrelated, which causes a reduction of the fringe contrast.

The experimental results shown in Fig. 4.5 were obtained by a rotation $\Delta\gamma$ of the hologram around the normal axis to the hologram plate. The two reference waves had a large angular separation of $\Delta k_y/k = 0.19$, which is necessary to keep the reconstructions separated in a typical holographic setup (Figs. 4.1a and 2a). A one-to-one image of the reconstructed object was formed by an objective of $f = 300$ mm ($d_1 = 600$ mm) and of $D = 9$ mm ($f/32$) effective aperture. The allowed repositioning error for a reduction of the fringe contrast to 0.5, as obtained from (4.12, 13 and 21), is found to be only $\Delta\gamma = 0.013°$. In a typical setup with reference sources close together (Fig. 4.1b) (0.5 mrad angular separation) and an imaging geometry as described above ($D = 9$ mm, $d_1 = 600$ mm), the acceptable misalignment is increased to about $\Delta\gamma = 5°$ [4.9].

4.3.1.2 Cross-Talk from Holographic Recording

Cross-talk between the two recorded holograms can be caused by nonlinear hologram recording [4.7, 15] or by improper switching between the two reference beams. The effects are in both cases the same, namely a systematic phase error which is correlated with the recorded interference pattern.

The cross-talk can be described by the two modified reconstructions

$$O'_1 = O_1 + \beta_1 O_2 \quad \text{and} \quad O'_2 = O_2 + \beta_2 O_1. \tag{4.38}$$

The complex cross-talk parameters β_1 and β_2 are in general assumed to be small ($|\beta| \ll 1$). The magnitude of β_1 and β_2 can be experimentally determined from the contrast of the interference fringes which are visible when the hologram is reconstructed with only one reference beam at the time. From (4.38) one gets with $O_2 = O_1 e^{i\phi}$, where ϕ is the interference phase,

$$|O'_1|^2 \cong |O_1|^2 [1 + 2|\beta_1|\cos(\phi + \delta_1)] \quad \text{and}$$

$$|O'_2|^2 \cong |O_2|^2 [1 + 2|\beta_2|\cos(\phi + \delta_2)]. \tag{4.39}$$

Following (4.8 and 9) one gets now with (4.38) for the intensity of the superposition of the reconstructions

$$I(x) = |O_1(x)|^2 \{[1 + \beta_1 e^{i\phi(x)}] e^{i\phi_1} + [e^{i\phi(x)} + \beta_2] e^{i\phi_2}\}, \tag{4.40}$$

where ϕ_1 and ϕ_2 are the phases of the two reconstructing reference waves. For phase-shifting and heterodyne detection, the difference $\Delta\phi = \phi_2 - \phi_1$ is controlled during the reconstruction. The corresponding variation of the intensity becomes then

$$I_{ac}(x) \cong 2|O_1(x)|^2 \, \text{Re}\left\{[1 + (\beta_1^* + \beta_2)e^{-i\phi(x)}]e^{i[\phi(x) + \Delta\phi]}\right\}. \tag{4.41}$$

The cross-talk introduces a systematic error with the same periodicity as the interference phase $\phi(x)$. From (4.41) the maximum value of this error is found to be

$$\delta\phi \cong |\beta_1^* + \beta_2| \leq |\beta_1| + |\beta_2|. \tag{4.42}$$

As an example, the extinction ratio of a Pockels cell used in pulsed holography to switch between the two reference beams is typically 1/50, which means $|\beta|^2 = 1/50$, and causes thus already an error of $\delta\phi = 16°$. For higher accuracy it is therefore recommended to employ other switching elements, e.g. acousto-optic modulators. The importance of the cross-talk from nonlinear hologram recording depends on the presence of visible parts in the recorded scene which remain unchanged between the two exposures and act as spurious reference sources [4.7]. Experimental results show that the corresponding phase error can be kept below $\delta\phi = 0.2°$, if the recorded scene is carefully cleared from spurious reference sources.

4.3.1.3 Spurious Fringes Due to Overlapping Reconstructions

The overlapping of the cross-reconstruction images (Fig. 4.2b) can produce extraneous interference in the interferogram. The effect depends on the speckle size in the image, i.e., on the aperture of the image forming lens. If the aperture of the lens is too small to resolve completely the reference sources close together, the speckle patterns of the cross-reconstructions are not fully decorrelated (the higher order maxima of the Airy function (Fig. 4.5) give still rise to some correlation with the undesired reconstructions) and reconstruction with only one reference produces already a fringe pattern in the interferogram. This will cause systematic phase errors, similar to the cross-talk fringes described in Sect. 4.3.1.2. To avoid such spurious fringe patterns and phase deviations, the angular separation between the two reference sources must be several times the diffraction limit of the lens. In other words, the interference between the two reference beams must produce many fringes (at least 6 fringes for $\delta\phi \leq 2.5°$) across the lens aperture.

4.3.2 Statistical Errors Due to Speckle Noise

The presence of speckles in the image of objects with diffusely scattering surfaces gives rise to a statistical error for the measured interference phase. The interference of non-correlated speckles (or non-correlated parts of speckles) introduces a random contribution to the phase of the mutual intensity of the interfering wave fields. Partial decorrelation of the speckle patterns may be caused by the transverse component of the object displacement. In addition, uncorrelated speckle fields (e.g., due to overlapping cross-reconstructions or diffusely scattered light from the hologram) may be superimposed to the desired interfering reconstructions. This source of phase error is of particular importance in phase-shifting holographic interferometry with reference sources close together (overlapping reconstructions).

In this subsection the influence of speckle decorrelation on the interference phase measurement is described. The fundamental statistical properties of the speckle pattern in an imaging system and the detection of intensity in a speckle field with a detector of finite size have been discussed in Sect. 4.2. The statistical errors due to the speckle noise in high-resolution interference phase measurement has been rigorously calculated for two different situations, namely non-overlapping and overlapping reconstructions [4.10]. In both cases, the resulting phase error depends essentially on the number of speckles within the detector area. The number of detected speckles and the statistical phase measurement error have been experimentally determined with the help of a heterodyne system and then compared with theory.

4.3.2.1 Non-Overlapping Reconstructions

First, the situation of two interfering wave fields shall be considered, as it is the case in two-reference-beam holographic interferometry with well separated reference sources. The mutual intensity I_{12} of (4.35) is detected by an ac-coupled photo-detector at the heterodyne frequency or calculated from phase-shifted samples (Sect. 4.1.1). The detector is averaging over many speckles, i.e. its surface is much larger than the average speckle size $(\Delta x)_s$. The detected optical power is then

$$P_{12} = \int d^2 x_I D(x_I) I_{12}(x_I). \tag{4.43}$$

Note that P_{12} is a complex quantity. It is proportional to the interference term in the detector current, which is described by its amplitude and phase.

The interference of non-correlated speckles (or non-correlated parts of speckles) gives rise to a random contribution to the phase of the complex mutual intensity I_{12}. The decorrelation of the two speckle patterns is due to the in-plane object displacement u_I. Since the detector averages over only a finite number of speckles, P_{12} differs from the (ideal) ensemble average $\langle P_{12} \rangle$ (Fig. 4.6). The statistical phase measurement error due to the speckle noise can be calculated from the fluctuations of the detected optical power P_{12}. Assuming an average interference phase of $\langle \phi(x_I) \rangle = 0$, $\delta\phi$ can be approximated by

$$\delta\phi \simeq \tan \delta\phi = \frac{\text{Im}\{P_{12}\}}{\text{Re}\{P_{12}\}} = \frac{\text{Im}\{P_{12}\}}{\langle P_{12} \rangle + \text{Re}\{P_{12} - \langle P_{12} \rangle\}} \simeq \frac{\text{Im}\{P_{12}\}}{\langle P_{12} \rangle}. \tag{4.44}$$

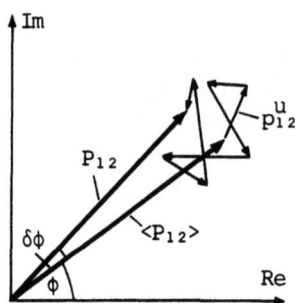

Fig. 4.6. Interference phase measurement error resulting from speckle noise. The power $p_{12}^s = p_{12}^c + p_{12}^u$ within each speckle is composed of a correlated and an uncorrelated part. The total detected power $P_{12} = \Sigma p_{12}$ differs from the (true) ensemble average $\langle P_{12} \rangle = \Sigma p_{12}^c$ and thus gives rise to the phase error $\delta\phi$

Note that this approach is only valid if the correlated part in P_{12} is much greater than the uncorrelated part, i.e., if there are enough speckles within the detector area.

For the mean square value of $\delta\phi$ one gets

$$\langle \delta\phi^2 \rangle = \frac{\langle \text{Im}^2\{P_{12}\} \rangle}{\langle P_{12} \rangle^2} = \frac{\langle |P_{12}|^2 \rangle - \langle P_{12}^2 \rangle}{2 \langle P_{12} \rangle^2}, \tag{4.45}$$

where we have assumed that $\langle I_{12}^2 \rangle = \langle I_{12}^{*2} \rangle$, which is true for a real valued auto-correlation function C_h.

The averages $\langle P_{12} \rangle$, $\langle P_{12}^2 \rangle$ and $\langle |P_{12}|^2 \rangle$ can now be evaluated with the help of the fundamentals presented in Chap. 3. $\langle P_{12} \rangle$ is obtained with (4.25 and 36) as

$$\langle P_{12} \rangle = \int d^2 x_I \, D(x_I) \langle I_{12}(x_I) \rangle = A_d \langle I \rangle C_h(u_I). \tag{4.46}$$

Similarly to (4.26 and 36) one gets

$$\langle P_{12}^2 \rangle = \int d^2 x_I \int d^2 x_I' \, D(x_I) D(x_I') \langle I_{12}(x_I) I_{12}(x_I') \rangle$$
$$= \int d^2 X \, C_d(X) [C_h^2(u_I) + C_h(X + u_I) C_h(X - u_I)], \tag{4.47}$$

and

$$\langle |P_{12}|^2 \rangle = \int d^2 X \, C_d(X) [C_h^2(u_I) + C_h(X) C_h(-X)]. \tag{4.48}$$

From (4.45, 47 and 48) one finds with (4.28)

$$\langle \text{Im}^2\{P_{12}\} \rangle = (1/2) \int d^2 X \, C_d(X) [|C_h(X)|^2 - C_h(X + u_I) C_h(X - u_I)], \tag{4.49}$$

and, finally, for the rms phase error that is due to the speckle noise

$$\langle \delta\phi^2 \rangle = \frac{1}{2N} \frac{1 - C_h(2u_I)}{C_h^2(u_I)}. \tag{4.50}$$

Figure 4.7 exhibits the statistical phase error $\delta\phi$ for two interfering speckle fields as a function of the speckle decorrelation (due to transverse displacement)

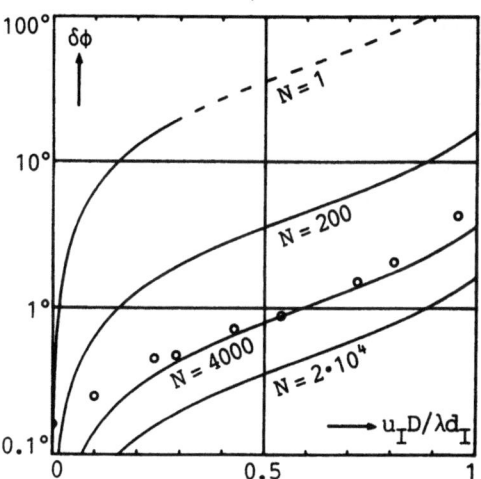

Fig. 4.7. Statistical phase error $\delta\phi$ for two interfering speckle fields versus normalized in-plane speckle decorrelation $u = u_1 D/\lambda d_1$, for different number N of speckles within the detector area. Experimental results are shown for $N = 4000$

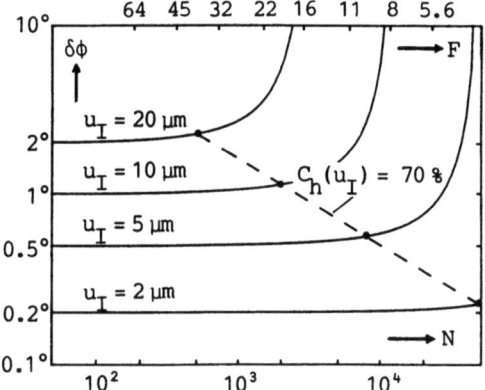

Fig. 4.8. Statistical phase error $\delta\phi$ versus number N of detected speckles for different in-plane displacement u_1. (Detector area $A_d = 1$ mm^2)

for different number of speckles N within the detector area. For small in-plane displacements u_1, $\delta\phi$ tends to zero.

The theoretical phase error of (4.50) versus speckle decorrelation has been experimentally verified with the help of a heterodyne system [4.10]. In Fig. 4.7, experimental results are shown for $N = 4000$ speckles within the detector surface. The speckle decorrelation was realized by hologram misalignment, which produces a shift between the object reconstructions and thus between the two (identical) speckle fields.

In Fig. 4.8, the statistical phase error $\delta\phi$ is plotted versus the number N of detected speckles for different in-plane displacements u_1. Through (4.31), the number of speckles is related to the F-number of the imaging lens. The figure reveals that over a wide range the phase error that is due to the speckle noise for constant u_1 does not depend on the lens aperture. The reason is that, when the lens aperture is closed, the reduction of the number of speckles, which increases the phase error, see (4.50), is compensated by the increase of the average speckle size, and thus the fringe contrast, which reduces the phase error. For a given maximum in-plane displacement u_I, the optimum choice for the F-number (minimum phase error and brightest possible image) is determined by the bend of the curves (dashed line in Fig. 4.8). The overall fringe contrast should therefore be better than about 70% of its maximum value.

4.3.2.2 Overlapping Reconstructions

For the calculation of the statistical phase error in the case of reference sources very close together, the speckle patterns of the overlapping reconstructions cannot be considered to be completely uncorrelated, as it was assumed in the preceding section. Each of the four wave fields V_n is described by (4.33) and the relations

$$O_2(x_0) = O_1(x_0 - u_0), \tag{4.51a}$$

$$O_3(x_0) = O_1(x_0 - v_0), \tag{4.51b}$$

$$O_4(x_0) = O_2(x_0 + v_0) = O_1(x_0 - u_0 + v_0), \tag{4.51c}$$

where v_0 denotes the lateral shift of the image, as seen in the object plane x_0, that is due to reconstruction with the wrong reference. The mutual intensity of the four wave fields, which is relevant for the detected interference phase, is now given by

$$I_{12} = (V_1 + V_3)(V_2 + V_4)^* = V_1 V_2^* + V_1 V_4^* + V_3 V_2^* + V_3 V_4^*. \tag{4.52}$$

After the calculation of the numerous cross-correlating terms, one finds for the statistical phase error of the detected signal containing N speckles that

$$\langle \delta\phi^2 \rangle = [6 + 8C_h(v_I) + 2C_h(2v_I) - 4C_h(2u_I + v_I) - 6C_h(2u_I + 2v_I)$$
$$- 4C_h(2u_I + 3v_I) - C_h(2u_I + 4v_I) - C_h(2u_I)]$$
$$\times \{2N[C_h(u_I) + 2C_h(u_I + v_I) + C_h(u_I + 2v_I)]\}^{-1}, \tag{4.53}$$

where u_I and v_I are now the displacements in the image plane x_I. Assuming that v_I is much larger than the speckle size $(\Delta x)_s$, all terms with C_h containing v_I can be neglected, and (4.53) can be rewritten as

$$\langle \delta\phi^2 \rangle = \frac{1}{2N} \frac{6 - C_h(2u_I)}{C_h^2(u_I)}. \tag{4.54}$$

The statistical phase error of (4.54) varies only slowly with the speckle decorrelation u_I of the interfering wave fields; this behavior is therefore not relevant. Note, however, that in the case of overlapping reconstructions the minimum error for $u_I = 0$ has a finite value, namely

$$\langle \delta\phi^2 \rangle = 5/2N. \tag{4.55}$$

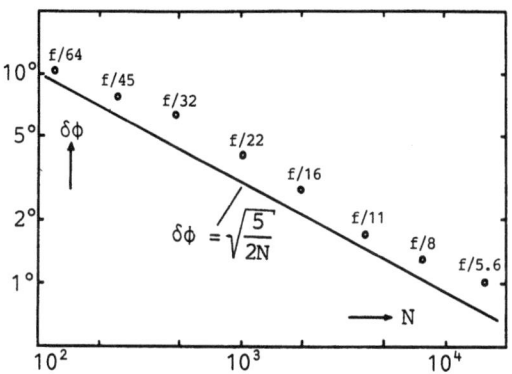

Fig. 4.9. Minimum statistical phase error $\delta\phi$ due to speckle noise from overlapping cross-reconstructions, measured for different number N of detected speckles, i.e., for different F-numbers

The minimum phase error has been measured for different number of speckles N, i.e., different lens apertures F [4.10]. The results are shown in Fig. 4.9. The measured phase errors are systematically larger than the theoretical ones, but the $1/N$ dependence is well established.

The statistical phase errors (Fig. 4.9) are relevant for the most commonly used arrangement, namely phase-shifting holographic interferometry of diffusely scattering objects, using two reference beams close together and video-electronic detection. In this case, the detector diameter is given by the pixel size of the TV-camera or the CCD array.

4.4 Applications

With the help of two-reference-beam holography, heterodyne and phase-shifting techniques for fringe analysis have been applied successfully to double-exposure and double-pulse holographic interferometry for the measurement of surface displacement, strain and vibration, as well as for the investigation of phase objects [4.16]. In the following, the most important applications and some of the most spectacular results will be presented.

4.4.1 Double-Exposure Holography with Phase-Shift Fringe Evaluation

Quasi-heterodyne holographic interferometry with TV-detection is nearly as simple as standard double-exposure holography, and it does not require any special instrumentation apart from a video-electronic data acquisition system (Fig. 4.10). It offers an accuracy of typically 1/100 of a fringe. The required two-reference-beam holography can be operated as easily as classical double-exposure holography by using two reference beams close together (Sect. 4.3.1). The method combines the simplicity of standard double-exposure holography, video-electronic processing, and the power of heterodyne holographic

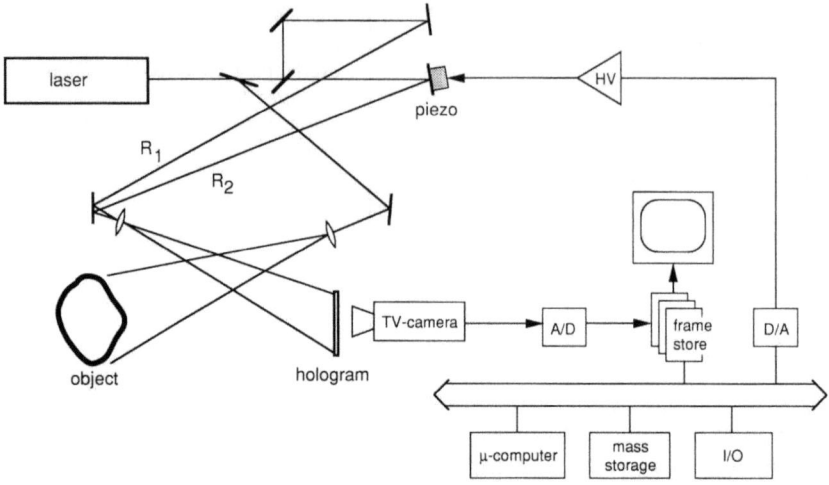

Fig. 4.10. Two-reference-beam holography with video-electronic data acquisition system

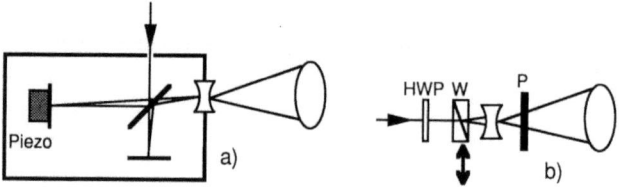

Fig. 4.11a,b. Two-reference-beam modules based on: (a) Michelson interferometer, (b) Wollaston prism (W) with $\lambda/2$-plate (HWP) and polarizer (P)

interferometry. Phase-shifting fringe processing is very well suited for industrial applications, where high speed and medium accuracy is required. Corresponding equipment for recording and reconstruction, as well as software for fringe analysis and data processing, are commercially available.

The optical arrangement shown in Fig. 4.10 is best suited for recording and reconstructing double-exposure holograms with the same cw laser in the same setup. Shutters are helpful to switch the reference beams between the two exposures. The phase shift for the fringe evaluation is obtained by the computer controlled piezo element. More compact optical modules for the generation of the two reference beams are shown in Fig. 4.11. A slightly misaligned Michelson interferometer (Fig. 4.11a) is very convenient to adjust the angle between the two beams. The shutters will be placed in the two arms of the interferometer. A Wollaston prism (Fig. 4.11b) generates two orthogonally polarized beams at a fixed angle. The polarizer at the output makes them interfere again. By rotating the input polarization with the help of the $\lambda/2$-plate the beams can be switched for the recording and adjusted to equal intensity during reconstruction.

The phase shift is obtained in a very reproducible manner by linear displacement of the Wollaston prism. This module is extremely stable and the direction of the light propagation is not altered, so that it can be inserted directly into existing holographic arrangements.

When the holograms are recorded with pulsed lasers (e.g., ruby laser), a separate setup with a cw laser (e.g., NeNe laser) is employed for the reconstruction [4.8]. In general, the wavelengths of these lasers will be different and therefore the reference beams for the reconstruction have to be readjusted to get minimum distortion and maximum fringe contrast (Sect. 4.1.2). The fringe pattern generated by the two reference beams in the hologram plane must be identical for recording and reconstruction. For the investigation of dynamic events, double-exposure holograms are recorded with Q-switched double-pulse lasers. The pulse separations are typically of the order of microseconds. Switching between the two reference beams is then accomplished either by electro-optic (Pockels cell) or by acousto-optic (Bragg cell) modulators [4.17] (see also Sects. 4.4.2, 3). A typical result of a computer aided fringe analysis of a vibrating object, recorded with a double-pulse Q-switched ruby laser, is shown in Fig. 4.12.

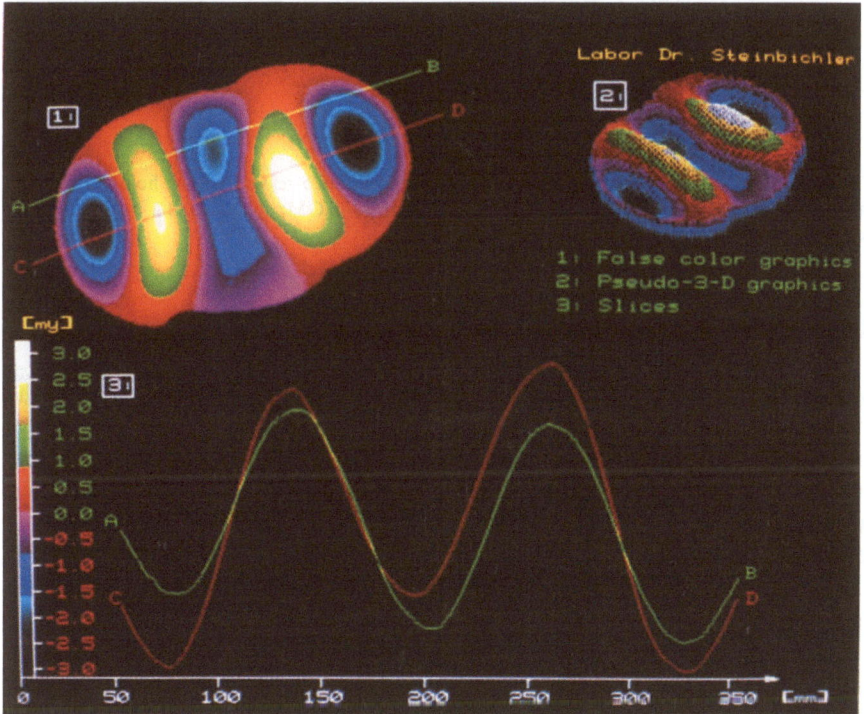

Fig. 4.12. Computer aided fringe analysis from a two-reference-beam hologram: 3-D plot of the digitized interference phase (top right), color representation of the vibration amplitude (top left), displacement [µm] versus position along the lines A–B and C–D (bottom). [Labor Dr. Steinbichler, Neubeuern, Germany]

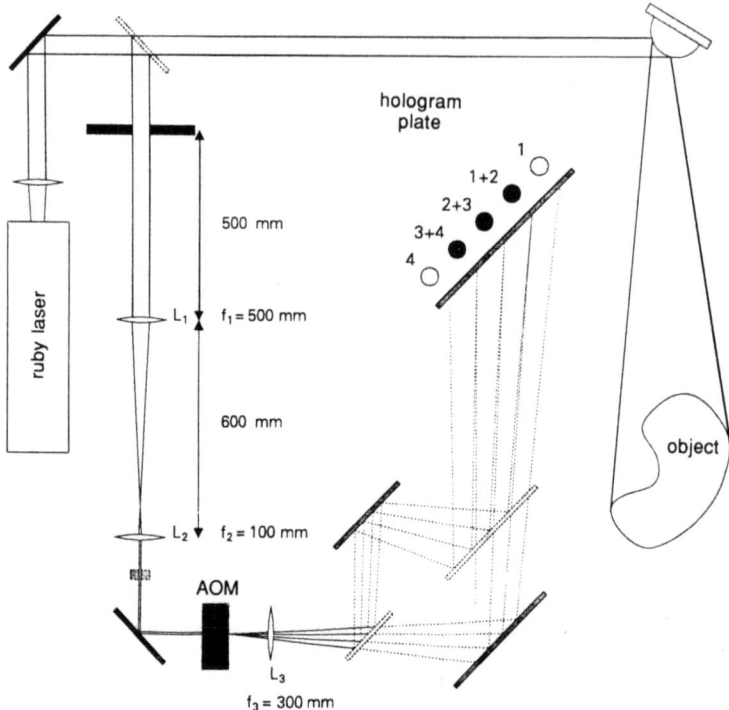

Fig. 4.13. Arrangement to record a series of double-exposure two-reference beam holograms on the same holographic plate using Acousto-Optic Modulators (AOM) to deflect the incoming beam [4.18]

4.4.2 Multiple-Exposure Holographic Interferometry

Multiple-exposure holographic interferometry is a very powerful tool for the analysis of dynamic events in mechanics and flow research. Figure 4.13 depicts an arrangement which allows to record a series of double-exposure two-reference beam holograms on the same holographic plate [4.18, 19]. The Q-switched ruby laser (Luminics HLS2) allows to produce four pulses with adjustable separations between 5 and 250 µs within one pump cycle of 800 µs duration. Each laser pulse produces two different spots of reference beams on the hologram plate. Position and angular separation of these spots are adjusted with a combination of mirrors and beam-splitters. Acousto-optic modulators are used to deflect the incoming beam of subsequent pulses by one spot position on the hologram plate. This gives a sequence of three double-exposure two-reference beam holograms, which can be analyzed off-line by standard phase-shifting methods. With this arrangement, cross-talk from recording (Sect. 4.3.1.2) is extremely small. To compensate for the frequency shift in the diffracted beams, two acousto-optic modulators in cascade must be used [Ref. 4.7, pp. 50–52].

$\Delta t_1 = 5\ \mu s$

$\Delta t_2 = 10\ \mu s$

$\Delta t_3 = 20\ \mu s$

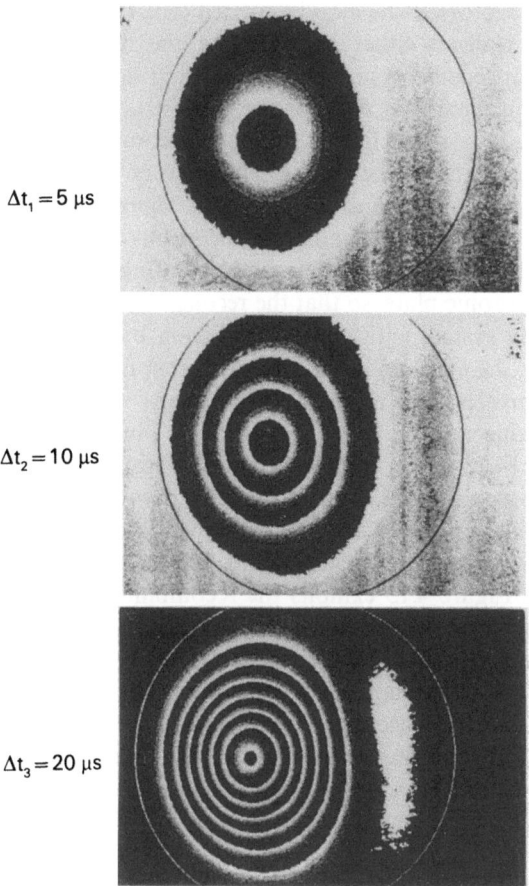

Fig. 4.14. Computer evaluated phase images for the transient behavior of a clamped steel plate after the impact of a hammer. (The phases are represented by a linear gray scale between 0 to 2π) [4.18]

The described system has been applied to analyze the transient behavior of a clamped steel plate after the impact of a hammer [4.18, 19]. The four laser pulses are triggered a few microseconds after the impact with intervals of 5, 10 and 20 μs, respectively. Figure 4.14 shows the results of the corresponding computer evaluated phase images. The phases are represented by a linear gray scale between 0 to 2π. Multiple-exposure holographic interferometry with quantitative phase evaluation is of great interest for modal analysis of transient vibrations and wave propagation.

4.4.3 Double-Pulse Holography for 3-D Displacement Measurement

The interference phase ϕ measures only the component u_g of the displacement vector L in the direction of the sensitivity vector $K = K_1 - K_2$, which points along the bisector of the illumination wave vector K_1 and the observation wave vector K_2 (Chap. 3). To determine all three components of the displacement vector L, at least three interference phase measurements ϕ_n ($n = 1, \ldots, N$;

$N \geq 3$) with different sensitivity vectors K^n have to be carried out. The different sensitivity vectors K^n can be realized either by changing the direction of observation K_2 during hologram reconstruction or by recording several holograms with different illumination directions K_1. For the reconstruction of the displacement vector L it is very important, that the measured components L^n correspond to exactly the same point of the object. This is very difficult to achieve, when different observation directions K_2^n are used. Therefore, three double-exposure holograms with different illumination directions K_1^n are recorded on the same hologram plate, spatially multiplexed by rotating a symmetric aperture in front of the holographic plate, so that the reconstructions can be observed with the same imaging system without any distortion by change of perspective [4.20] (see also Sect. 4.4.4). In addition, using different illumination directions allows to get high sensitivity for in-plane deformations.

It is possible to combine this approach with double- or multiple-pulse two-reference beam holography [4.18, 21]. Three holograms with three different illumination directions are recorded independently by three temporally separated pulses, which are obtained from one laser pulse using optical delay lines [4.22]. The lengths of the delay lines are typically 4 and 8 m for Q-switched pulses of about 10 ns duration. Figure 4.15 exhibits the recombination of the three delayed pulses to produce three reference beams which are coded with three aperture masks. The following Pockels cell switches the two reference beams between the two exposures of the hologram (Sect. 4.4.1). The aperture masks are imaged through the two-reference-beam module onto the hologram plate. Using the corresponding masks during reconstruction, the three spatially multiplexed double-exposure two-reference beam holograms can be analyzed independently by standard phase-shifting methods.

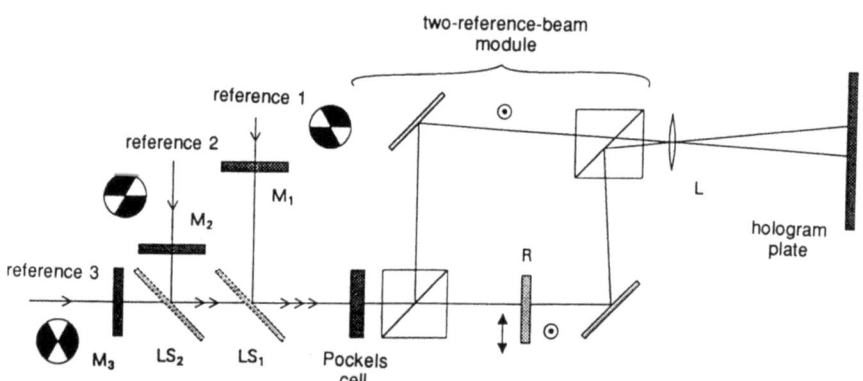

Fig. 4.15. Arrangement to recombine three delayed pulses to produce three reference beams coded with different aperture masks. The Pockels cell switches the two reference beams between the two exposures of the hologram [4.18]

The described system has been successfully applied to 3-D vibration analysis [4.18, 21]. Previously it was tested for the case of a rotating disk. The disk had a diameter of 150 mm and was rotating at an angular speed of about 0.1 rad/s. The two pulses of the Q-switched ruby laser were fired at an interval of 600 μs. Figure 4.14 shows the results of three computer evaluated holograms, corresponding to the three different illumination vectors. The measured phases are represented by a linear gray scale between 0 and 2π. Taking into account the geometry of the experimental setup, the three Cartesian components of the displacement vector L have been calculated. The results are given in Fig. 4.16b, still represented by their phase values, similar to Fig. 4.16a. The theoretical prediction is zero fringes for the z-component (out-of plane) and parallel, equidistant fringes for both the x- and y-components (in-plane). A statistical analysis of the results shows an rms phase error of the order of 10° with respect to the theoretical values for the in-plane components. The residual fringe in the z-direction can be explained by out-of plane vibrations induced by the driving motor.

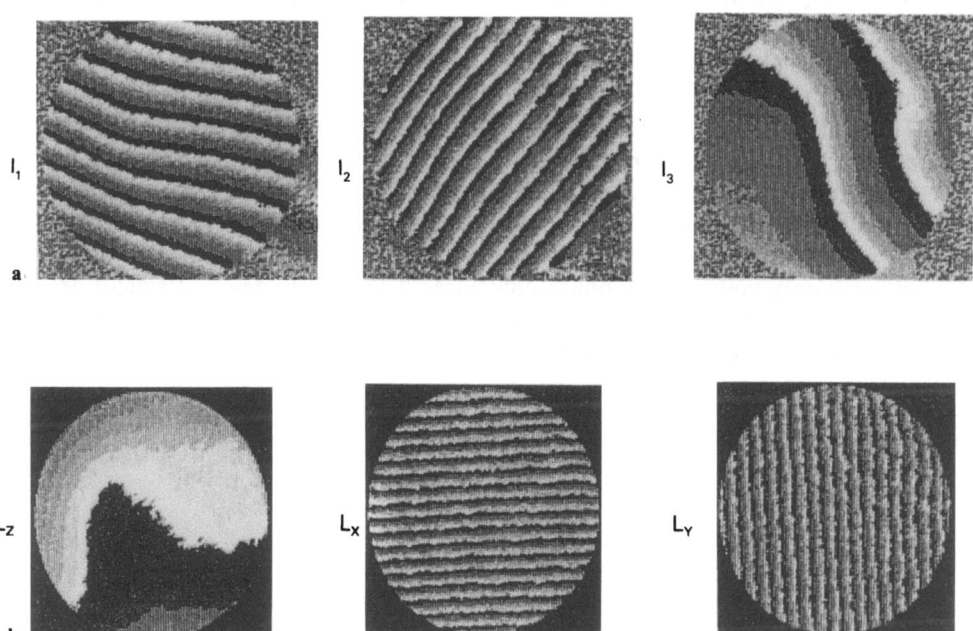

Fig. 4.16a,b. Measuring the 3-D displacement of a rotating disk by double-pulse holography. (a) Computer evaluated holograms, corresponding to the three different illumination vectors. (b) Calculated Cartesian components L_x, L_y, L_z of the displacement vector L. (Phases representation with a linear gray scale between 0 to 2π) [4.18]

4.4.4 Strain Measurement by Heterodyne Holographic Interferometry

Strain and rotation of the object surface can be determined by numerical differentiation of the displacement vector field $L(x)$ in the tangential plane at each point. This requires highly accurate measurement of the 3-D displacement vector L. The accuracy for the determination of surface strain by holographic interferometry depends therefore essentially on the error in the interference phase measurement [4.23]. For this reason, heterodyne holographic interferometry with two well separated reference sources is the best choice.

The experimental arrangement for double-exposure heterodyne holographic interferometry, as it has been used in the experiments reported here [4.20], is sketched in Fig. 4.17. The setup is essentially based on the one described in [4.7]. The frequency difference $\Omega/2\pi$ of 100 kHz between the two reference beams is realized by two commercially available Acousto-Optical Modulators (AOM) in cascade to give opposite frequency shifts. During recording, both modulators are driven with 40 MHz, so that the net shift is zero. During reconstruction, one modulator is driven with 40 MHz and the other with 40.1 MHz, so that the net shift is the desired beat frequency of 100 kHz.

The interferogram is observed in the image of the virtual object. To avoid any error from overlapping reconstructions (Sect. 4.3.2.2), the two reference sources are placed on the same side of the object with a mutual separation larger than the angular size of the object in the corresponding direction (Fig. 4.2a). The interference phase is determined by scanning an array of three detectors which measures the phase difference in two orthogonal directions in the image plane. The total phase $\phi(x)$ can be obtained afterwards by appropriate integration. The detector array is realized by the ends of three fiber bundles, which feed the light to photomultiplier tubes. The beat-frequency signals at 100 kHz are filtered with a bandwidth of 10 kHz, and the amplitudes are kept constant, independent of the intensity variations across the image, by a feedback control of the photomultiplier supply voltage. The phase differences between the detected signals are measured with zero-crossing phasemeters, which interpolate the phase angles to 0.1° and also count the multiples of 360° (fringe number). Note, that all

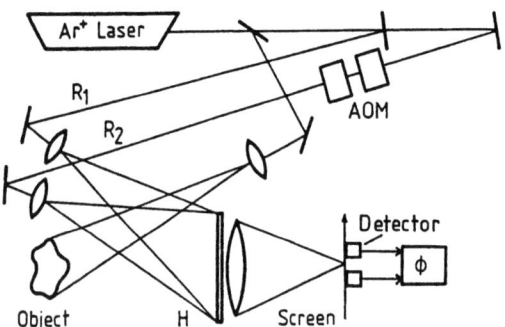

Fig. 4.17. Arrangement for heterodyne holographic interferometry using two acousto-optical modulators to generate the frequency offset

electronic amplifiers and filters in the signal path should be carefully designed to avoid phase distortion that could reduce the accuracy of the phase measurement [4.24]. The detector array is mounted on a stepper motor driven stage to scan the image. Scanning and data acquisition is automated and computer controlled. The measuring time for one position, including displacement of the detector head, is about 1 s.

For the determination of vector displacement and strain, the interference phase caused by the object deformation must be measured for different sensitivity vectors, by changing either the observation or the illumination directions (Sect. 4.4.3 and Chap. 3). In order to have a fixed imaging system and changes of the sensitivity vector which are large enough, we used a setup with three illumination sources [4.25]. The deformation is recorded with three double-exposure holograms, each of them with another object illumination. For reasons of simplicity in the recording procedure and sensitivity to misalignment errors of the holograms, only one holographic plate is used. The different holograms are spatially multiplexed by rotating a symmetric aperture in front of the holographic plate, as shown in Fig. 4.18. The three illumination sources are arranged around the optical axis. The angle between the illumination directions and the

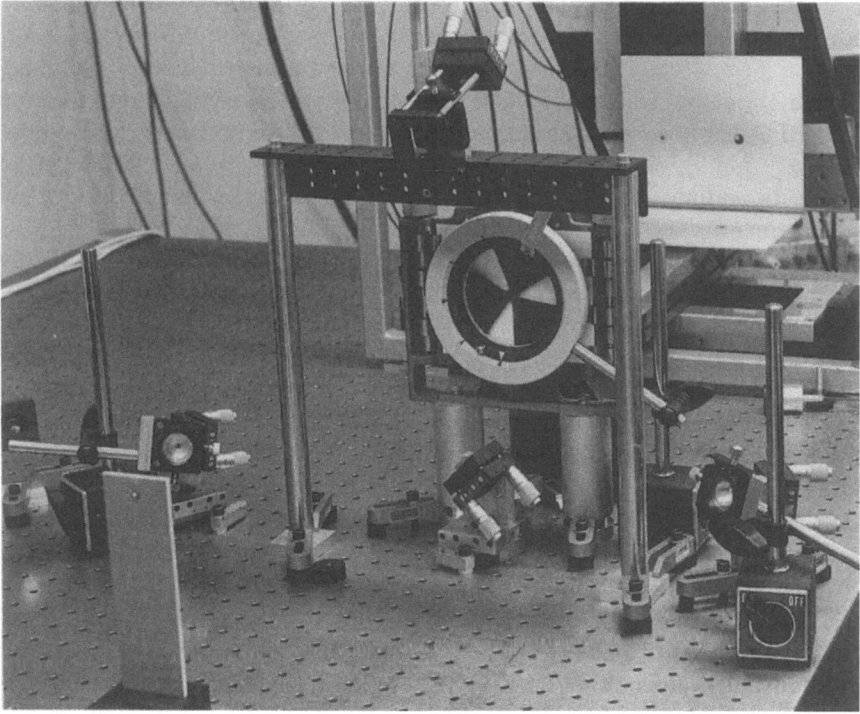

Fig. 4.18. Optical arrangement for the spatial multiplexing of three holograms with a rotating aperture. The three illumination sources are arranged around the optical axis

optical axis is typically about 26°. First, the three holograms of the undeformed object state are recorded with the first reference wave on the appropriate part of the hologram plate. Then, the corresponding holograms of the deformed state are recorded with the second reference wave. It is extremely important, that the optical setup as well as the object in its undeformed and its deformed state are stable during the whole recording procedure. The heterodyne evaluation of the interference phase is performed in the same optical arrangement after careful realignment of the hologram plate.

The sensitivity and reliability of experimental strain determination has been tested by measuring an undeformed object and an object undergoing only a rigid body motion [4.20]. An aluminum plate, which could be rotated about its normal axis, was chosen for that purpose. The interference phase has been measured on a grid of 15×15 sample points, 3 mm apart from each other. After the evaluation of the displacement vector field, the three surface strain components and the three rotation components have been calculated by numerical derivation of the displacement field. For the undeformed object, the mean values and the standard deviations of strain and local rotation for the evaluated sample points are smaller than 0.3 µm/m. For the rotated disk, all components, except the rotation ω_z, should be equal to zero. In the experiment, the rotation was $\omega_z = 38 \times 10^{-6}$. This is small enough to have even at the border of the evaluated surface area in-plane displacement components < 1.5 µm, which do not yet introduce a detectable statistical phase error due to the speckle decorrelation (Sect. 4.3.2.1). The measured rms errors are larger than for the undeformed object, but still smaller than one micro-strain for a spatial resolution of 6 mm. Furthermore, it has been observed that the two normal strain components ε_{xx} and ε_{yy} show deviations from zero to about 2µm/m, which are of systematic nature and not due to statistical errors in the interference phase measurement.

The systematic deviations observed in the experiments above are mainly caused by repositioning errors of the hologram plate and atmospheric turbulences during hologram recording. The statistical interference phase measurement accuracy has been shown to be much better, namely of the order of 1/1000 fringe or 0.3° rms. This proves clearly that limitations for strain determination in heterodyne holographic interferometry do not arise from limitations in the interference phase measurement technique, but rather from the interferometry itself, that is to say from the stability of the optical setup, air turbulences, zero order fringe errors, and inaccurately known geometrical data of the setup.

Another investigated test object was a cylindrical vessel loaded by internal pressure [4.20, 23]. In this case, the strain has to be evaluated on a curved surface. The cylindrical vessel was an aluminum tube with a radius of 10 cm and a wall thickness of 5 mm. In the experiment [4.20], the difference of the internal pressure between the two exposures was 2.6 bar. Figure 4.19 exhibits the interference fringe patterns of the three holograms, recorded with different illumination directions. The indicated rectangular area corresponds to the field

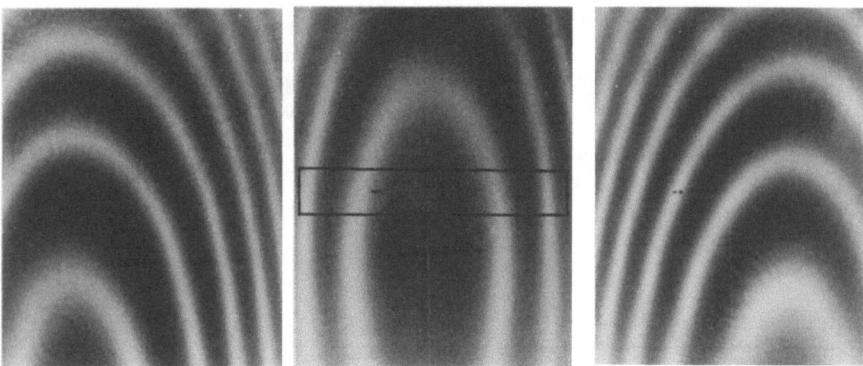

Fig. 4.19. Interference fringe patterns on a cylindrical vessel deformed by internal pressure for three illumination directions. The rectangular field indicates the area where the interferogram has been evaluated

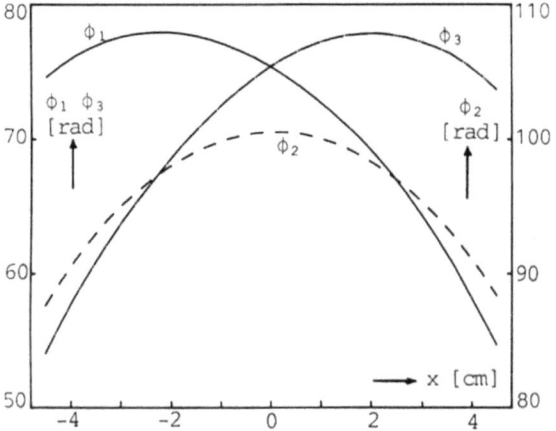

Fig. 4.20. Measured phase function along a horizontal line through the interferogram of Fig. 4.19

of 31 × 5 sample points, for which the interferogram has been analyzed. In Fig. 4.20, the measured phase functions along a horizontal line through the three interferograms are plotted. Figure 4.21 shows the experimentally determined displacement components along a radial section on the cylinder surface. The corresponding strain components are displayed in Fig. 4.22. The reliability of the measurements has been checked by the evaluation of the interferogram for the same holographic recording on two interlaced grids of sample points (o and ∗, respectively). The standard deviation of the difference between the two measurements is of the order of 1 μm/m.

These results illustrate impressively the power of heterodyne holographic interferometry. It is possible to determine all components of surface strain and

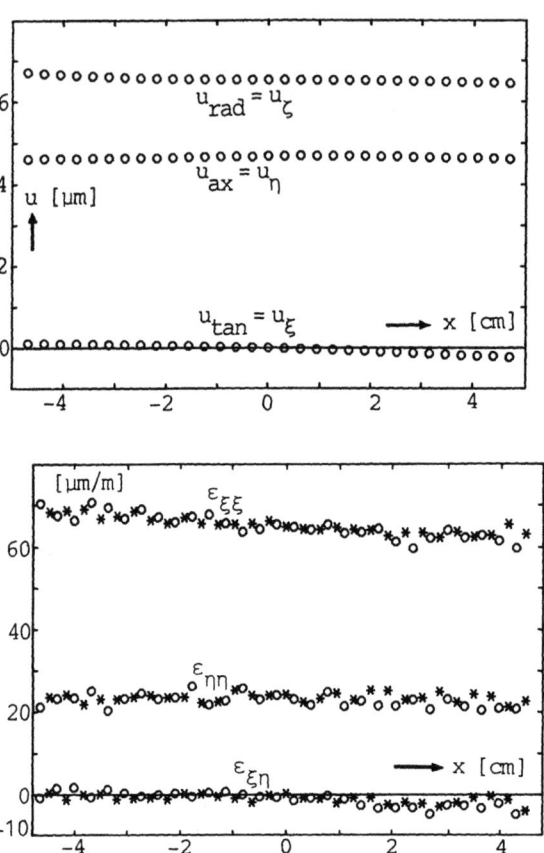

Fig. 4.21. Experimentally determined radial, axial and tangential displacement components on the (curved) cylinder surface

Fig. 4.22. Experimentally determined surface strain components $\varepsilon_{\xi\xi}$, $\varepsilon_{\eta\eta}$, $\varepsilon_{\xi\eta}$, of the cylindrical vessel evaluated from interference phases measured on two interlaced grids of sample points (o and *)

rotation on a curved surface with an accuracy of about one micro-strain for a spatial resolution of a few millimeters.

4.5 Conclusions

Heterodyne holographic interferometry offers very high accuracy for the interference phase measurement (better than 1/1000 of a fringe). An arrangement with reference sources close together, which would introduce additional errors due to the speckle decorrelation, is therefore not adequate. Well separated reference sources, on the other hand, impose strong requirements to the stability of the optical setup and to the repositioning of the hologram. For recording and reconstruction, one and the same experimental setup has to be used. Surface-strain measurement of extremely high resolution is possible, when three two-reference-beam holograms for different illumination directions are recorded by spatial multiplexing on the same hologram plate. The experimental results show that it is not the heterodyne fringe interpolation technique that limits the measurement accuracy, but rather air turbulence, hologram repositioning and speckle decorrelation due to large in-plane displacement.

Quasi-heterodyne holographic interferometry with TV-detection offers a moderate accuracy for the fringe interpolation (about 1/100 of a fringe), with all the advantages of phase shifting interferometry. An optical arrangement with two reference sources close together is nearly as simple and uncritical for the hologram recording and reconstruction as standard double-exposure holography. It is also possible to record the holograms with a double-pulse Q-switched laser (e.g., ruby laser) and to analyze the interferograms in a separate setup with a cw laser (e.g., HeNe laser). Finally, the 3-D displacement vector can be determined even in an industrial environment from two-reference-beam holograms recorded with a double-pulse Q-switched laser. For that purpose, three two-reference-beam holograms with three different illumination directions are recorded independently on the same plate using optical delay lines and appropriate aperture masks to code the reference beams. The three spatially multiplexed double-exposure, two-reference-beam holograms are then observed with the same camera position and analyzed by standard phase-shifting methods to get the three components of the displacement vector for any point on the object.

References

4.1   See, e.g., J. Schwider: *Progress in Optics* **28**, 273–359 (North Holland, Amsterdam 1980)
4.2   R. Dändliker, E. Marom, F.M. Mottier: J. Opt. Soc. Am. **66**, 23–30 (1976)
4.3   J.E. Greivenkamp: Opt. Eng. **23**, 350–352 (1984)
4.4   C.J. Morgan: Opt. Lett. **7**, 368–370 (1982)
4.5   K. Creath: *Progress in Optics* **26**, 350–393 (Elsevier Science Publishers, Amsterdam 1988)
4.6   J.H. Bruning, D.R. Herriott, J.E. Gallagher, D.P. Rosenfeld, A.D. White, D.J. Brangaccio: Appl. Opt. **13**, 2693–2703 (1974)
4.7   R. Dändliker: *Progress in Optics* **17**, 1–84 (North Holland, Amsterdam 1980)
4.8   B. Breuckmann, W. Thieme: Appl. Opt. **24**, 2145–2149 (1985)
4.9   R. Dändliker, R. Thalmann, J.-F. Willemin: Opt. Commun. **42**, 301–306 (1982)
4.10 R. Thalmann, R. Dändliker: J. Opt. Soc. Am. A, **3**, 972–981 (1986)
4.11 S. Lowenthal, H. Arsenault: J. Opt. Soc. Am. **60**, 1478–1483 (1970)
4.12 See, e.g., J.W. Goodman: *Introduction to Fourier Optics* (McGraw-Hill, New York 1979)
       B.R. Frieden: *Probability, statistical optics and Data Testing*, 2nd edn., Springer Ser. Int. Sci., Vol. 10 (Springer, Berlin, Heidelberg 1992)
4.13 J.W. Goodman: In *Laser Speckle and Related Phenomena*, ed. J.C. Dainty, 2nd edn. (Springer, Berlin, Heidelberg 1984) pp. 9–75
4.14 C.M. Vest: *Holographic Interferometry* (Wiley, New York 1979)
4.15 R. Dändliker, B. Ineichen: Opt. Commun. **19**, 365–369 (1976)
4.16 D.W. Watt, C.M. Vest: Exp. in Fluids, **5**, 401–406 (1987)
4.17 G. Lai, T. Yatagai: Appl. Opt. **27**, 3855–3858 (1988)

4.18 V. Linet: "Développement d'une méthode d'interférométrie holographique appliquée à l'analyse quantitative 3D du comportement dynamique de structures", Thèse de Doctorat, Orsay (1992)
4.19 X. Bohineust, V. Linet, F. Dupuy: "Dynamic analysis of structures by holographic interferometry: strategy and developments for vehicle", Troisième Colloque Franco-Allemand sur les Applications de l'Holographie, St.-Louis (1991)
4.20 R. Thalmann, R. Dändliker: Appl. Opt. **26**, 1964–1971 (1987)
4.21 V. Linet, X. Bohineust, F. Dupuy: "Three dimensional dynamic analysis of parts of automobile body by holographic interferometry", Troisième Colloque Franco-Allemand sur les applications de l'Holographie, St.-Louis (1991)
4.22 Z. Füzessy: SPIE Proc. **398**, 17–21 (1983)
4.23 R. Dändliker, B. Eliasson: Exp. Mech. **19**, 93–101 (1979)
4.24 J. Mastner and V. Masek: Rev. Sci. Instrum. **51**, 926–931 (1980)
4.25 P. Hariharan, B.F. Oreb, N. Brown: Appl. Opt. **22**, 876–880 (1983)

5. Phase-Shifting Holographic Interferometry

K. Creath

Optical Sciences Center, University of Arizona, Tucson, AZ 85721, USA

Quantitative data can be extracted from holographic interference fringes using Phase-Measurement Interferometry (PMI) techniques. These techniques are used to determine the phase of the secondary interference fringe pattern and can be divided broadly into spatial and temporal techniques. Temporal techniques introduce a known phase shift between the object and reference beams in an interferometer and take a series of data over time as the phase shift is varied. Spatial techniques rely on encoding the phase shift information spatially in a single interferogram by using a large number of fringes as a carrier for the phase information which are generated by tilting the reference wavefront relative to the test wave front. Temporal techniques which process data in electronics are known as heterodyne techniques and were discussed in the last chapter. The processing of spatial phase-measurement data will be discussed in the next chapter. Temporal techniques which process the data analytically on a point-by-point basis will be the concentration of this chapter.

This chapter will begin by providing some background on holographic secondary interference fringes and the basics of PMI. It then continues by outlining the various algorithms which can be used with real-time holographic interferometry to measure the difference between two states of an object. The phase difference may correspond to a static object displacement or to the shape of an object. Other topics to be discussed are calibration, major error sources, and equipment and experimental considerations. Specialized technqiues for electronic holography and time-average vibration analysis will be covered at the end of this chapter.

5.1 Background

5.1.1 Real-Time Holographic Secondary Interference Fringes

Phase-measurement techniques in holography are most often applied to real-time holographic interferometry. This technique uses a hologram to store the wave front scattered from an object. After the state of the object is changed, the wave front scattered from the object will be changed by a small amount. The reference beam will playback the object wave front stored in the hologram. This

110   5. Phase-Shifting Holographic Interferometry

stored wave front will interfere with the live wave front scattered by the object to produce secondary interference fringes. These secondary interference fringes correspond to the differences in the object position and shape between making and reading out of the hologram. As long as the change in the object is small, the secondary interference fringes will be localized at the object. It is the phase of the difference wave front of the secondary interference fringes which is measured using PMI. This difference can correspond to a displacement of the object due to applied stress. Or it can correspond to the shape of the object if two angles of incidence, two wavelengths, or two indices of refraction are used. When the object is in motion, a pulsed laser or a high-speed shutter may be used to freeze the motion. More detail about the specific techniques for holographic interferometry is given in (Chap. 7).

The secondary interference fringes for a real-time holographic measurement can be written mathematically as

$$I(x,y) = I_0(x,y)\,[1 + \gamma(x,y)\cos\Delta\phi(x,y)], \tag{5.1}$$

where I_0 is the unknown dc intensity of the interferogram, γ is the unknown fringe visibility, and $\Delta\phi$ is the phase of the secondary interference fringes to be determined (Fig. 5.1). This equation assumes a single point x, y in the the viewing plane. All of the variables may change from point to point across the interferogram.

5.1.2 From Wavefront to Object Displacement

The measurement phase difference $\Delta\phi$ can be written as

$$\Delta\phi = \mathbf{K}\cdot\mathbf{L}, \tag{5.2}$$

where $\mathbf{L}$ is a vector corresponding to the object displacement, and $\mathbf{K}$ is a vector corresponding to the measurement direction bisecting the illumination and viewing directions of a single object point. It is also known as the sensitivity

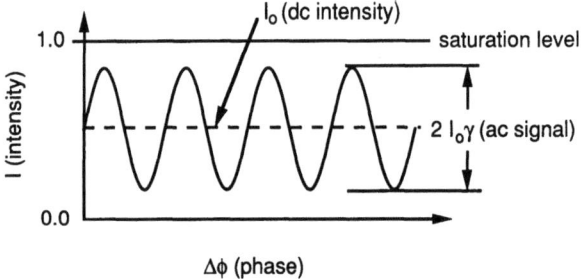

Fig. 5.1. Quantities related to secondary interference fringes

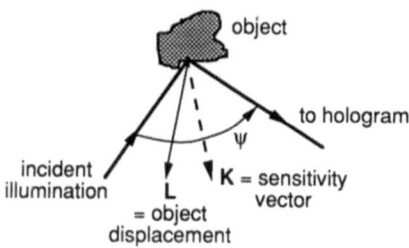

Fig. 5.2. Relationship between illumination and viewing directions and object displacement

vector. These vectors will change orientation across an object which is not illuminated and viewed in collimated light. They will also vary if the object is not flat. Figure 5.2 presents a drawing indicating these vectors.

The displacement of the object at a point x, y in the direction of the sensitivity vector is given by

$$D(x, y) = \frac{\Delta\phi(x, y)\lambda}{4\pi \cos(\psi/2)}, \tag{5.3}$$

where λ is the illumination wavelength, and ψ is the angle between the illumination and viewing directions. The measured displacment is usually a combination of in-plane (along the surface of object) and out-of-plane (perpendicular to the object surface) displacements. Specific setups to measure only one of these components are described elsewhere in this book. If all three components of displacement (x, y, and z) are desired, at least three measurements must be made [5.1].

5.2 Phase-Measurement Basics

PMI has been used to measure wave front phase in interferometers for 25 years [5.2-7] and in holographic interferometry since the early 1980's [5.8-11]. Generally, a number of data frames are recorded as the reference-beam phase is changed in a known manner. The data are shipped to a computer where the phase at each detector point is calculated. Information about the test surface is geometrically related to the calculated wave front phase.

The direct measurement of phase information has many advantages over simply tracing fringes in interferograms. The precision of PMI techniques is a factor of ten to a hundred greater than tracing fringes, and it is simple. The only necessary modifications to an interferometer are a detector array placed at plane conjugate to the object (assuming the fringes are localized at the object) and a phase-shifting device placed in one beam. With state-of-the-art frame grabbers and desktop computers data can be taken rapidly and phase determined in a matter of seconds over hundreds of thousands of data points.

112   5. Phase-Shifting Holographic Interferometry

5.2.1 General Phase-Measurement Theory

After adding the induced phase shift to (5.1), a frame of measured intensity data can be written as

$$I(x,y) = I_0(x,y)\{1 + \gamma_0(x,y)\cos[\Delta\phi(x,y) + \alpha]\}, \tag{5.4}$$

where α is the induced relative phase shift between the test and reference beams of the interferometer. It is assumed to be constant across the entire interferogram. In most cases α is known; however, there are algorithms where the phase shift does not need to be known. The induced phase shift can be anything between 0 and 180° (0 to π radians). The three unknowns (I_0, γ_0, and ϕ) require a minimum of three measurements to determine the phase.

For the temporal phase-shifting class of techniques, N frames of intensity data are recorded as the phase is shifted. In general, the phase-shift is assumed to be constant from frame to frame and may be changing during the detector's integration time. (If the phase shift α changes from frame to frame, it must be determined by calibration or calculation.) One frame of integrated recorded intensity data is written as [5.12]

$$I_i(x,y) = \frac{1}{\delta}\int_{\alpha_i-\delta/2}^{\alpha_i+\delta/2} I_0(x,y)\{1 + \gamma_0(x,y)\cos[\Delta\phi(x,y) + \alpha(t)]\}\,d\alpha(t), \tag{5.5}$$

where $I_0(x,y)$ is the average intensity, $\gamma_0(x,y)$ is the visibility of the incident fringe pattern, $\Delta\phi(x,y)$ is the phase of the wavefront being measured, $\alpha(t)$ is the relative phase shift between test and reference beams as a function of time, α_i is the average value of the phase shift for the ith exposure, and δ is the integrated phase shift (or detector integration time). The integration over a phase shift δ enables this expression to apply to any phase-shifting technique. The relationship between δ and α is shown in Fig. 5.3.

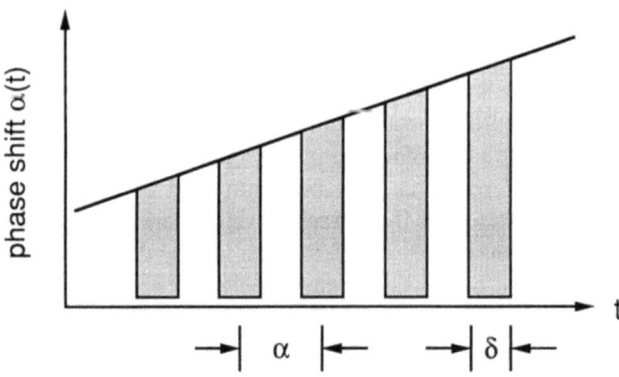

Fig. 5.3. Relationship between the phase shift between data frames α and the integration interval δ

5.2 Phase-Measurement Basics

After integrating (5.5), the recorded intensity is

$$I_i(x, y) = I_0(x, y) \{1 + \gamma_0(x, y) \operatorname{sinc}(\delta/2) \cos[\Delta\phi(x, y) + \alpha_i]\}$$
$$= I_0(x, y) \{1 + \gamma(x, y) \cos[\Delta\phi(x, y) + \alpha_i]\}, \tag{5.6}$$

where $\operatorname{sinc}(\delta/2) = \sin(\delta/2)/(\delta/2)$ and γ is the detected fringe visibility. When the phase shift is stepped ($\delta = 0$), the sinc function has a value of one. When $\delta = \alpha$, the phase shift is ramped continuously and the value of the sinc function is reduced. For $\delta = \pi/2$ (90°), the value of the since function is 0.9. The only difference between ramping and stepping the phase shift is a reduction in fringe visibility after detection. At the extreme of $\delta = 2\pi$, there is no intensity modulation.

For N intensity measurements, the most general phase calculation uses a least squares technique. This approach is outlined in detail by the *Greivenkamp* [5.12] and *Morgan* [5.13]. Equation (5.6) is first rewritten in the form

$$I_i(x, y) = a_0(x, y) + a_1(x, y) \cos\alpha_i + a_2(x, y) \sin\alpha_i, \tag{5.7}$$

where

$$a_0(x, y) = I_0(x, y), \tag{5.8}$$

$$a_1(x, y) = I_0(x, y)\gamma_0(x, y) \operatorname{sinc}(\delta/2) \cos[\Delta\phi(x, y)], \tag{5.9}$$

$$a_2(x, y) = -I_0(x, y)\gamma_0(x, y) \operatorname{sinc}(\delta/2) \sin[\Delta\phi(x, y)]. \tag{5.10}$$

The unknowns in these equations are represented by $a_0(x, y)$, $a_1(x, y)$, and $a_2(x, y)$. The least squares solution to these equations is

$$\boldsymbol{a}(x, y) = \begin{bmatrix} a_0(x, y) \\ a_1(x, y) \\ a_2(x, y) \end{bmatrix} = \boldsymbol{A}^{-1}(\alpha_i) \boldsymbol{B}(x, y, \alpha_i), \tag{5.11}$$

where

$$\boldsymbol{A}(\alpha_i) = \begin{bmatrix} N & \sum \cos\alpha_i & \sum \sin\alpha_i \\ \sum \cos\alpha_i & \sum \cos^2\alpha_i & \sum (\cos\alpha_i)\sin\alpha_i \\ \sum \sin\alpha_i & \sum (\cos\alpha_i)\sin\alpha_i & \sum \sin^2\alpha_i \end{bmatrix}, \tag{5.12}$$

and

$$\boldsymbol{B}(\alpha_i) = \begin{bmatrix} \sum I_i(x, y) \\ \sum I_i(x, y)\cos\alpha_i \\ \sum I_i(x, y)\sin\alpha_i \end{bmatrix}. \tag{5.13}$$

The matrix A only needs to be calculated and inverted once because it is dependent just on the phase shift. The phase at each point in the interferogram is determined by evaluating the value of B at each point and then solving for the a_1 and a_2 coefficients:

$$\Delta\phi(x, y) = \tan^{-1}\left(\frac{-a_2(x, y)}{a_1(x, y)}\right)$$

$$= \tan^{-1}\left(\frac{I_0(x, y)\gamma_0(x, y) \operatorname{sinc}(\delta/2) \sin[\Delta\phi(x, y)]}{I_0(x, y)\gamma_0(x, y) \operatorname{sinc}(\delta/2) \cos[\Delta\phi(x, y)]}\right). \tag{5.14}$$

This phase calculation assumes that the phase shifts between measurements are known, and that the integration period δ is constant for each data frame. A generalized least-squares phase calculation has also been developed which will work with any phase shifts [5.14]. Because of the arctangent function, (5.14) (and most of the techniques described in this chapter) determines the phase modulo π. To determine the phase modulo 2π, either you have a function which does this in your compiler or the signs of quantities proportional to $\sin \Delta\phi$ and $\cos \Delta\phi$ must be examined [5.5].

The fringe visibility will not only be reduced by ramping the phase shift, but it will also be reduced by averaging over the finite size of the detector element and could be affected by other less quantifiable effects such as scattered light. To make reliable phase measurements, the intensity recorded by the detector must modulate as the phase is shifted between frames of data. In fact, the larger the fringe modulation, the more accurate the phase calculation will be because the data will cover a larger number of levels of the Analog-to-Digital Converter (ADC). The recorded fringe visibility can be calculated from the intensity data using

$$\gamma(x, y) = \gamma_0(x, y) \operatorname{sinc}(\delta/2) = \frac{\sqrt{a_1(x, y)^2 + a_2(x, y)^2}}{a_0(x, y)}. \tag{5.15}$$

This expression can be used to determine if a data point will yield an accurate phase measurement or if it should be ignored. Expressions for the detected fringe visibility are given for each of the specific techniques described in Sect. 5.3.

5.2.2 Phase-Modulation Techniques

A phase shift (modulation) in an interferometer can be induced by moving a mirror, tilting a glass plate, moving a grating, rotating a half-wave plate or analyzer, using an acousto-optic or electro-optic modulator, or using a Zeeman laser [5.15–23]. Phase shifters such as moving mirrors, gratings, tilted glass plates, or polarization components can produce continuous as well as discrete phase shifts between the object and reference beams. All of these methods effectively shift the frequency of one beam in the interferometer with respect to

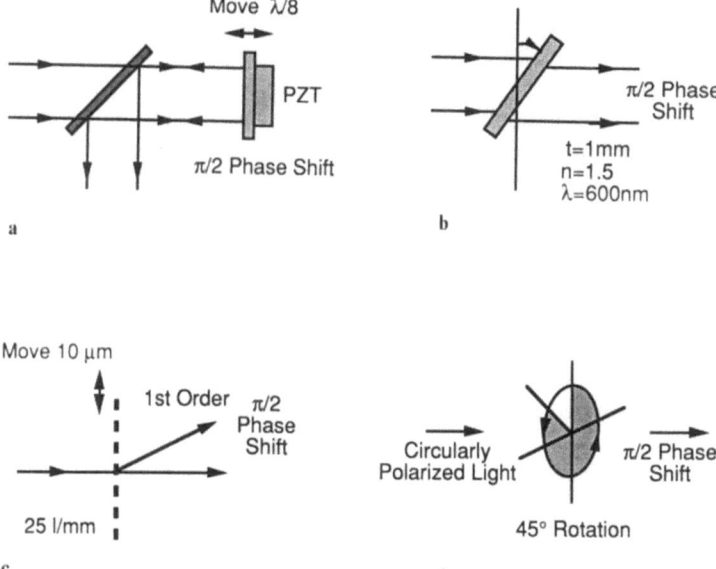

Fig. 5.4. Means of modulating or shifting the phase in an interferometer

the other to introduce a phase difference between beams. Phase shifters may either be placed in one arm of the interferometer or positioned so that they shift the phase of one of two orthogonally polarized beams. In the latter case, the interferometer must use a polarizing beamsplitter to separate the two beams and then recombine them in the output arm with an analyzer to get interference fringes.

A very common phase-shifting technique used in commercial interferometers is to push a mirror in the reference beam with a piezo-electric transducer (PZT) (Fig. 5.4a) [5.24, 25]. Many brands of PZTs are available to linearly move a mirror over a 1 μm range. A high-voltage amplifier is used to produce a linear ramping signal from 0 to several hundred volts. Care needs to be taken if the phase shifter is placed at an angle other than normal incidence or in a converging beam to make sure that the phase shift is constant over the entire beam and to ensure the beam is not translated as it is shifted.

A tilted glass plate placed in one beam can provide a relative phase shift between object and reference beams (Fig. 5.4b) [5.19]. However, this plate must have high optical quality, and care must be taken to have an equal optical path over the entire beam diameter. To minimize aberrations introduced by the plate, it should be placed in a collimated beam. Different amounts of phase shift are achieved by tilting the plate to different angles. Large amounts of tilt will displace the beam.

Another phase-modulation technique continuously moves a diffraction grating in one arm of the interferometer (Fig. 5.4c). Diffraction gratings produce a wavelength-independent frequency shift in the nth diffracted order of nvf when a grating of spatial frequency f is moved with a velocity v [5.16, 19].

In an interferometer with polarization isolation, the object and reference beams have orthogonal linear or circular polarizations. A rotating half-wave plate (or a quarter-wave plate in double pass) in the output of an interferometer will produce a frequency shift at twice its rotation frequency (Fig. 5.4d) [5.15, 18, 19, 22, 23]. (A rotation of 45° will yield a $\pi/2$ phase shift.) Likewise, a rotating analyzer will produce a phase modulation at twice the rotation frequency [5.20].

A phase shift can also be produced using an acousto-optic Bragg cell. In the Bragg cell, a traveling acoustic wave serves as a grating, and the frequency shift obtained in the first diffracted order is equal to the device's driving frequency [5.19, 21]. A Zeeman laser which outputs two different frequencies in two different orthogonal polarizations can also produce the phase shift [5.19].

5.2.3 Phase Unwrapping

The phase calculated modulo 2π using a PMI algorithm is shown in Fig. 5.5a. Note the sharp discontinuities caused when the phase rolls over the 2π boundary. These phase ambiguities need to be removed to reconstruct the measured wave front.

Removing phase ambiguities is known as phase unwrapping or *integrating* the phase. The phase ambiguities can be removed by comparing phase differences between adjacent pixels. When the phase difference between adjacent pixels is greater than π, a multiple of 2π is added or subtracted to make the difference less than π. For the reliable removal of discontinuities, the phase must not change by more than π between adjacent pixels. As long as the data are sampled as described in the sampling requirements, the wave front can be reconstructed (Fig. 5.5b). Specific algorithms can be found in other references [5.26–37]. Techniques have also been developed which enables sampling beyond the Nyquist limit when the surface is either continuous or has known step heights [5.38, 39]

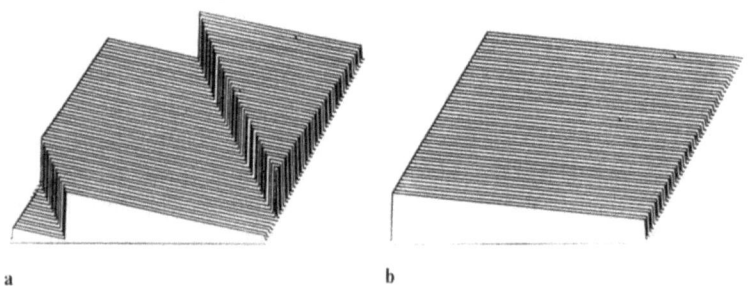

Fig. 5.5. (a) Modulo 2π phase calculation and (b) unwrapped phase

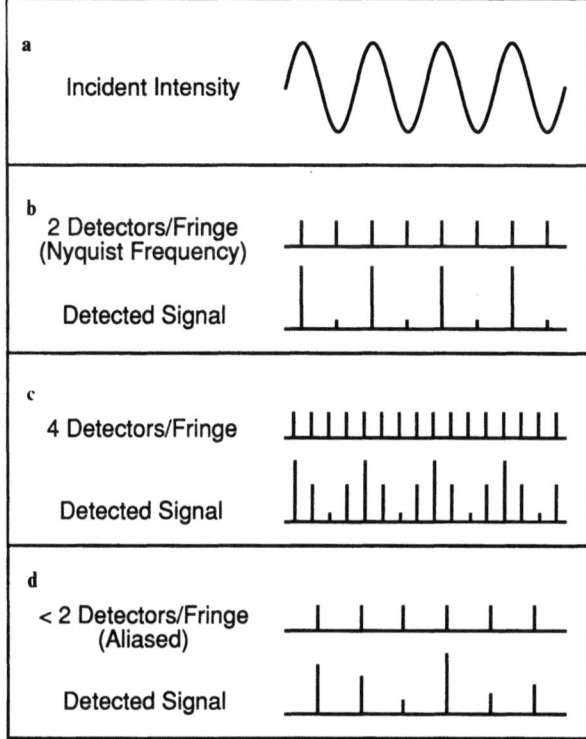

Fig. 5.6. (a) Incident fringe intensity, and detected signal with (b) 2 detectors per fringe, (c) 4 detectors per fringe, and (d) aliasing with less than 2 detectors per fringe

5.2.4 Sampling Requirements

To reconstruct a wave front, the Nyquist sampling theorem states that the highest frequency present must be sampled at least twice per period [5.40]. This has two implications. The smallest feature on the object to be reconstructed has to have at least two sample points across it or else it will not be seen. And the interference fringes must be sampled at least twice per fringe in order to reconstruct the object. Figure 5.6b illustrates sampling a wave front at the Nyquist frequency with two detectors per fringe. Sampling with four detectors per fringe is shown in Fig. 5.6c, and aliasing by sampling with fewer than two detectors per fringe is depicted in Fig. 5.6d.

Because of the sampling theorem, the closest fringe spacing (highest wave front slope) in the interferogram will dictate the number of detector elements necessary to reconstruct the wave front producing that interferogram. Since detector arrays have a finite number of detector elments, the range of measurable surfaces is limited.

118   5. Phase-Shifting Holographic Interferometry

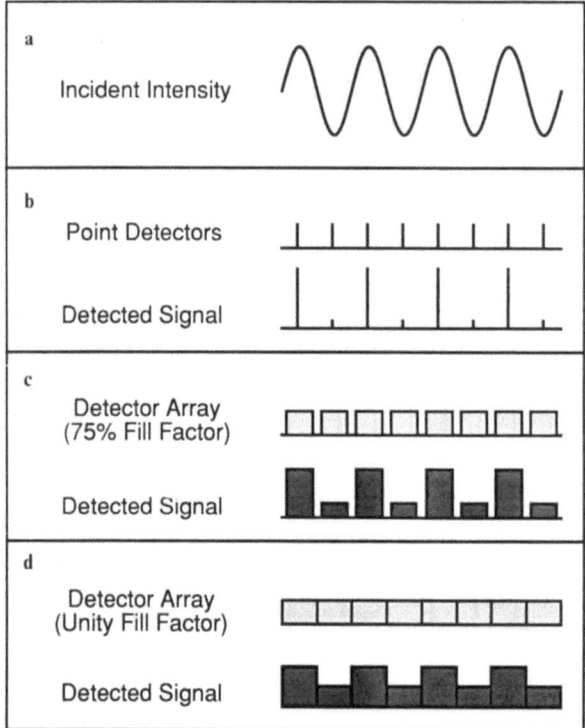

Fig. 5.7. (a) Incident fringe intensity, and detected signal from (b) point detectors, (c) 75% fill factor, and (d) unity fill factor

Another sampling consideration is related to the size of the individual detector elements. If detector elements in an array were point detectors as assumed in the Nyquist theorem, then detecting a high frequency signal with two detectors per period would be no problem. In reality, detector elements have a finite size and therefore average the signal over the sensing area of the element. This smoothes the signal, thereby reducing the detected fringe visibility and intensity modulation. This effect is illustrated by Fig. 5.7. A signal which has a fringe spacing with a period two times the detector spacing is shown in Fig. 5.7a and the detected signal assuming point detectors is shown in Fig. 5.7b. Figure 5.7c depicts the detected signal when the detector width is three-quarters of the detector spacing. Note that the signal amplitude (modulation) is reduced. When the detector width is equal to the detector spacing, a significantly smaller signal is seen as shown in Fig. 5.7d. The modulation of the signal will depend upon the initial phase and when there is no space between detector elements (unity fill factor), there may be no signal modulation at all. About three detector elements per fringe are generally required to provide sufficient phase modulation for typical detector arrays or TV cameras. A good rule of thumb is four detector elements per fringe for most measurement situations.

5.2.5 Intensity-Modulation Requirements

The intensity at a single detector point must modulate as the phase is shifted through a 2π phase shift to make a measurement with PMI. If the intensity at a single detector point does not modulate as the phase is shifted, the phase of the wave front cannot be calculated. Besides the obvious reductions in modulation due to the detector sampling, scattered light within the interferometer and defects or dirt on the test object can also reduce the modulation of the signal. Figure 5.8 exhibits the fringes for a number of data frames with 90° ($\pi/2$) phase shifts along with the calculated ac fringe modulation. Good fringe modulation is illustrated in Fig. 5.8a and a falloff in modulation is illustrated in Fig. 5.8b. Because intensity modulation is important for making a good measurement, it can

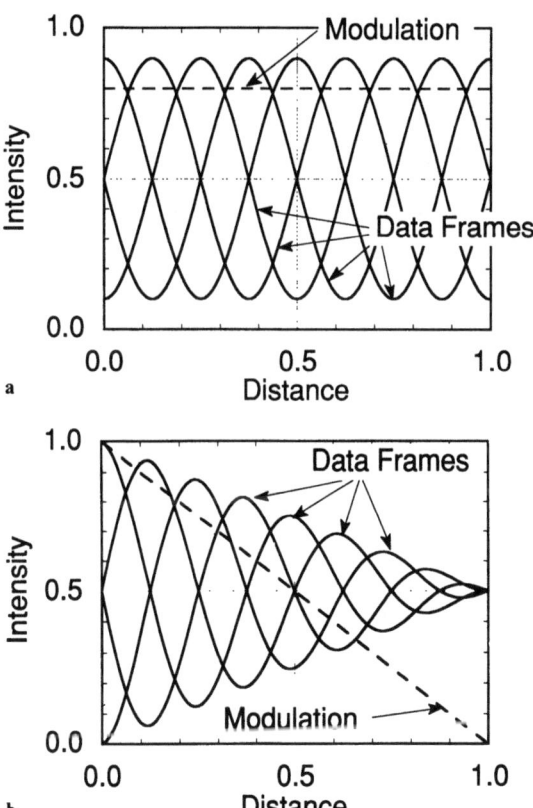

Fig. 5.8. (a) Good fringe modulation, and (b) fringe modulation decreasing

120   5. Phase-Shifting Holographic Interferometry

be calculated at every point in the interferogram and used as a test to determine if a data point is good or bad. The Peak-to-Valley (P–V) of a sinusoidal intensity pattern is $2I_0\gamma$. The maximum value corresponds to the saturation level of the detector (Fig. 5.1). To have a good data point, the ac amplitude of the intensity at a given point has to be greater than some minimum value which is a fraction of the detector's dynamic range. This criterion can be written as

$$2I_0\gamma > ac_{min}. \tag{5.16}$$

For the intensity modulation using the general least-squares phase calculation, this test can be written as

$$2I_0\gamma = \sqrt{a_1(x, y)^2 + a_2(x, y)^2} > ac_{min}; \tag{5.17}$$

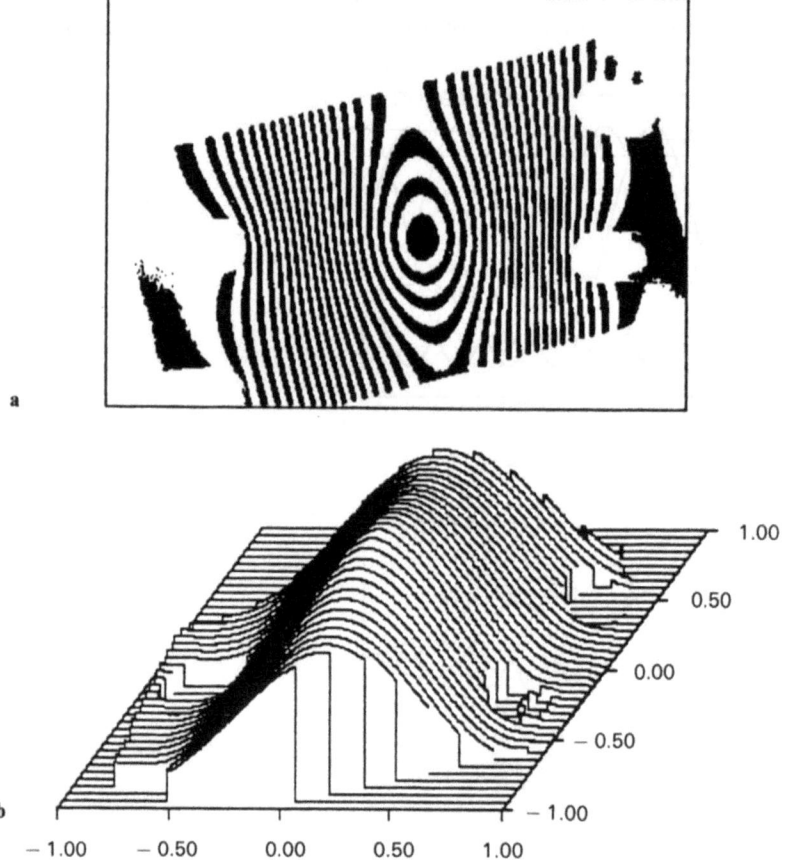

Fig. 5.9. (a) Interference fringe contours (contour interval = 0.3653 μm) and (b) isometric plot (peak-to-valley = 7.3 μm) showing out-of-plane displacement of a metal plate

ac_{min} is typically set between 0.05 and 0.1. The larger its value, the less noisy the data are. Measurements can be made with ac_{min} as small as 0.01.

5.2.6 Measurement Examples

There are many applications of phase-shifting methods in holographic interferometry. The two most common are out-of-plane displacement measurement and shape measurement. Figure 5.9 illustrates a map of the out-of-plane displacement of a metal plate bolted on all four corners when a screw pushing on the back of the plate has been tightened after exposing the holographic plate. The contour interval for Fig. 5.9a is 0.3653 µm, and the angle between the illumination and viewing directions is 60° yielding a peak-to-valley displacement of 7.3 µm. Figure 5.10 shows a contour map of a foot casting using

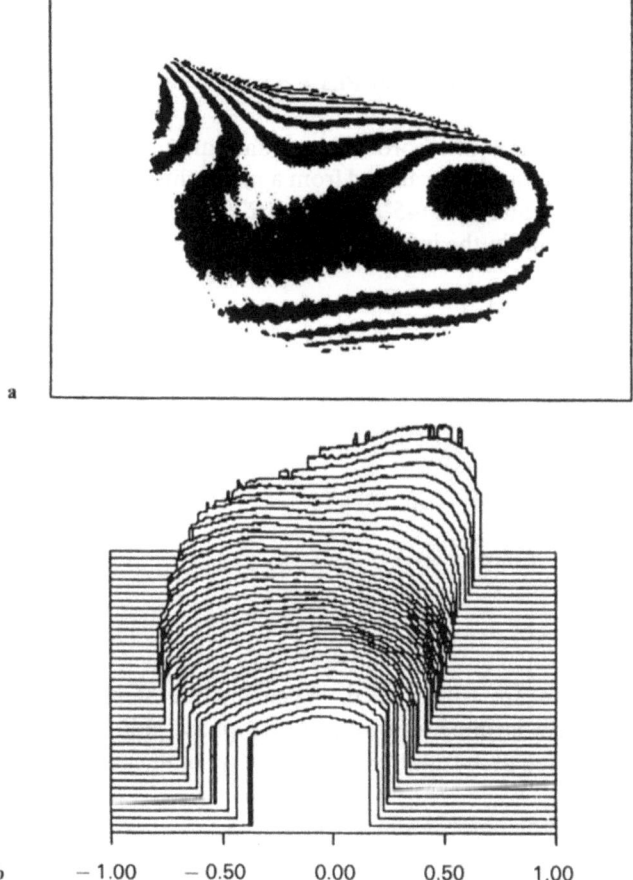

Fig. 5.10. (a) Contour plot (contour interval = 1 mm) and (b) isometric plot of a foot casting with tilt removed

two-angle holographic contouring. This enables the shape of the object to be measured rather than a change in the object between exposures. The contour interval in Fig. 5.10a is 1 mm yielding a peak-to-valley departure of 10 mm over the measured area of the casting. For these examples, holograms were recorded using a thermoplastic plate and secondary interference fringe patterns were recorded utilizing 1320×900 pixels of a Kodak (Videk) Megaplus camera [5.41]. The specific experimental techniques for obtaining these measurements are discussed in more detail in Chap. 7.

5.3 Temporal Phase-Measurement (Phase-Shifting) Algorithms

This section lists the most popular PMI algorithms along with some of the newer techniques which show promise for specific applications.

5.3.1 Three Frames (90°, 120°, and General Phase Shifts)

Since a minimum of three frames of recorded fringe data are needed to reconstruct a wave front, the phase can be calculated from a phase shift of $\pi/2$ (90°) per exposure with $\alpha_i = \pi/4$, $3\pi/4$, and $5\pi/4$ (45°, 135°, and 225°). The three intensity measurements at a single point in the interferogram may be expressed as [5.42]

$$I_1 = I_0[1 + \gamma \cos(\Delta\phi + \pi/4)]$$
$$= I_0[1 + (\sqrt{2}/2)\gamma(\cos\Delta\phi - \sin\Delta\phi)], \qquad (5.18)$$

$$I_2 = I_0[1 + \gamma \cos(\Delta\phi + 3\pi/4)]$$
$$= I_0[1 + (\sqrt{2}/2)\gamma(-\cos\Delta\phi - \sin\Delta\phi)], \qquad (5.19)$$

$$I_3 = I_0[1 + \gamma \cos(\Delta\phi + 5\pi/4)]$$
$$= I_0[1 + (\sqrt{2}/2)\gamma(-\cos\Delta\phi + \sin\Delta\phi)], \qquad (5.20)$$

where the x, y dependencies are still implied. When discrete steps are used, $\gamma = \gamma_0$, and when the phase is integrated over a $\pi/2$ phase shift per frame, $\gamma = 0.9\gamma_0$. Integrating over the phase shift produces a very small effect for a $\pi/2$ phase shift per exposure.

The phase at each detector point is simply

$$\Delta\phi = \tan^{-1}\left(\frac{I_3 - I_2}{I_1 - I_2}\right), \qquad (5.21)$$

where the detected fringe visibility is

$$\gamma = \frac{\sqrt{(I_3 - I_2)^2 + (I_1 - I_2)^2}}{\sqrt{2}I_0}. \tag{5.22}$$

If a general phase shift of α is used, the three intensity measurements become

$$I_1 = I_0[1 + \gamma \cos(\Delta\phi - \alpha)], \tag{5.23}$$
$$I_2 = I_0[1 + \gamma \cos \Delta\phi], \tag{5.24}$$
$$I_3 = I_0[1 + \gamma \cos(\Delta\phi + \alpha)]. \tag{5.25}$$

The phase can be calculated using

$$\Delta\phi = \tan^{-1}\left[\left(\frac{1 - \cos\alpha}{\sin\alpha}\right)\frac{I_1 - I_3}{2I_2 - I_1 - I_3}\right], \tag{5.26}$$

where the detected fringe visibility is

$$\gamma = \frac{\sqrt{[(1 - \cos\alpha)(I_1 - I_3)]^2 + [\sin\alpha(2I_2 - I_1 - I_3)]^2}}{2I_0 \sin\alpha(1 - \cos\alpha)}. \tag{5.27}$$

when $\alpha = 2\pi/3$ (120°), $\gamma = 0.83\gamma_0$ for integration over a $2\pi/3$ phase shift, and Eq. (5.26) becomes

$$\Delta\phi = \tan^{-1}\left(\frac{\sqrt{3}(I_1 - I_3)}{2I_2 - I_1 - I_3}\right). \tag{5.28}$$

5.3.2 Four Frames (90° Phase Shifts)

A common algorithm for phase calculations is the four-frame method [5.24]. In this case, the four recorded sets of intensity measurements can be written as

$$I_1 = I_0[1 + \gamma \cos \Delta\phi], \tag{5.29}$$
$$I_2 = I_0[1 + \gamma \cos(\Delta\phi + \pi/2)] = I_0[1 - \gamma \sin \Delta\phi], \tag{5.30}$$
$$I_3 = I_0[1 + \gamma \cos(\Delta\phi + \pi)] = I_0[1 - \gamma \cos \Delta\phi], \tag{5.31}$$
$$I_4 = I_0[1 + \gamma \cos(\Delta\phi + 3\pi/2)] = I_0[1 + \gamma \sin \Delta\phi], \tag{5.32}$$

where $\alpha_i = 0, \pi/2, \pi$, and $3\pi/2$ (0°, 90°, 180°, and 270°). The phase at each point is

$$\Delta\phi = \tan^{-1}\left(\frac{I_4 - I_2}{I_1 - I_3}\right), \tag{5.33}$$

and the detected fringe visibility is calculated using

$$\gamma = \frac{\sqrt{(I_4 - I_2)^2 + (I_1 - I_3)^2}}{2I_0}. \tag{5.34}$$

5.3.3 Five Frames (90° Phase Shifts)

Another technique which uses $\pi/2$ (90°) phase shifts to minimize phase calibration errors was developed by *Hariharan* et al. [5.43]. This algorithm is designed to reduce the possibility of having the numerator and denominator tend to zero, and thereby reduce the uncertainty in the calculation. This algorithm uses five frames of intensity with relative phase shifts of α

$$I_1 = I_0[1 + \gamma \cos(\Delta\phi - 2\alpha)], \tag{5.35}$$

$$I_2 = I_0[1 + \gamma \cos(\Delta\phi - \alpha)], \tag{5.36}$$

$$I_3 = I_0[1 + \gamma \cos(\Delta\phi)], \tag{5.37}$$

$$I_4 = I_0[1 + \gamma \cos(\Delta\phi + \alpha)], \tag{5.38}$$

$$I_5 = I_0[1 + \gamma \cos(\Delta\phi + 2\alpha)]. \tag{5.39}$$

These equations can be combined to yield

$$\frac{I_2 - I_4}{2I_3 - I_5 - I_1} = \frac{\sin\alpha \sin\Delta\phi}{(1 - \cos 2\alpha)\cos\Delta\phi}. \tag{5.40}$$

When $\alpha = \pi/2$ (90°), the right-hand side of (5.40) reduces to $(1/2)\tan\Delta\phi$. In this situation, the five frames of intensity may be written as

$$I_1 = I_0[1 + \gamma \cos(\Delta\phi - \pi)] = I_0[1 - \gamma \cos\Delta\phi], \tag{5.41}$$

$$I_2 = I_0[1 + \gamma \cos(\Delta\phi - \pi/2)] = I_0[1 + \gamma \sin\Delta\phi], \tag{5.42}$$

$$I_3 = I_0[1 + \gamma \cos(\Delta\phi)], \tag{5.43}$$

$$I_4 = I_0[1 + \gamma \cos(\Delta\phi + \pi/2)] = I_0[1 - \gamma \sin\Delta\phi], \tag{5.44}$$

$$I_5 = I_0[1 + \gamma \cos(\Delta\phi + \pi)] = I_0[1 - \gamma \cos\Delta\phi], \tag{5.45}$$

where the phase shifts are $-\pi$, $\pi/2$, 0, $\pi/2$, and π ($-180°$, $-90°$, 0°, 90°, and 180°). The phase calculated from this set of intensities is given by

$$\Delta\phi = \tan^{-1}\left(\frac{2(I_2 - I_4)}{2I_3 - I_5 - I_1}\right). \tag{5.46}$$

This is a very simple calculation and has a large tolerance to miscalibration of the phase shift. For this technique, the detected fringe visibility is given by

$$\gamma = \frac{\sqrt{[2(I_2 - I_4)]^2 + (2I_3 - I_5 - I_1)^2}}{4I_0}. \tag{5.47}$$

5.3.4 Carré (Four Frames with General Phase Shifts)

In the previous subsections, the phase shift has been known by calibrating the phase shifter. *Carré* [5.2] presented a technique of phase measurement which is independent of the amount of phase shift. It assumes that the phase is shifted by α between consecutive intensity measurements to yield four equations

$$I_1 = I_0[1 + \gamma\cos(\Delta\phi - 3\alpha/2)], \tag{5.48}$$

$$I_2 = I_0[1 + \gamma\cos(\Delta\phi - \alpha/2)], \tag{5.49}$$

$$I_3 = I_0[1 + \gamma\cos(\Delta\phi + \alpha/2)], \tag{5.50}$$

$$I_4 = I_0[1 + \gamma\cos(\Delta\phi + 3\alpha/2)], \tag{5.51}$$

where the phase shift is assumed to be linear. From these equations, the phase shift can be calculating using

$$\alpha = 2\tan^{-1}\left(\sqrt{\frac{3(I_2 - I_3) - (I_1 - I_4)}{(I_2 - I_3) + (I_1 - I_4)}}\right), \tag{5.52}$$

and the phase at each point is

$$\Delta\phi = \tan^{-1}\left[\tan(\alpha/2)\left(\frac{(I_1 - I_4) + (I_2 - I_3)}{(I_2 + I_3) - (I_1 + I_4)}\right)\right]. \tag{5.53}$$

The above two equations are combined to yield

$$\Delta\phi = \tan^{-1}\left(\frac{\sqrt{[(I_1 - I_4) + (I_2 - I_3)][3(I_2 - I_3) - (I_1 - I_4)]}}{(I_2 + I_3) - (I_1 + I_4)}\right). \tag{5.54}$$

The detected fringe visibility is

$$\gamma = \frac{1}{2I_0}\sqrt{\frac{[(I_1 - I_4) + (I_2 - I_3)]^2 + [(I_2 + I_3) - (I_1 + I_4)]^2}{2}}, \tag{5.55}$$

where this equation assumes that α is near $\pi/2$. If the phase shift is off by $\pm 10°$, the estimation of γ will be off by $\pm 10\%$. An obvious advantage of the Carré technique is that the phase shift does not need to be calibrated. It also has the advantage of working when a linear phase shift is introduced in a converging or diverging beam where the amount of phase shift varies across the beam.

5.3.5 Synchronous Detection ($2\pi/N$ Phase Shifts)

A technique for phase measurement based on the generalized least-squares method of Sect. 5.2.1 utilizes techniques of communication theory [5.3, 44]. To synchronously detect a noisy signal, it is correlated (or multiplied) with sinusoidal and cosinusoidal signals of the same frequency and averaged over many periods of oscillation. With phase shifts

$$\alpha_i = 2\pi i/N, \quad \text{with } i = 1, \ldots, N, \tag{5.56}$$

such that N measurements are equally spaced over one modulation period, (5.14) reduces to that given by both *Bruning* et al. [5.3] and *Morgan* [5.13] (see also [5.12, 44])

$$\Delta\phi(x, y) = \tan^{-1}\left(\frac{\sum I_i(x, y) \sin \alpha_i}{\sum I_i(x, y) \cos \alpha_i}\right). \tag{5.57}$$

For the least-squares estimation of (5.14), (5.57) is the special case in which the matrix A, see (5.9), is diagonal.

5.3.6 $(N + 1)$ Frames $(2\pi/N$ Phase Shifts)

A new class of algorithms has recently been introduced by *Larkin* and *Oreb* [5.45, 46] which have reduced sensitivity to phase shifter miscalibration and detector nonlinearities. This class of algorithms have phase shifts $\alpha_i = 2\pi i/N$ similar to the synchronous-detection technique, see (5.56). However, they have an extra data frame which has a 2π phase shift relative to the first data frame. Ideally, the first and last frames should be identical. However, phase-shifter errors may cause differences between them and the extra data frame helps to reduce the error sensitivity. The five-frame algorithm of Sect. 5.4.3. with $\pi/2$ phase shifts is an example of an $(N + 1)$-frame algorithm. The specific algorithm which *Larkin* and *Oreb* [5.46] developed for application to projected fringe contouring utilizes seven frames of intensity data with relative phase shifts of $\pi/3$ (60°). The phase calculated using seven frames has multiple solutions. Using Fourier analysis, a solution which minimizes second- and third-order harmonics is given by

$$\Delta\phi = \tan^{-1}\left(\frac{\sqrt{3}(I_2 + I_3 - I_5 - I_6) + (I_7 - I_1)/\sqrt{3}}{-I_1 - I_2 + I_3 + 2I_4 + I_5 - I_6 - I_7}\right). \tag{5.58}$$

For these methods (and synchronous detection), a larger number of data frames will reduce error sensitivity [5.46, 47]. The reason for this will be discussed in the section on major error sources.

5.3.7 $2 + 1$ (90° Phase Shifts)

A three-frame technique for use in the presence of vibration and air turbulence to take data very quickly has been developed by *Angel* and *Wizinowich* [5.48, 49]. Fast data taking is important when large mirrors (or long path lengths) are tested which cannot be mounted on the same table as the interferometer. With this technique, vibration isolation is not as critical as for the previous techniques described in this chapter. The three frames of data required by this technique are written as

$$I_1 = I_0[1 + \gamma \cos(\Delta\phi)], \tag{5.59}$$

$$I_2 = I_0[1 + \gamma \cos(\Delta\phi - \pi/2)] = I_0[1 + \gamma \sin(\Delta\phi)], \tag{5.60}$$

$$I_3 = \tfrac{1}{2}\{I_0[1 + \gamma \cos(\Delta\phi)]\} + \tfrac{1}{2}\{I_0[1 + \gamma \cos(\Delta\phi + \pi)]\} = I_0, \tag{5.61}$$

where the first two frames of data I_1 and I_2 are taken very quickly with a $\pi/2$ (90°) phase shift between them, and the third frame of data is the dc intensity (the average of two frames with a π (180°) phase shift between them) which can be acquired at any time. This means that there are only two frames of data which can be affected by vibration and air turbulence. If these two frames are taken on either side of the interline transfer in a standard CCD video camera, 1 ms exposures can taken as quickly as 1 μs apart. This will freeze most vibrations and air turbulence which may affect the measurement. These three frames of data can then be combined to calculate the wavefront phase using

$$\Delta\phi = \tan^{-1}\left(\frac{I_2 - I_3}{I_1 - I_3}\right). \tag{5.62}$$

The detected fringe visibility can be calculated using

$$\gamma = \frac{\sqrt{(I_2 - I_3)^2 + (I_1 - I_3)^2}}{I_0}. \tag{5.63}$$

5.3.8 Scanning Phase Shift (Random Phase Shifts)

Another technique developed by *Vikhagen* [5.50] for use in the presence of vibration and air turbulence utilizes a large number of data frames with random phase shifts. This technique was originally developed for TV holography to look at large structures which could not be isolated; however, it can be very useful for other applications. It requires the collection of many frames of intensity data with random phase shifts. The intensity for a single data frame is given by (5.6) with the phase shift α_i having numerous random values between 0 and 2π. Every time a new data frame is recorded, the maximum and minimum intensity values at each detector point are determined. When the number of frames becomes large, these values approach the maximum and minimum fringe intensities I_{max} and I_{min}. These values can be used to determine the dc intensity I_0 and the detected fringe visibility γ at each data point. Once the dc intensity and detected fringe visibility are known, the phase is the only unknown left. The phase is then calculated using

$$\Delta\phi = \cos^{-1}\left(\frac{I_i - I_0}{\gamma I_0}\right), \tag{5.64}$$

where I_i is the intensity data frame with a phase shift of α_i, and I_0 and γ are calculated using

$$I_0 = (I_{max} + I_{min})/2, \tag{5.65}$$

$$\gamma = \frac{I_{max} - I_{min}}{I_{max} + I_{min}} = \frac{I_{max} - I_{min}}{2I_0}. \tag{5.66}$$

Once a large number of frames (about 20) has been recorded, the phase can be calculated as each new intensity frame is recorded. This technique works well as long as the vibration and air turbulence do not cause the object to move more than a fraction of a pixel from frame to frame. It is most useful when random phase values are present; however, a computer-controlled phase shifter making small phase shifts can be used. This algorithm can easily be programmed in an array processor to enable calculation of the phase modulo 2π at video frame rates.

5.4 Phase-Shifter Calibration

It is important to make sure that the phase shifter is calibrated to minimize errors in the measurement. Calibration should be performed on a regular basis. A phase shifter can be calibrated by determining the value of the phase shift α. If either the *Carré* method or the scanning phase shift method are used, the phase shift does not need to be calibrated

The simplest technique for phase shifter cailbration involves looking at sequential frames of intensity data. Figure 5.11 shows five frames of intensity data with phase shifts of $\pi/2$ (90°). Frames 1 and 5 should look the same because there is a 2π phase shift (360° or 1 fringe) between them. Frames 1 and 3 as well as frames 2 and 4 should have complementary intensity patterns because they are shifted by π (180°). When $N + 1$ intensity data frames are recorded with phase shift increments of $2\pi/N$, the $(N + 1)$th measurement should overlap the first measurement.

A more exact determination of the phase shift α is obtained by direct calculation from four frames of intensity data using (5.52). When five frames of intensity data are taken with relative phase shifts of α, a simpler equation can be used [5.43, 47, 51]:

$$\alpha = \cos^{-1}\left(\frac{I_5 - I_1}{2(I_4 - I_2)}\right). \tag{5.67}$$

After applying (5.67) to each data point in the interferogram, a histogram of the phase shifts can be plotted, as shown in Fig. 5.12. The voltage applied to the phase shifter needs to be adjusted to provide the desired value. Skewness or spread in the histogram indicates a variation in the phase shift across the interferogram which may be due to tilting or uneven motion as the shifter is moved.

5.4 Phase-Shifter Calibration   129

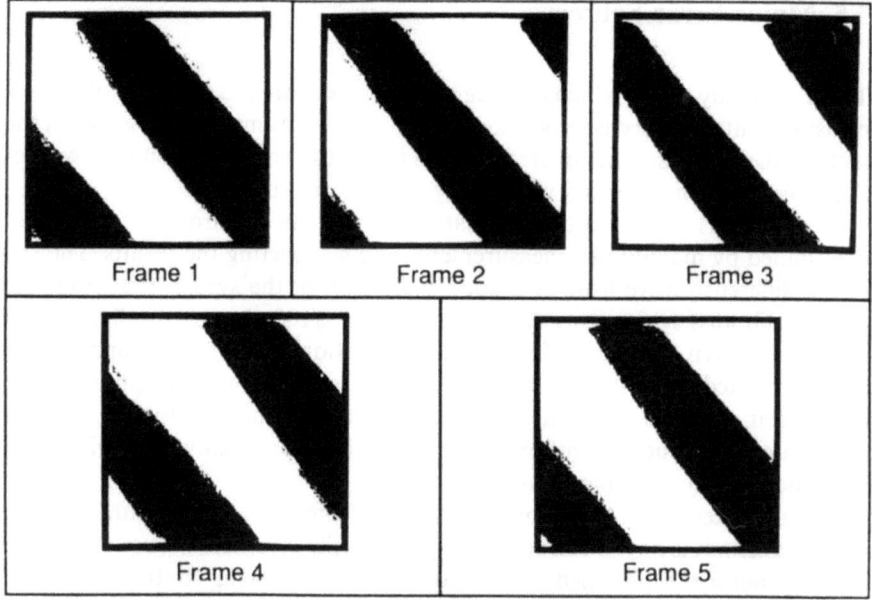

Fig. 5.11. Five frames of intensity data with $\pi/2$ (90°) phase shifts

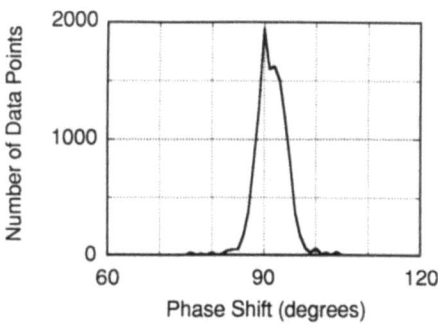

Fig. 5.12. Histogram showing spread in phase shifts for a calibrated phase-shifter

If the phase shifter motion is nonlinear, it is possible to use a reference signal to determine the wave form necessary to drive the phase shifter linearly [5.52]. It can be done either by calculating the proper shifter motion, or by doing a least squares fit as shown by *Ai* and *Wyant* [5.53]. A generalized least-squares phase calculation can also be used which will work with any phase shifts and does not require calibration [5.14].

5.5 Major Error Sources

There are many errors which can degrade the precision and accuracy of a phase measurement. Care needs to be taken to reduce these as much as possible. In this section, the major sources of error will be reviewed. The user should be warned that these are not all errors which may affect a measurement.

Precision is a day-to-day indication of system performance. It can be determined by making two measurements and subtracting the results. The rms of the difference wave front is a good indication of the system precision. For a well-calibrated system, this should be less than $\lambda/100$. Precision will mostly be affected by environmental effects such as vibration and air turbulence. It can also be affected by systematic errors such as miscalibration of the phase shifter, nonlinearities in the detected fringe intensities, and quantization of the measured intensity values. Monitoring system precision on a daily basis will let the user know when something has changed.

Measurement with precision does not indicate the actual accuracy of the measurement. It is an indication of the spread in the measured values. As long as the same number is obtained, then whatever process is being controlled will still provide usable products. Accuracy tells the user whether the measured value is the correct one. It is usually determined relative to a known standard. (Caution: standards may not be as accurate as you think they are.) If a blessed standard (from a national standards lab) is not available, an in-house standard which is known to provide acceptable results can be used. Many interferometric instruments have scale factors so that the measurements on different instruments can be made to provide the same value. This is not to say that the value is correct; only that it is consistent. For most users, consistency is the most important consideration so they can compare measurements with other instruments or labs.

Error sources in PMI have been studied by a large number of researchers namely, *Koliopoulos* [5.54], *Schwider* et al. [5.47], *Hayes* [5.52], *Cheng* and *Wyant* [5.55], *Creath* [5.56], *Hariharan* et al. [5.43], *Ai* [5.57], *Ai* and *Wyant* [5.58], *Ohyama* et al. [5.59], *Creath* [5.5], *Kinnstaetter* et al. [5.60], *Schwider* [5.61], *Brophy* [5.62], *Freischlad* and *Koliopoulos* [5.63], *van Wingerden* et al. [5.64], and *Larkin* and *Oreb* [5.65]. Measurement errors can be due to many things such as miscalibration of the phase shifter, nonlinearities due to the detector, quantization of the detector signal, spurious reflections, quality of the reference beam (surface), aberrations in the optics of the interferometer, air turbulence, and vibrations. Vibration and air turbulence are dynamic, random variables which reduce both precision and accuracy. Averaging can help reduce the affect of these significantly. The quality of the reference beam (surface) and aberrations in the interferometer are systematic errors which degrade accuracy. Other error sources listed affect both precision and accuracy. They are deterministic and systematic, but the actual form of the error will depend upon the initial phase difference between the object and reference beam. Their affect can be reduced (but not eliminated) by averaging a number of frames of data. All of these errors are affected by the algorithm used to calculate the phase. The choice of an algorithm to reduce one error may exacerbate the effect of another.

A lot of the effects due to the error sources covered in this section can be traced to the behavior of the arctangent function in the presence of errors [5.65]. *Larkin* and *Oreb* have shown that the error in a phase calculation utilizing an arctangent can be written as

$$\Delta\Phi = \tan^{-1}\left(\frac{\varepsilon_s \cos\Delta\phi - \varepsilon_c \sin\Delta\phi}{1 + \varepsilon_c \cos\Delta\phi + \varepsilon_s \sin\Delta\phi}\right), \tag{5.68}$$

where $\Delta\Phi$ is the phase error in the estimation of the phase $\Delta\phi$, and ε_s and ε_c are the normalized errors in the sine term (numerator) and cosine term (denominator) within the arctangent. This can be simplified for normalized errors less than about $\pm 10\%$ to become

$$\Delta\Phi \cong \varepsilon_s \cos\Delta\phi - \varepsilon_c \sin\Delta\phi. \tag{5.69}$$

After expanding (5.69) as a Fourier series and rearranging, the approximated error function is now represented by

$$\Delta\Phi \cong \frac{1}{2}\sum_n \varepsilon_{sn}\{\cos[(n+1)\Delta\phi + \beta_{sn}] + \cos[(n-1)\Delta\phi + \beta_{sn}]\}$$

$$-\frac{1}{2}\sum_n \varepsilon_{cn}\{\sin[(n+1)\Delta\phi + \beta_{cn}] - \sin[(n-1)\Delta\phi + \beta_{cn}]\}, \tag{5.70}$$

where n represents harmonic orders in the error functions, and β_{sn} and β_{cn} are constant phase offsets. ε_{sn} and ε_{cn} are normalized errors for different harmonics. $n = 0$ corresponds to piston offsets, $n = 1$ corresponds to phase dependent errors such as phase shift miscalibration, $n = 2$ corresponds to second-order nonlinearities in the measured intensities which can be caused by detector nonlinearities, and $n = 3$ or higher, corresponds to higher-order nonlinearities in the measured intensities. The result of (5.70) shows that an error which goes as $n\Delta\phi$ in phase will generate error terms at frequencies of $(n+1)\Delta\phi$ and $(n-1)\Delta\phi$. It is possible for some algorithms to have the either or both of these orders cancel out.

The rest of this section concentrates on presenting the effects of phase-shifter errors, detection nonlinearities, quantization, and vibration and air turbulence for four of the most popular algorithms. Results for phase-shifter and detector errors have been generated by computer simulations, whereas the random errors of quantization and vibration and air turbulence are treated analytically.

5.5.1 Phase-Shifter Errors

Phase errors caused by inaccurate phase-shifter calibration can be minimized by adjusting the interferometer for a single fringe (fluffing out the fringes). However, with large amounts of phase deviation across the measured wave front, it will not be possible to obtain a single fringe. If a constant calibration error is present, the phase shift may be written as

$$\alpha' = \alpha(1 + \varepsilon), \tag{5.71}$$

where α is the desired phase shift, α' is the actual phase shift, and ε is the normalized error. It has been shown by *Schwider* et al. [5.47] that the errors in phase due to a calibration error or nonlinearity in the phase shifter will decrease as the number of measurements increases. For a phase-shift miscalibration error, the calculated phase has a periodic error with a spatial frequency of twice the fringe frequency as shown in Fig. 5.13. The 10% error indicated in the figure means that the phase shift is 81° or 99° instead of 90°. The amount of phase error depends upon the algorithm used to calculate the phase as well as the magnitude of the error. The *Carré* algorithm shows no error because it is independent of the actual phase shift as long as it is linear. The three-frame and four-frame algorithms yield the largest amount of error and the five-frame algorithm is relatively insensitive to this error. Figure 5.14 depicts a plot of the peak-to-valley (P–V) phase error versus amount of phase calibration error for the same algorithm shown in Fig. 5.13.

Nonlinear phase-shift errors are not as easy to detect or remove [5.56, 59–61]. In normal operation, a nonlinear phase shifter with a quadratic error will be partially compensated in the calibration of the interferometer by adding a linear bias to its movement. The most straightforward approach to calibration is to make sure that the phase shifter actually moves 2π over a 2π desired change in phase. This function minimizes the error due to nonlinear phase-shifter motion. The phase shift then becomes

$$\alpha' = \alpha(1 + \varepsilon\alpha - \varepsilon). \tag{5.72}$$

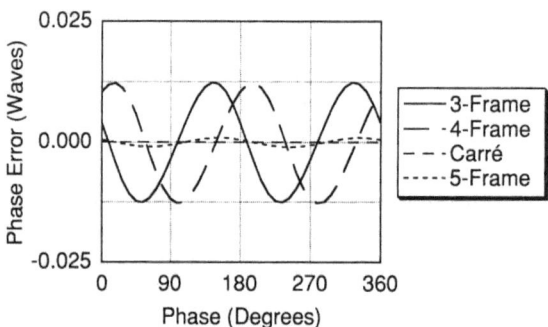

Fig. 5.13. Phase error for 10% phase-shift miscalibration

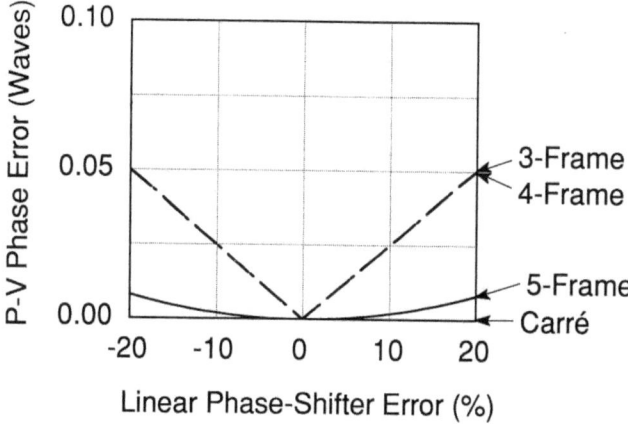

Fig. 5.14. P–V phase error versus percent phase-shifter miscalibration

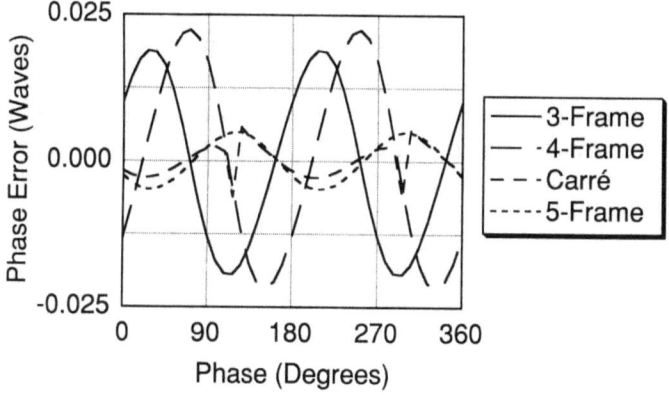

Fig. 5.15. Phase error for 10% quadratic phase-shift nonlinearity

Nonlinear phase-shifter errors can be reduced by applying certain algorithms such as the *Carré* technique and the five-frame technique; however, these errors can only be eliminated by determining the voltage necessary to produce the correct phase shift or by using an algorithm which is insensitive to variations in the phase shift [5.14, 52, 53]. The nonlinear phase-shifter also causes periodic errors at twice the fringe spatial frequency (Fig. 5.15). Again the three-frame and four-frame algorithms are the most sensitive and the five-frame and *Carré* algorithms are the least sensitive. Figure 5.16 presents a plot of P–V phase error versus amount of nonlinearity for these algorithms.

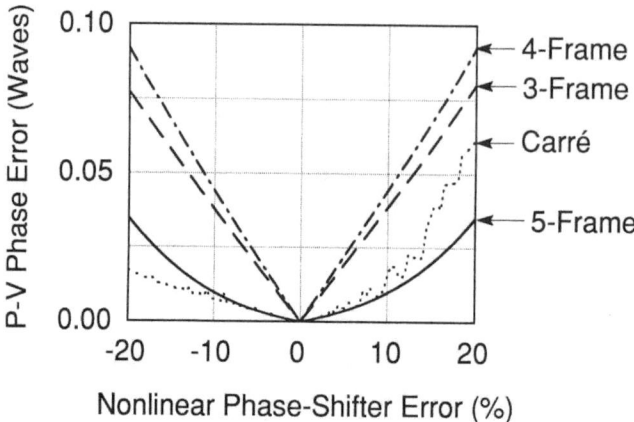

Fig. 5.16. P–V phase error versus percent quadratic nonlinear phase-shifter error

5.5.2 Detection (Intensity) Nonlinearities

Creath [5.56], *Ai* [5.57], and *Kinnstaetter* et al. [5.60] have shown that a nonlinear response from a detector can introduce phase errors. These errors are espcially noticeable if they are not consistent from detector to detector in an array. Most detector arrays have gain adjustments so that these errors can be minimized. The effects of these errors are also reduced by minimizing the number of fringes in the interferogram.

When a detector has a second-order nonlinear response or if the fringe intensities have a second-order nonlinearity, the measured intensity I' can be written in terms of the incident intensity I as

$$I' = I + \varepsilon I^2, \tag{5.73}$$

where ε is a normalized coefficient for the nonlinear term. It can be demonstrated that effects due to this error will cancel out if four or five frames of intensity data are used in the phase calculation [5.5, 64, 66]. Figure 5.17 exhibits a plot of the phase error for a 10% second-order nonlinearity. There is no error for the four-frame and five-frame algorithms. The three-frame and *Carré* algorithms have an error function which contains harmonics at both the fringe frequency and three times the fringe frequency. The P–V error is shown in Fig. 5.18. It is noticeable that the amount of error depends upon the sign of the error as well as the algorithm.

For most applications, the largest nonlinearities affecting the measurement will be due to third-order harmonics where the detected intensity is given by

$$I' = I + \varepsilon I^3, \tag{5.74}$$

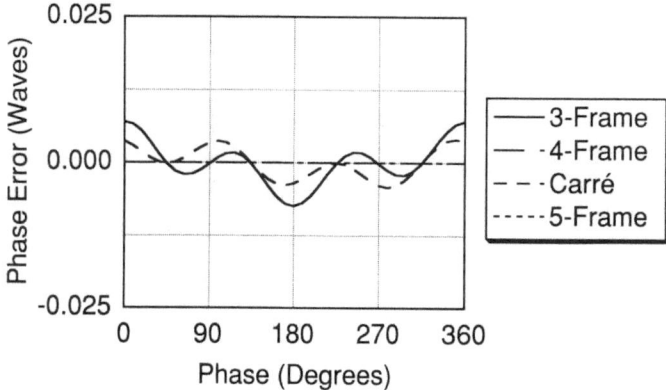

Fig. 5.17. Phase error for − 10% second-order detection nonlinearity error

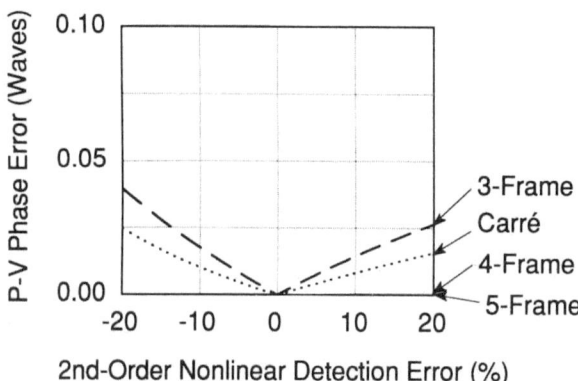

Fig. 5.18. P–V phase error versus percent second-order detection nonlinearity error

The phase error due to third-order nonlinearities are displayed in Fig. 5.19. The functional form is similar to the second-order nonlinearities. The four-frame and five-frame algorithms show a small error at four times the fringe frequency. The errors for the three-frame and *Carré* algorithms are much larger and contain harmonics at the fringe frequency and three times the fringe frequency with an amplitude about 50% larger than the second-order nonlinearities. The P–V phase errors for these algorithms as a function of percent nonlinearity are illustrated in Fig. 5.20. For most applications, the five-frame algorithm is sufficient to reduce effects of detector nonlinearities; however, some applications such as those employing projected fringe illumination may need to utilize an algorithm with a larger number of frames of data such as the seven-frame algorithm of (5.58). In general, the larger number of frames used in the calculation, the smaller the effects of nonlinearities.

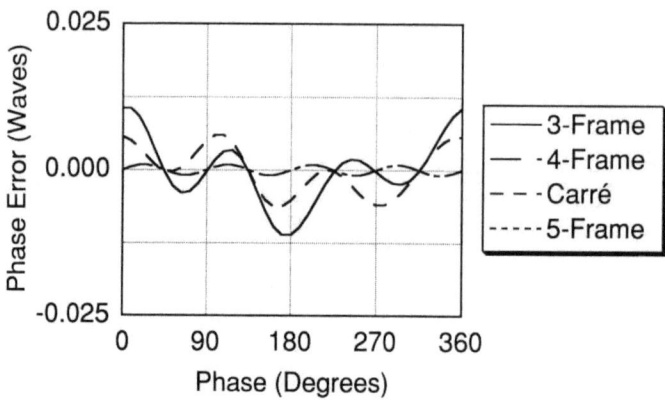

Fig. 5.19. Phase error for − 10% third-order detection nonlinearity error

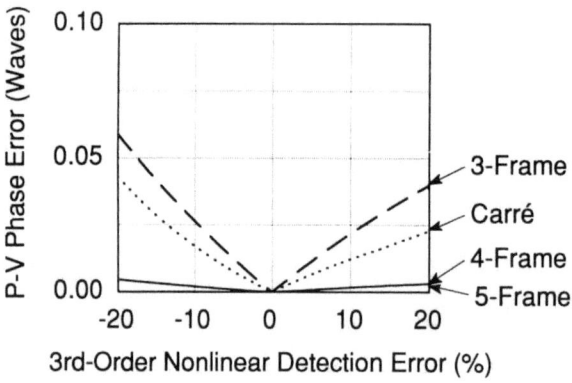

Fig. 5.20. P–V phase error versus percent third-order detection nonlinearity error.

If both phase-shifter errors and detection nonlinearities are present, then the five-frame algorithm works the best of the four algorithms compared in these simulations. It is possible to tailor an algorithm to reduce a specific error [5.46, 65]. Better error compensation is always available if more frames of data are used. However, the time to take data as well as storage considerations are trade-offs which need to be made when deciding upon a particular algorithm.

5.5.3 Quantization

When data frames are acquired by a camera, an analog video signal needs to be converted to a digital signal with an Analog-to-Digital Converter (ADC).

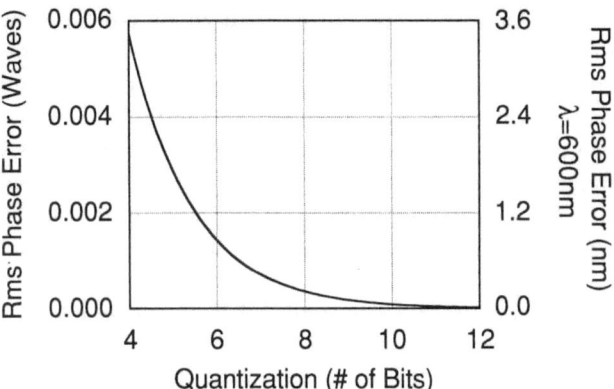

Fig. 5.21. Rms phase error due to intensity quantization versus number of bits digitized

Brophy [5.62] has shown that there is some correlation between frames of intensity data when $\pi/2$ (90°) phase shifts are used, and that for the class of phase measurement algorithms with $\pi/2$ (90°) phase shifts, the rms phase error due to quantizing the intensity into Q levels is given by

$$\sigma_\phi = \frac{1}{\sqrt{3}\gamma Q}, \tag{5.75}$$

where γ is the fringe visibility. An intensity signal digitized to 8 bits ($Q = 2^8 = 256$) has an rms phase error due to quantization of the signal of 2.26×10^{-3} radians or 3.59×10^{-4} waves. For a wavelength of 600 nm, this corresponds to an rms error of 0.2 nm. A plot of rms phase error in waves (for $\gamma = 1$) versus number of bits the signal is digitized to is shown in Fig. 5.21. This error is very small, and 8 bits of ADC resolution should be sufficient for most measurement applications.

Equation (5.75) shows that the modulation of the signal will affect the rms phase error due to quantization. The fringe visibility times the number of quantization levels will yield an effective number of quantization levels. Therefore, the lower the modulation, the fewer the number of quantization levels which span the signal P–V, and the greater the rms phase error. To minimize this effect, the fringe intensity should cover as much of the detector's dynamic range as possible and the sampling of the interference fringes should be sufficient so that the modulation is not significantly reduced.

5.5.4 Vibration and Air Turbulence

Vibration and air turbulence cause the most problematic errors in PMI. To get a good measurement, the optical system must be isolated from vibration and shielded from air turbulence. Air turbulence can be reduced by covering the optical paths and positioning the system away from air ducts and fans. Vibration can be significantly reduced by placing the system on an isolation table and removing any equipment including motors and fans from the table. Acoustic coupling of vibrations from loud fans, machinery, or stereos can also cause noticeable phase measurement errors. These effects can be reduced by placing the system in a quiet room. If the system happens to be next door to a machine shop, there is not much which can be done except to wait until there is a quiet time to take data.

Vibration effects in PMI data cause a sinusoidal phase error which also has a frequency of two times the fringe spatial frequency. Once the system has been calibrated and the detector adjusted, the most likely reason for a double frequency phase error is vibration. Ai [5.57] has simulated a number of simple vibration errors due to sinusoidal vibrations. For high-frequency vibrations where a large number of periods are contained in a single intensity exposure, the intensity fluctuations caused by the vibrations are uncorrelated from frame to frame, and the fluctuations tend to average out. The relationship between the rms of the phase averaged over all possible phase values $\sigma_{\bar{\phi}}$ ($\bar{\phi}$ indicates the phase averaged over all possible phase values) and the rms of the intensity fluctuations σ_I can be written in a simple form [5.62],

$$\sigma_{\bar{\phi}} = \frac{\sqrt{k}\,\sigma_I}{\gamma I_0}, \qquad (5.76)$$

where k is a constant determined by the measurement algorithm. For the three-frame algorithm of (5.21) with $\pi/2$ phase shifts, $k = 1$. If $\gamma = 1$, and $I_0 = 1/2$, then the average rms phase error equals one-half of the rms intensity fluctuations. Values of k for a number of different PMI algorithms are listed in Table 5.1. As the number of frames of data taken increases, the average rms phase error will be reduced. Equation (5.76) can also be employed to determine the average rms phase error due to shot noise or thermal noise in the detector.

Table 5.1. Proportionality constant for rms phase error due to high-frequency vibration

Algorithm	Equation	k
Three-Frame, $\pi/2$	(5.21)	1
Three-Frame, $2\pi/3$	(5.28)	2/3
Four-Frame	(5.33)	1/2
Five-Frame	(5.46)	7/16
N-Frame, (Synchronous Detection)	(5.57)	$2/N$

5.6 Equipment and Experimental Consideration

To build a phase-measurement system, there are many things which need to be considered when choosing components. This chapter has already discussed various methods of providing a phase modulation and explained the importance of proper sampling and intensity modulation. Cameras to record interferograms and frame grabbers to transfer data into the computer system for analysis are very important components in phase-measurement systems. Since the interferogram recording process needs to be synchronized with a phase shifter, some possible methods of doing this will be outlined. Finally, alignment considerations for the optical system will be discussed. Good measurements can not be obtained unless the system is properly aligned.

5.6.1 Cameras and Frame Grabbers

To analyze data frame for phase-measurement techniques, interferogram intensity data needs to be recorded and processed in a computer. This requires a camera capable of resolving the interference fringes and surface features of interest as well as a frame grabber to transfer the images into computer memory for analysis. There are many trade-offs which must be considered when choosing these components.

Camera signals can be divided into three major categories: RS-170, CCIR, and slow-scan. RS-170 is a USA standard. It provides an analog output of 30 frames per second where each frame includes 2 interlaced fields. The pixels are not square and the format has a 4:3 aspect ratio. A typical RS-170 sensor will have 754×488 pixels. CCIR is a European standard. It provides an analog output of 25 frames (50 fields) per second. The pixels are sometimes square and a typical sensor contains 756×581 pixels. Neither RS-170 or CCIR specifies a specific pixel size or number of pixels. These standards were created for television signals and have been adopted for image processing uses. Everything else is usually lumped together as slow-scan cameras. There is no standard for these cameras. They generally have at least 512×512 pixels, clock rates of about 10–14 MHz (10–14 million pixels per second), and can have either analog or digital output. In addition to the many different signal types, there are many different sensor architectures [5.67].

Issues which need to be considered when choosing a camera are numerous. Does the camera provide an interlaced or non-interlaced signal? Are the pixels square? Does the camera require shuttering? How noisy is it? Does it sample the signal sufficiently? And how large are the pixels relative to their spacing? For most phase shifting applications, interlaced signals are not desirable. This is because the second field either will access different rows of pixels or it may be generated by averaging adjacent rows. In either case this does not provide usable information. Square pixels provide a uniform spatial frequency response in the x and y directions. Non-square pixels mean there is a preferred fringe orientation and must be compensated for in graphics routines. Many cameras need to be covered when reading out so that the signal does not smear. If the camera requires shuttering, make sure that shutter motion does not reduce fringe visibility by inducing vibrations. The signal from the sensor will be digitized either by the camera or the frame grabber. The noise level in a sensor needs to be less than 1 bit rms to provide the best results. The number of pixels and spacing is important to be able to resolve the interference fringes and provide signal modulation as the phase is shifted along with being able to resolve necessary features on the surface of the test object. For many sensors the number of pixels is not the same as the number of lines of resolution. This can be explained by noting that the effective size of a pixel is usually larger than the pixel spacing. This effect is usually greater along a row than a column. It is best to have some dead space between pixels to get the maximum resolution from a sensor. Sparse array sensors with pixels much smaller in size than the pixel spacing are useful for many applications in phase-measurement interferometry because they allow measurement of much smaller fringe spacings [5.39, 68].

A frame grabber takes the signal from the camera and transfers it to a data array accessible by a computer program. The frame grabber needs to be compatible with the specific camera being used. It can have either analog or digital input and may follow the RS-170 or CCIR standards. The number of lines of resolution is fixed by the sensor. The number of samples along a line is variable. It is best for the frame grabber to sample every pixel from the camera signal. Otherwise, aliasing is likely to occur. It is also important to sample each pixel at its center to get a good signal. The best way to sample the signal is to couple the clocks of the camera and frame grabber. A single clock with the correct phase delay will ensure that every pixel is sampled at its center. Beware that many frame grabbers use an ac signal from either the computer clock or from the input power to provide a clock. This can lead to synchronization problems.

There are simple tests for checking a sensor. An easy way to measure the camera and frame grabber noise is to cover the sensor and capture two consecutive frames. The rms of the difference between the two frames is a measure of the noise in the signal. A good camera/frame grabber combination will have less than 1 bit of rms noise. The frequency response of a sensor is also simple to determine. High-frequency straight fringes are oriented either horizontally or vertically on the sensor and the modulation of the fringes determined. When the modulation goes to zero, the cutoff frequency of the camera has been determined. It is possible that the modulation will increase again when the fringe spacing is increased because of aliasing. Another way to measure the cutoff frequency is to put a large number of fringes across the array and see how many fringes it takes before the phase unwrapping algorithm cannot reconstruct the wavefront. The cutoff will be the spatial frequency determined by the maximum number of fringes which can be measured without a discontinuity. Because the effective size of a pixel tends to be larger than the pixel-to-pixel spacing, the theoretical Nyquist limit can not usually be reached.

The ideal camera for phase-measurement applications is non-interlaced, has square pixels, low noise, some sort of standard video signal to make interfacing easy, and is coupled to the frame grabber using a single clock signal.

5.6.2 Phase Shifters: Ramping Versus Stepping

When providing a phase shift for phase measurement, the phase shifter can either be ramped or stepped. Ramping provides a continuous motion for the phase shifter without jerking it. This is preferred because no time delay is necessary to wait for vibrations to settle down after moving the phase shifter. However, ramping requires good synchronization of the frame grabber and the phase shifter to get the correct phase shifts between data frames. Stepping is easier to coordinate with a frame grabber. The phase shift is adjusted, a delay allows for settling, then a data frame is recorded, and the process is repeated. Timing can be worked out so that every other field is recorded and the phase shifter moved during the unrecorded field. This limits how fast data can be taken. Ramping allows faster data taking but requires more sophisticated electronics.

Another consideration is the fact that it takes a finite time for a mass to move linearly. When ramping the phase shift, the first frame or two of data usually needs to be discarded because the phase shift is not correct until the movement is linear. This is illustrated in Fig. 5.22. When stepping the phase shift, it is best to change the phase shift gradually to reduce vibrations caused by abrupt movement (Fig. 5.23).

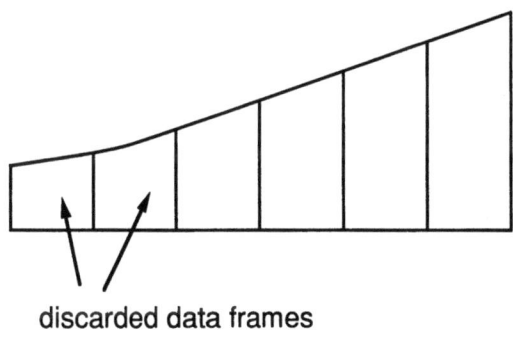

Fig. 5.22. First frame or two of data is discarded because of nonlinear phase shift.

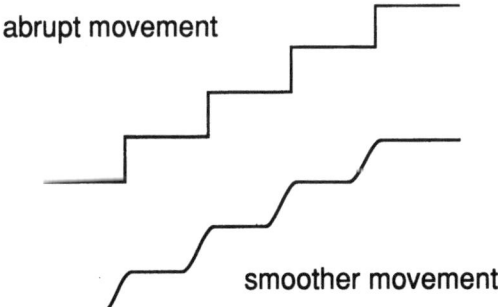

Fig. 5.23. Abrupt movement causes vibration whereas a smoother motion will reduce vibration effects.

5.6.3 Alignment Considerations

To provide high-contrast fringes which correspond to the test surface, an interferometer needs to be correctly aligned. If the source needs to be collimated, the accuracy of the measurement will depend upon the wave front quality of the collimated beam. This wave fronts needs to be better than the required accuracy of the measurements. When a laser source is used, the difference in the path lengths of the object and reference beams needs to be within the coherence length of the laser or a multiple of twice the cavity length of the laser. To get the best signal, the object needs to be uniformly illuminated. Otherwise, there will be more bits of information for some areas of the object than others. Ideally the interference fringes should cover the entire dynamic range of the camera and be uniform over the entire image. The relative intensities of the test and reference beams need to be adjusted if possible to provide the maximum fringe contrast. Another consideration is the polarization of the object and reference beams. Interference only occurs when the polarizations of the beams are the same; it is maximum when this polarization is in the plane of intersection. When the polarizations are not correct, fringe visibility will be low and there will be a lot of background light. When imaging the object, it is important to make sure that the entire object is in focus. To easily interpret the fringe pattern, the fringes need to be localized at the test object and the object needs to be imaged onto the camera. Telecentric imaging (putting the pupil at infinity) helps to reduce distortions in the determination of depths on the object.

5.7 Specialized Techniques

These techniques encompass research areas of interest for specialized applications. In recent years, electronic holography techniques which do not need an intermediate holographic recording, have been developed. Analysis of time-average holographic fringes is another special area of interest to investigate objects which are excited with a periodic vibration. The section ends with a discussion on the application of phase-shifting techniques to holographic moiré.

5.7.1 Electronic Holography

Electronic holography encompasses a number of techniques which can be applied to quantitative displacement analysis without the use of an intermediate holographic recording. Rather than recording secondary interference fringes, a camera is employed to directly record the primary interference fringes, and secondary fringes are generated either using electronic circuitry or numerically in an array processor or computer. Because of the limited resolution of cameras relative to holographic film, the resolution of these systems is much less than real-time holography. Two different types of processing have been developed to look at primary interferograms. Even though both techniques have stemmed from speckle interferometry techniques, they both essentially generate electronic holograms. One technique has been referred to as Digital Speckle-Pattern Interferometry (DSPI), Phase-Shifting Speckle Interferometry (PSSI), or TV holography [5.69–73]. It has evolved from Electronic Speckle-Pattern Interferometry (ESPI). The other type of processing involves more calculation but reduces the amount of speckle noise in the secondary interference fringes. It has been referred to as Electro-Optic Holography (EOH) [5.36, 66, 74–76].

Electronic holography utilizes phase-measurement techniques to measure the phase of primary interference fringes before and after the object state is changed by applying stress. At the simplest level, the difference between these phases is the phase of the secondary interference pattern.

$$\Delta\phi = \phi' - \phi, \tag{5.77}$$

where ϕ and ϕ' are the modulo 2π phases before and after object displacement, and $\Delta\phi$ is the modulo 2π phase of object displacement. The final step is to unwrap the phase $\Delta\phi$. Because of speckle noise, the raw phase usually needs to be filtered before phase unwrapping. Speckle averaging can be utilized to reduce the noise in the individual frames of secondary interference fringe data used to calculate the phase.

An alternative way to calculate the phase of the object displacement employes a single equation with the phase-shifted intensity data frames before and after displacement.

$$\Delta\phi = \tan^{-1}\left(\frac{\sin(\phi - \phi')}{\cos(\phi - \phi')}\right) = \tan^{-1}\left(\frac{\sin\phi\cos\phi' - \cos\phi\sin\phi'}{\cos\phi\cos\phi' + \sin\phi\sin\phi'}\right), \tag{5.78}$$

which for the four-frame method becomes

$$\Delta\phi = \tan^{-1}\left(\frac{(I_4 - I_2)(I'_1 - I'_3) - (I_1 - I_3)(I'_4 - I'_2)}{(I_1 - I_3)(I'_1 - I'_3) - (I_4 - I_2)(I'_4 - I'_2)}\right), \tag{5.79}$$

where I_i and I'_i are the intensity data frames recorded before and after displacement.

The fewest number of data frames required to determine the phase of the object displacement is three (one before and two after). This technique has been developed by *Kerr* et al. [5.77].

Electronic holography techniques assume that the difference between the wave fronts measured before and after applied force can be reconstructed with two detector points per secondary interference fringe. The phase distribution for each of the measurements may not be reconstructible. Each phase distribution may contain high-frequency phase data such as speckle or high-frequency fringe data. The detector size should be small enough to resolve the high-frequency data. A sparse array sensor where the detector spacing is much larger than the detector size can be used for this purpose.

A major disadvantage of the electronic holography techniques discussed above is that after processing the interference fringe data, there are terms proportional to the phase of the speckle produced by the interference between the reference beam and the speckle in the object beam [5.78]. These terms cause noisy-looking fringes. By processing the data differently, *Stetson* and *Brohinsky* [5.74] have shown that the speckle term can be removed. Eight frames of data taken for a static measurement can be used to calculate the phase change $\Delta\phi$ due to the displacement of the object,

$$\Delta\phi = \tan^{-1}\left(\frac{C_1 - C_2}{C_3 - C_4}\right), \tag{5.80}$$

where the quantities C_i are given by

$$C_1 = [(I_1 - I_3) + (I'_1 - I'_3)]^2 + [(I_2 - I_4) + (I'_2 - I'_4)]^2, \tag{5.81}$$

$$C_2 = [(I_1 - I_3) - (I'_1 - I'_3)]^2 + [(I_2 - I_4) - (I'_2 - I'_4)]^2, \tag{5.82}$$

$$C_3 = [(I_1 - I_3) + (I'_2 - I'_4)]^2 + [(I_2 - I_4) + (I'_1 - I'_3)]^2, \tag{5.83}$$

$$C_4 = [(I_1 - I_3) - (I'_2 - I'_4)]^2 + [(I_2 - I_4) - (I'_1 - I'_3)]^2. \tag{5.84}$$

These calculations involve simple calculations and the use of look-up tables which can be done in an array processor. It is possible to calculate modulo 2π phase in real time at video rates [5.36]. In the near future it will be possible to calculate the unwrapped phase at video rates using array processors.

5.7.2 Time-Average Vibration Analysis

Quantitative data can also be extracted from time-average vibration fringes via phase-shifting techniques [5.69, 79–83]. One way to do this utilizes three separate phase measurements [5.82]. The first one is made with the object vibrating while the relative phase between the object and reference beams is shifted. The second measurement is made by applying a vibration of the same frequency as the object vibration to a PZT in the reference beam of the interferometer. A bias phase between the object and reference vibration is added such that the relative phase difference between the vibrations is $+\pi/3$. Relative phase shifts between the object and reference beams are then applied and standard phase-shifting methods are used to calculate the phase. A third measurement is taken such that the relative phase difference between the object and reference vibrations is $-\pi/3$. Assuming a sinusoidal object vibration and 90° relative phase shifts for the phase calculations, one of the total of twelve frames of data recorded can be written as

$$I_{ji} = I_0[1 + \gamma\cos(\phi + \delta_i)J_0(\Omega + \beta_j)], \tag{5.85}$$

where $\delta_i = 0$, $\pi/2$, π and $3\pi/4$, and $\beta_j = -\pi/3, 0$, and $\pi/3$. The amplitude of the vibration can then be calculated using

$$\Omega = \tan^{-1}\left[\frac{1}{\sqrt{3}}\left(\frac{H_1 - H_3}{2H_2 - H_1 - H_3}\right)\right], \tag{5.86}$$

where

$$H_1 = (I_{11} - I_{13})^2 + (I_{12} - I_{14})^2 = 4I_0^2 \gamma^2 J_0^2(\Omega - \pi/3), \tag{5.87}$$

$$H_2 = (I_{21} - I_{23})^2 + (I_{22} - I_{24})^2 = 4I_0^2 \gamma^2 J_0^2(\Omega), \tag{5.88}$$

$$H_3 = (I_{31} - I_{33})^2 + (I_{32} - I_{34})^2 = 4I_0^2 \gamma^2 J_0^2(\Omega + \pi/3). \tag{5.89}$$

Equation (5.86) assumes the form of $\cos^2$ for the fringes. Because of this, a look-up table is necessary to find the difference between the $J_0^2(\Omega)$ and $\cos^2(\Omega)$ functions. The error due to the difference between the $J_0^2(\Omega)$ and $\cos^2(\Omega)$ functions is dependent upon the fringe order, which can be determined from the H_2 measurement. *Pryputniewicz* and *Stetson* [5.83] have implemented this technique using an array processor to determine the vibration amplitude.

5.7.3 Phase-Shifting Holographic Moiré

The use of phase-shifting techniques has been extended recently to include holographic moiré (four-wave) interferometers [5.84–89]. Four-wave interference has more flexibility than two-wave interference because it permits manipulation of the measured phase differences to isolate specific deformation components. The addition of a second illumination beam in the interferometer setup enables two displacement components to be measured simultaneously. Likewise, a third illumination beam would enable all three displacement components to be measured simultaneously [5.1]. The components measurable using holographic moiré are in-plane displacement, out-of-plane displacement, and the difference of displacements. The specific measurement techniques for holographic moiré are discussed in Chap. 7.

A holographic moiré arrangement consists of two object illumination beams with equal and opposite angles of incidence. This provides two sets of interference fringes at the hologram with phases ϕ_1 and ϕ_2 which interfere to form a moiré pattern. After the object is deformed, another two sets of fringes are formed with phases ϕ'_1 and ϕ'_2, and a second moiré pattern is generated. The phase-shifting techniques enable the measurement of either the difference or the sum of the phases of the two moiré patterns. The intensity distribution in a double-exposure holographic moiré pattern can be written as

$$I = 2I_0\left[1 + \frac{\gamma}{2}(\cos \Delta \phi_1 + \cos \Delta \phi_2)\right], \tag{5.90}$$

where $\Delta\phi_1 = \phi_1' - \phi_1$ and $\Delta\phi_2 = \phi_2' - \phi_2$ are the phases of the secondary interference fringes produced for each illumination beam. The information concerning the in-plane and out-of-plane displacements are contained in the expressions corresponding to the difference of phases $\Delta\phi_1 - \Delta\phi_2$ and sum of phases $\Delta\phi_1 + \Delta\phi_2$.

Rastogi [5.84] has presented a phase-stepping procedure which consists of introducing a number of pairs of phase shifts in the two illumination beams. Three successive phase shifts in pairs of $(-2\pi/3, -4\pi/3)$, $(0, 0)$, and $(2\pi/3, 4\pi/3)$ give rise to three sets of intensity measurements

$$I_1 = 2I_0 \left\{ 1 + \frac{\gamma}{2}\left[\cos\left(\Delta\phi_1 - \frac{2\pi}{3}\right) + \cos\left(\Delta\phi_2 - \frac{4\pi}{3}\right) \right] \right\}, \tag{5.91}$$

$$I_2 = 2I_0 \left[1 + \frac{\gamma}{2}(\cos\Delta\phi_1 + \cos\Delta\phi_2) \right], \tag{5.92}$$

$$I_3 = 2I_0 \left\{ 1 + \frac{\gamma}{2}\left[\cos\left(\Delta\phi_1 + \frac{2\pi}{3}\right) + \cos\left(\Delta\phi_2 + \frac{4\pi}{3}\right) \right] \right\}. \tag{5.93}$$

The phase difference corresponding to $\Delta\phi_1 - \Delta\phi_2$ at each point in the interferogram is given by

$$\Delta\phi_1 - \Delta\phi_2 = 2\tan^{-1}\left(\frac{\sqrt{3}(I_1 - I_3)}{2I_2 - I_1 - I_3} \right). \tag{5.94}$$

Rastogi [5.88] also has applied the Carré method [5.2] to holographic moiré. Four frames of intensity are recorded corresponding to four pairs of phase steps $(-3\alpha/2, 3\alpha/2)$, $(-\alpha/2, \alpha/2)$, $(\alpha/2, -\alpha/2)$, and $(3\alpha/2, -3\alpha/2)$ in the two illumination beams yielding the set of equations

$$I_1 = 2I_0 \left\{ 1 + \frac{\gamma}{2}\left[\cos(\Delta\phi_1 - 3\alpha/2) + \cos(\Delta\phi_2 + 3\alpha/2) \right] \right\}, \tag{5.95}$$

$$I_2 = 2I_0 \left\{ 1 + \frac{\gamma}{2}\left[\cos(\Delta\phi_1 - \alpha/2) + \cos(\Delta\phi_2 + \alpha/2) \right] \right\}, \tag{5.96}$$

$$I_3 = 2I_0 \left\{ 1 + \frac{\gamma}{2}\left[\cos(\Delta\phi_1 + \alpha/2) + \cos(\Delta\phi_2 - \alpha/2) \right] \right\}, \tag{5.97}$$

$$I_4 = 2I_0 \left\{ 1 + \frac{\gamma}{2}\left[\cos(\Delta\phi_1 + 3\alpha/2) + \cos(\Delta\phi_2 - 3\alpha/2) \right] \right\}, \tag{5.98}$$

The phase shift α at each point in the interferogram can be extracted from this set of equations with

$$\alpha = 2\tan^{-1}\left(\sqrt{\frac{3(I_2 - I_3) - (I_1 - I_4)}{I_2 - I_3 + I_1 - I_4}} \right), \tag{5.99}$$

and the phase difference $\Delta\phi_1 - \Delta\phi_2$ with

$$\Delta\phi_1 - \Delta\phi_2 = 2\tan^{-1}\left(\frac{\sqrt{[3(I_2 - I_3) - (I_1 - I_4)][(I_2 - I_3) + (I_1 - I_4)]}}{(I_2 + I_3) - (I_1 + I_4)}\right). \tag{5.100}$$

When a different set of phase steps $(-3\alpha/2, -3\alpha/2)$, $(-\alpha/2, -\alpha/2)$, $(\alpha/2, \alpha/2)$, and $(3\alpha/2, 3\alpha/2)$ is applied to the same configuration, the sum of phases $\Delta\phi_1 + \Delta\phi_2$ at each point in the interferogram can easily be determined. If I'_1, I'_2, I'_3, and I'_4 correspond to the four intensity measurements, then

$$\Delta\phi_1 + \Delta\phi_2 = 2\tan^{-1}\left(\frac{\sqrt{[3(I'_2 - I'_3) - (I'_1 - I'_4)][(I'_2 - I'_3) + (I'_1 - I'_4)]}}{(I'_2 + I'_3) - (I'_1 + I'_4)}\right). \tag{5.101}$$

Many other algorithms have been presented [5.85–87, 89] which have the potential to simultaneously determine information about both the sum and difference phases. Phase-shifting procedures have also been reported [5.87, 89] which are relatively insensitive to phase-shifter inaccuracies.

5.8 Emerging Trends

As computers and image processing systems become faster and more powerful, PMI techniques will utilize larger detector arrays and faster processors. It typically takes a few seconds to calculate and unwrap phase at 256 × 256 data points. Detector arrays are available with 2048 × 2048 detector elements. Larger detector arrays enable the measurement of more complex surfaces; however, processing time is much longer. Commercial systems which calculate the phase modulo 2π over areas of 512 × 512 at TV frame rates are available [5.36, 50, 84]. Using dedicated image processing systems, the unwrapped phase (without phase ambiguities) can be calculated very rapidly in array processors [5.32].

Besides faster processing, new advances are being made in many areas of holographic phase-shifting interferometry. For many years, the major limitation of phase-measurement techniques has been phase unwrapping techniques which are insensitive to noise. The latest work in this area was referenced in Sect. 5.2.3. Another area of development is algorithms which are insensitive to errors [5.45, 46, 65] and algorithms which combine techniques such as spatiotemporal methods [5.91]. More complex measurement techniques enable the measurement of in-plane displacements with holographic moiré [5.84, 89]. Whole-field analysis to measure three-dimensional displacements is another area of recent research [5.92–94]. Holographic interferometry is also being used to look at less stable objects with shearing interferometry techniques [5.97, 98]. And it is even being applied to investigate objects underwater [5.99]. Although this is not a complete list of current research, it shows the breadth of topics being discussed in the literature.

References

5.1   K.A. Stetson: Appl. Opt. **29**, 502–504 (1990)
5.2   P. Carré: Metrologia **2**, 13–23 (1966)
5.3   J.H. Bruning, D.R. Herriott, J.E. Gallagher, D.P. Rosenfeld, A.D. White, D.J. Brangaccio: Appl. Opt. **13**, 2693–2703 (1974)
5.4   G.T. Reid: Opt. Lasers Eng. **7**, 37–68 (1986)
5.5   K. Creath: *Progress in Optics*, **26**, 349–393 (Elsevier, Amsterdam 1988)
5.6   J. Schwider: *Progress in Optics*, **29**, 271–359 (North Holland, Amsterdam 1990)
5.7   J.E. Greivenkamp, J.H. Bruning: In *Optical Shop Testing* 2nd ed. by D. Malacara (Wiley, New York 1992) pp. 501–598
5.8   P. Hariharan, B.F. Oreb, N. Brown: Opt. Commun. **41**, 393–396 (1982)
5.9   B.F. Oreb, N. Brown, P. Hariharan: Rev. Sci. Instrum. **53**, 697–699 (1982)
5.10 P. Hariharan: Opt. Eng. **24**, 632–638 (1985)
5.11 R. Thalmann, R. Dändliker: Opt. Eng. **24**, 930–935 (1985)
5.12 J.E. Greivenkamp: Opt. Eng. **23**, 350–352 (1984)
5.13 C.J. Morgan: Opt. Lett. **7**, 368–370 (1982)
5.14 G.-M. Lai, T. Yatagai: J. Opt. Soc. Am. A **8**, 822–827 (1991)
5.15 R. Crane: Appl. Opt. **8**, 538–542 (1969)
5.16 W.H. Stevenson: Appl. Opt. **9**, 649–652 (1970)
5.17 J.C. Wyant: Appl. Opt. **14**, 2622–2626 (1975)
5.18 R.N. Shagam, J. C. Wyant: Appl. Opt. **17**, 3034–3035 (1978)
5.19 J.C. Wyant, R.N. Shagam: Use of electronic phase measurement techniques in optical testing, Proc. ICO-11, Madrid (1978) pp. 659–662
5.20 H.Z. Hu: Appl. Opt. **22**, 2052–2056 (1983)
5.21 R.N. Shagam: SPIE Proc. **429**, 35–42 (1983)
5.22 M.P. Kothiyal, C. Delisle: Opt. Lett. **9**, 319–321 (1984)
5.23 M.P. Kothiyal, C. Delisle: Appl. Opt. **24**, 2288–2290 (1985)
5.24 J.C. Wyant: Laser Focus 65–71 (May 1982)
5.25 J.C. Wyant, K. Creath: Laser Focus/Electro-Optics 118–132 (Nov. 1985)
5.26 K. Itoh: Appl. Opt. **21**, 2470 (1982)
5.27 D.G. Ghiglia, G.A. Mastin, L.A. Romero: J. Opt. Soc. Am. A **4**, 267–280 (1987)
5.28 R.M. Goldstein, H.A. Zebker, C.L. Werner: Radio Sci. **23**, 712–720 (1988)
5.29 J.M. Huntley: Appl. Opt. **28**, 3268–3270 (1989)
5.30 D.J. Bone: Appl. Opt. **30**, 3627–3632 (1991)
5.31 A.A.M. Maas: Phase shifting speckle interferometry, Dissertation, Delft Technical University (1991)
5.32 A. Spik, D.W. Robinson: Opt. Lasers Eng. **14**, 25–37 (1991)
5.33 D.P. Towers, T.R. Judge, P.J. Bryanston-Cross: Opt. Lasers Eng. **14**, 239–281 (1991)
5.34 H.A. Vrooman, A.A.M. Mass: Appl. Opt. **30**, 1636–1641 (1991)
5.35 T.R. Judge, C. Quan, P.J. Bryanston-Cross: Opt. Eng. **31**, 533–543 (1992)
5.36 K.A. Stetson: Appl. Opt. **31**, 5320–5325 (1992)
5.37 D.W. Robinson, G.T. Reid (eds): *Interferogram Analysis: Digital Processing Techniques for Fringe Pattern Measurement* (IOP, London 1993)
5.38 J.E. Greivenkamp: Appl. Opt. **26**, 5245–5258 (1987)
5.39 R.J. Palum, J.E. Greivenkamp: SPIE Proc. **1162**, 378–388 (1989)

5.40 R.N. Bracewell: *The Fourier Transform and Its Applications*, 2nd edn. (McGraw-Hill, New York 1986)
5.41 K. Creath: Appl. Opt. **28**, 2170–2175 (1989)
5.42 J.C. Wyant, C.L. Koliopoulos, B. Bhushan, O.E. George: ASLE Trans. **27**, 101–113 (1984)
5.43 P. Hariharan, B.F. Oreb, T. Eiju: Appl. Opt. **26**, 2504–2505 (1987)
5.44 J.H. Bruning: In *Optical Shop Testing*, ed. by D. Malacara (Wiley, New York 1978) pp. 409–438
5.45 K.G. Larkin, B.F. Oreb: In *Interferometry: Techniques and Analysis* SPIE Proc. **1755** (1992)
5.46 K.G. Larkin, B.F. Oreb: J. Opt. Soc. Am. A **9**, 1740–1748 (1992)
5.47 J. Schwider, R. Burow, K.-E. Elssner, J. Grzanna, R. Spolaczyk, K. Merkel: Appl. Opt. **22**, 3421–3432 (1983)
5.48 P.L. Wizinowich: Appl. Opt. **29**, 3271–3279 (1990)
5.49 J.R.P. Angel, P.L. Wizinowich: Europ. Southern Observatory Conf. Proc. **30**, 561–567 (1988)
5.50 E. Vikhagen: Appl. Opt. **29**, 137–144 (1990)
5.51 Y.-Y. Cheng, J.C. Wyant: Appl. Opt. **24**, 804–807 (1985)
5.52 J.B. Hayes: Linear methods of computer controlled optical figuring, Ph.D. Thesis, Optical Sciences Center, University of Arizona, Tucson, AZ (University Microfilms, Ann Arbor, MI 1984)
5.53 C. Ai, J.C. Wyant: Appl. Opt. **26**, 1112–1116 (1987)
5.54 C.L. Koliopoulos: Interferometric optical phase measurement techniques, Ph.D. Thesis, Optical Sciences Center, University of Arizona, Tucson, AZ (University Microfilms, Ann Arbor, MI 1981)
5.55 Y.-Y. Cheng, J.C. Wyant: Appl. Opt. **24**, 3049–3052 (1985)
5.56 K. Creath: SPIE Proc. **680**, 19–28 (1986)
5.57 C. Ai: Phase measurement accuracy limitation in phase-shifting interferometry, Ph.D. Thesis, Optical Sciences Center, University of Arizona, Tucson, AZ (University Microfilms, Ann Arbor, MI 1987)
5.58 C. Ai, J.C. Wyant: Appl. Opt. **27**, 3039–3045 (1988)
5.59 N. Ohyama, S. Kinoshita, A. Cornejo-Rodriguez, T. Honda, J. Tsujiuchi: J. Opt. Soc. Am. A **5**, 2019–2025 (1988)
5.60 K. Kinnstaetter, A.W. Lohmann, J. Schwider, N. Streibl: Appl. Opt. **27**, 5082–5089 (1988)
5.61 J. Schwider: Appl. Opt. **28**, 3889–3892 (1989)
5.62 C.P. Brophy: J. Opt. Soc. Am. A **7**, 537–541 (1990)
5.63 K. Freischlad, C.L. Koliopoulos: J. Opt. Soc. Am. A **7**, 542–551 (1990)
5.64 J. van Wingerden, H.J. Frankena, C. Smorenburg: Appl. Opt. **30**, 2718–2729 (1991)
5.65 K.G. Larkin, B.F. Oreb: In *Interferometry: Techniques and Analysis*, SPIE Proc. **1755** (1992)
5.66 K.A. Stetson, W.R. Brohinsky: Appl. Opt. **24**, 3631–3637 (1985)
5.67 E.L. Dereniak, D.G. Crowe: *Optical Radiation Detectors* (Wiley, New York 1984)
5.68 K. Creath, J.C. Wyant: SPIE Proc. **645**, 101–106 (1986)
5.69 S. Nakadate, H. Saito: Appl. Opt. **24**, 2172–2180 (1985)
5.70 K. Creath: Appl. Opt. **24**, 3053–3058 (1985)
5.71 K. Creath: Digital speckle-pattern interferometry, Ph.D. Thesis, Optical Sciences Center, University of Arizona, Tucson, AZ (University Microfilms, Ann Arbor, MI 1985)
5.72 K. Creath: Direct measurement of deformations using digital speckle-pattern interferometry. Proc. SEM Spr. Conf. on Exp. Mech., New Orleans (1986) pp. 370–377
5.73 D.W. Robinson, D.C. Williams: Opt. Commun. **57**, 26–30 (1986)
5.74 K.A. Stetson, W.R. Brohinsky: Appl. Opt. **25**, 2643–2644 (1986)
5.75 K.A. Stetson, W.R. Brohinsky: Opt. Eng. **26**, 1234–1239 (1987)
5.76 K.A. Stetson, W.R. Brohinsky, J. Wahid, T. Bushman: J. Nondestructive Evaluation **8**(2), 69–76 (1989)
5.77 D. Kerr, F. Mendoza-Santoyo, J.R. Tyrer: J. Opt. Soc. Am. A **7**, 820–826 (1990)
5.78 K. Creath, J.C. Wyant: In *Optical Shop Testing*, 2nd ed., ed. by D. Malacara (Wiley, New York 1992) pp. 599–651
5.79 D.B. Neumann, C.F. Jacobson, G.M. Brown: Appl. Opt. **9**, 1357–1362 (1970)

5.80 K.A. Stetson: Opt. Lett. **7**, 233–234 (1982)
5.81 Y. Oshida, K. Iwata, R. Nagata, Opt. Lasers Eng. **4**(2), 67–79 (1983)
5.82 K.A. Stetson, W.R. Brohinsky: J. Opt. Soc. Am. A **5**, 1472–1476 (1988)
5.83 R. Pryputniewicz, K.A. Stetson: SPIE Proc. **1162**, 456–467 (1989)
5.84 P.K. Rastogi: Appl. Opt. **31**, 1680–1681 (1992)
5.85 P.K. Rastogi: Opt. Commun. **93**, 336–338 (1992)
5.86 E.S. Simova, K.N. Stoev: Appl. Opt. **31**, 2405–2408 (1992)
5.87 E.S. Simova, K.N. Stoev: Appl. Opt. **31**, 5965–5974 (1992)
5.88 P.K. Rastogi: Opt. Eng. **32**, 190 (1993)
5.89 P.K. Rastogi: Appl. Opt. **32**, 3669 (1993)
5.90 M. Küchel: SPIE Proc. **1332**, 655–663 (1990)
5.91 M. Takeda, M. Kitoh: J. Opt. Soc. Am. A **9**, 1607–1614 (1992)
5.92 P. Hariharan, B.F. Oreb, N. Brown: Appl. Opt. **22**, 876–880 (1983)
5.93 M.C. Shellabear, J.R. Tyrer: Opt. Lasers Eng. **15**, 43–56 (1991)
5.94 D.W. Watt, T.S. Gross, S.D. Hening: Appl. Opt. **30**, 1617–1623 (1991)
5.95 F. Mendoza-Santoyo, M.C. Shellbear, J.R. Tyrer: Appl. Opt. **30**, 717–721 (1991)
5.96 R. Spooren: Appl. Opt. **31**, 1000–1007 (1992)
5.97 K.A. Stetson: Electro-optic holography for real-time display and quantitative analysis of interference fringes. Fringe '89, Automatic Processing of Fringe Patterns, (E. Berlin, April 25–28, 1989) and Optical Sensing and Measurement, Proc. ICALEO, Laser Inst. Am. **70**, 78–85 (1989)
5.98 M. Owner-Petersen: Appl. Opt. **30**, 2730–2738 (1991)
5.99 J.T. Malmo, O.J. Lokberg, S. Ellingsrud: Opt. Lasers Eng. **15**, 25–41 (1991)

6. Computer-Aided Evaluation of Holographic Interferograms

T. Kreis

Bremer Institut für angewandte Strahltechnik (BIAS), Klagenfurter Str. 2, D28359 Bremen, Germany

This chapter is designed to provide to the reader an overview of digital image processing of holographic interferograms. It deals primarily with the quantitative determination of the interference phase from the digitally recorded and stored interferograms, and related aspects like suppression of systematic and statistical errors. Digital image processing and the physical theory of holographic interferometry is treated briefly and only as far as it is needed for the description of quantitative evaluation. The same holds for the treatment concerning further processing of the evaluated phase data like derivation of deformation fields, strain and stress analysis, structural analysis methods, or computer tomography for the determination of refractive-index fields. This brief introduction on fringe-pattern processing should stimulate further studies of the related literature and encourage developments in this field. A more detailed treatment would go beyond the scope of this chapter.

The main aim of this contribution is to provide a brief but thorough background of the computer-based and optoelectronic approaches currently in use in the evaluation of the holographic interference patterns utilized in the deformation analysis of diffuse object surfaces. The determination of refractive-index fields is treated only briefly, and other topics like holographic contour measurements are omitted. Nevertheless the problem of extracting the interference phase distribution, which is central to all these applications, is discussed in detail.

6.1 Background

6.1.1 Holographic Interferometry

Holographic interferometry has proved to be a powerful tool for measuring deformations and in non-destructive testing [6.1–6.6]. Deformations of an opaque rough surface or refractive-index variations of a transparent body are measured to an accuracy of better than the wavelength of the used laser light. The main advantages of this method are:

- the measurements are contactless and non-invasive,
- the measurements are made on rough, diffusely reflecting surfaces; no specular reflection is required,

- two-dimensional continuous information are obtained; local deformation extrema cannot go undetected,
- the surface may be of arbitrary shape,
- the areas of the examined surfaces range from a few mm^2 to several m^2,
- the measured displacements range roughly speaking from 0.05 to 500 μm,
- reliable deformation analysis can be performed at low loading intensities; the testing is non-destructive,
- the method is independent of the material properties: deformations of hard or soft materials are measured; refractive-index variations in solids, fluids, gases, and even plasmas can be determined,
- the achievable resolution and accuracy of the displacement measurement allows subsequent numerical strain and stress calculations,
- the measurements can be made on moving surfaces,
- the measurements can be made in closed box-like structures like pressure or vacuum chambers through transparent windows,
- the generation and the evaluation of the information can be separated temporally and locally.

Holographic interferometry does not measure directly the inside integrity of a specimen, like X-ray or ultrasonics, but records its behaviour under an applied load. Thus it enables a validation of the tested structure and provides the deformation field relative to the load [6.7]. A defect in a structure may be critical with regard to a loading type or loading direction but not to another (Fig. 6.1). There are a number of possible loading-types, some of which are given below:

- direct mechanical loading: bending moment, tensile stress, torsional stress, gravity,
- pressure loading: internal pressure in hollow vessels, pressure or vacuum chambers,
- vibrational loading: acoustic fields by loudspeakers, electrodynamic shakers in point contact,
- impulse loading: local impact,
- thermal loading: radiation sources, hot air jets, volatile fluids, high power DC in conductive material.

In the early years of holographic testing the evaluation was done by visual interpretation of the interferograms. Quantitative measurement was performed by manual fringe counting in photographs of the interferograms on the desktop. Later the quantitative evaluation was improved drastically by the utilization of computers. The first attempts to automate the fringe counting procedure relied

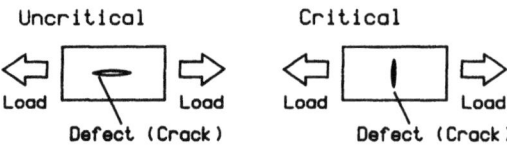

Fig. 6.1. Critical and uncritical defects with regard to applied tensile stress

on locating the fringe centres by computer, but furtheron refined techniques using precise control of experimental parameters, like the optical phase were developed [6.8].

Computer simulations of ideal interference patterns help to find the best optical arrangement, to optimize the loading, to develop improved evaluation methods, and to interpret the recorded interferograms. Computer-aided evaluation can be combined with techniques of structural analysis, like the finite element method [6.9, 10] or the boundary element method [6.11], to get further information. Complete computerized systems controlling the generation of the holographic interferograms and performing their quantitative evaluation are offered on the market today.

6.1.2 Digital Image Processing

The processing of multidimensional signals, especially two-dimensional information, is commonly called picture or image processing [6.12–6.15]. A typical image processing system consists of the modules shown in Fig. 6.2. The sensor, usually a TV-camera, converts the image to an electrical signal. Photoconductive sensors, like the vidicon have been almost totally replaced by solid-state sensors, especially the Charge Coupled Devices (CCD). The main advantages of solid-state sensors over photo-conductive sensors are their compactness and stability, their higher sensitivity and the lower lag factor. Monochrome sensors are sufficient for evaluating holographic interferograms due to the use of monochromatic laser light. If the interferogram varies rapidly – in the kHz-range – the two-dimensional image sensors cannot resolve this variation; instead pointwise detectors like phototransistors have to be scanned over a real image.

The digitizer samples the camera signal into an array of discrete picture elements, the pixels. Common digital image sizes are 256×256, 512×512, and 1024×1024 pixels. The amplitude at each pixel is quantized into a number of gray-levels. For evaluating holographic interferograms, 256 levels, corresponding to 8 bits, have proven sufficient. The gray-levels are denoted by integers with 0 belonging to black, the darkest level, and 255 corresponding to white, the brightest level. The output of the digitizer is a sequence of numbers. These

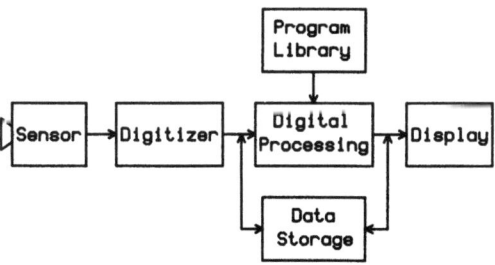

Fig. 6.2. Structure of typical image processing system

numbers are stored and processed by digital image processing algorithms, which may be implemented on a general purpose computer, a microprocessor, or special purpose hardware. Signal processors or transputer networks enable rapid measurements for real-time applications.

The results of digital processing range from single numbers to full digital images. They may be stored again or displayed. Images are displayed on TV-monitors with a Cathode-Ray Tube (CRT). The results may also be printed as numbers or be depicted as perspective plots, vector plots, false colour plots, etc.

The sampling theorem indicates the least number of pixels necessary to digitize reliably a bandlimited function. In the case of ideally cosine-shaped interferograms, one needs more than two pixels across one interference fringe. However, this is only a theoretical limitation; for practical applications at least five to eight pixels per fringe are necessary to produce reliable evaluation results.

Some typical operations in digital image processing, as they arise in the evaluation of holographic interferograms, are summarized as follows [6.16–6.18]:

- Point operations map a single input image to a single output image so that each output gray level depends only on the gray level of the corresponding input pixel [6.19]. A typical point operation may achieve a contrast stretching of low contrast interferograms or may perform a fixed threshold binarization.
- Algebraic operations produce an output image which is the pixel-by-pixel sum, difference, product, or quotient of two or more images. Combination of these operations are used extensively in phase sampling evaluation of holographic interferograms. Temporal filtering consists of averaging over several consecutively recorded interferograms to suppress time-variant intensity fluctuations.
- Geometric operations change the spatial relationships between the details within an interferogram. Typical examples are shifting, rotation, and magnification of patterns, correction of perspective distortion, or correction of geometric sensor aberrations.
- In local operations all pixels in the neighbourhood of an input pixel determine the gray value of the corresponding output pixel. Typical examples are the linear and nonlinear filtering techniques to enhance the interferogram. A smoothing of the interference pattern is obtained by low-pass filtering. Averaging over the neighbourhood of each pixel minimizes the speckle noise. The uneven brightness distribution of the background may be eliminated by high-pass filtering or shading correction. This is achieved by subtracting from the original image the result of low-pass filtering obtained with a large area averaging window [6.20].
- Global operations use the gray values of all pixels in an interference pattern to produce the gray value of the output pixel, if the result of the operation is again an image, or to produce a single number output. Single number outputs may be statistical features like minimal and maximal fringe densities, mean,

variance, histograms, etc. The Fourier-transform represents the input image in the spatial frequency domain, where the spectral contents of the pattern can be investigated.

Based on these fundamental types of operations a variety of digital image processing techniques to analyze interference patterns become possible. Image restoration by linear filtering restores the interferograms degraded by various distortions and noise. Pattern segmentation analyzes the image content. Skeletonizing of the fringes is useful for quantitative measurements as well as for a classification of the patterns into those stemming from sound and those stemming from defective specimens. Image compression, image coding and feature extraction are other processes which are used to process holographic interferograms.

6.2 Fringe Formation in Holographic Interferometry

A prerequisite for the computer-aided evaluation of holographic interferograms is the knowledge of the process which leads to the formation of the interference pattern. For quantitative evaluation the spatial distribution of the measured value is determined from the recorded intensity pattern or patterns. Therefore the quantitative relationship between the physical quantity to be measured and the interference pattern must be known (Chap. 3).

6.2.1 Double Exposure and Real-Time Holographic Interferometry

In double-exposure holographic interferometry two wave fronts reflected from the same surface or refracted by the same medium are recorded holographically onto the same holographic plate. The two wave fronts correspond to different deformation states – one before and one after a loading variation – or to different refractive index distributions in the medium.

Let the complex amplitude of the first wave front at a point P on the object surface be

$$U_1(P) = |U_1(P)| \exp[-i\phi(P)]; \tag{6.1}$$

$|U_1(P)|$ is the real amplitude and $\phi(P)$ the phase distribution. $\phi(P)$ varies spatially in a random manner due to the microstructure of the reflecting rough surface or refracting medium.

The change of the object shape, due to deformation, or the change of the refractive index distribution, have to be so small that only the phase distribution of the reflected or refracted wave front is modified. Let the phase change at point P be $\Delta\phi(P)$. The complex amplitude of the second wave front obtained after changing the load level can be written as

$$U_2(P) = |U_2(P)| \exp\{-i[\phi(P) + \Delta\phi(P)]\}. \tag{6.2}$$

The two wave fronts U_1 and U_2 are recorded consecutively onto the same holographic plate with identical reference waves and after development are reconstructed together. The two reconstructed wave fronts interfere and give rise to a stationary intensity distribution expressed by

$$I(P) = |U_1(P) + U_2(P)|^2$$
$$= I_1(P) + I_2(P) + 2\sqrt{I_1(P)I_2(P)} \cos[\Delta\phi(P)]. \tag{6.3}$$

For identical intensities, $I_1(P) = I_2(P)$, one gets

$$I(P) = 2I_1(P)\{1 + \cos[\Delta\phi(P)]\}. \tag{6.4}$$

The reconstructed image of the object appears covered with cosine-shaped fringes. Bright centres of fringes arise, where the interference phase is an even integer multiple of π

Bright $I(P)$: $\Delta\phi(P) = 2n\pi, \quad n \in \mathscr{Z}$ \hfill (6.5)

dark centres of fringes correspond to odd integer multiples of π

Dark $I(P)$: $\Delta\phi(P) = (2n + 1)\pi, \quad n \in \mathscr{Z}$. \hfill (6.6)

In real-time holographic interferometry only one wave front, corresponding to a reference state of the tested object, is holographically recorded. After processing, the hologram is replaced in its initial recording position. This replacement has to be done within sub-wavelength precision. During illumination of the hologram by the original reference wave, the reconstructed virtual-image wave front coincides with the wave front reflected directly from the object surface. A deformation of the surface now gives rise to an interference pattern, which can be observed in real-time. Changes in deformation lead to simultaneously observable changes in the interference pattern.

In both real-time and double-exposure holographic interferometry, the intensity variation in the fringe patterns has a cosine shape. However, in the real-time technique bright fringes correspond to odd multiples of π and dark fringes belong to even multiples of π. The intensity distribution is now expressed as

$$I(P) = 2I_1(P)\{1 - \cos[\Delta\phi(P)]\}. \tag{6.7}$$

6.2.2 Vibration Analysis by Holographic Interferometry

An in-depth discussion of vibration analysis is given in Chap. 8. A number of methods exist to record and display the amplitude distribution of a vibrating diffusely reflecting surface by holographic interferometry. Of these methods the double-exposure method is performed by triggering two short laser pulses at two different vibration states. Closely related to this is the stroboscopic method, where a sequence of light pulses is used for recording the hologram. These pulses, triggered at the frequency of the vibration of the object illuminate its two vibration states [6.21, 22]. The intensity is similar to (6.4).

In the real-time method, the holographically recorded and reconstructed waves represent the stationary state of the object. During vibration, at each instant of time, a cosine-shaped intensity pattern is formed:

$$I(P, t) = 2I_1(P)\{1 - \cos[\Delta\phi(P)\sin(\omega t)]\}, \quad (6.8)$$

where ω is the angular frequency of the vibration, and the interference phase $\Delta\phi(P)$ is related to the maximal amplitude of the vibration. If the frequency ω is higher than the temporal resolution of the eye or the applied sensor, a time-averaged intensity is seen

$$I(P) = 2I_1(P) \lim_{T\to\infty} \int_0^T \{1 - \cos[\Delta\phi(P)\sin(\omega t)]\}\,dt$$

$$= 2I_1(P)\{1 - J_0[\Delta\phi(P)]\} \quad (6.9)$$

where J_0 is the zero-order Bessel function of the first kind. These fringes have low contrast; nevertheless the method can be used for the identification of resonant frequencies by monitoring the fringe pattern while varying the excitation frequency.

The most frequently applied method for holographic vibration analysis is the time-average method [6.23]. Here the vibrating surface is recorded holographically with an exposure time which is long compared with the period of vibration. During reconstruction, we obtain an intensity pattern described by

$$I(P) = 2I_1(P)J_0^2[\Delta\phi(P)]. \quad (6.10)$$

The fringes are contours of equal vibration amplitudes of the spatial vibration modes. Maximal intensity corresponds to $\Delta\phi(P) = 0$, which occurs at the nodes of the vibration mode. Dark centres of the fringes refer to zeros of the Bessel-function J_0.

Further refinements of holographic vibration analysis use temporal modulation of the reference beam by varying its amplitude, phase, or frequency [6.6].

6.2.3 Interference Phase Variation due to Deformation

In holographic interferometric deformation measurement, the displacement of each surface point gives rise to an optical path difference $\delta(P)$. This path difference is related to the interference phase $\Delta\phi(P)$ by

$$\Delta\phi(P) = \frac{2\pi}{\lambda}\delta(P) \quad (6.11)$$

where λ denotes the wavelength of the used laser light. The observed intensity, which is to be evaluated, is related to $\Delta\phi$ by (6.3 to 7).

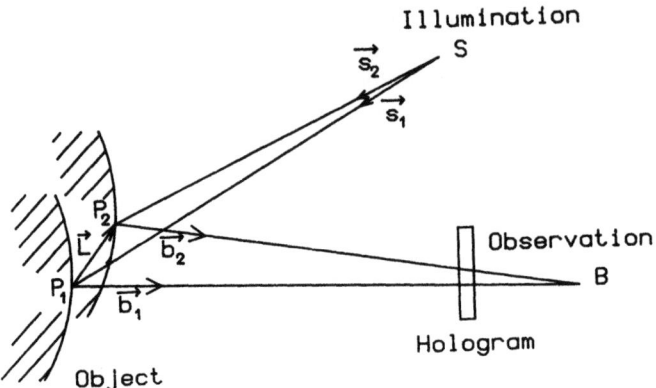

Fig. 6.3. Holographic arrangement

In a holographic arrangement with diverging illumination and observation waves, let $S = (x_S, y_S, z_S)$ be the illumination point and $B = (x_B, y_B, z_B)$ be the observation point, described in a Cartesian coordinate system (Fig. 6.3). When the object is deformed, the surface point $P_1 = (x_{P1}, y_{P1}, z_{P1})$ moves, by an amount defined by the displacement vector $L(P_1) = (L_x(P_1), L_y(P_1), L_z(P_1))$, to its new position P_2. The optical path difference can be expressed as

$$\delta(P) = \overline{SP_1} + \overline{P_1 B} - (\overline{SP_2} + \overline{P_2 B})$$
$$= s_1 \cdot SP_1 + b_1 \cdot P_1 B - s_2 \cdot SP_2 + b_2 \cdot P_2 B, \quad (6.12)$$

where s_1 and s_2 are unit vectors in the illumination directions, and b_1 and b_2 are unit vectors in the observation directions.

Let $K_1(P)$ be the bisector of the unit vectors in illumination direction and $\Delta K_1(P)$ half the difference. The same way $K_2(P)$ and $\Delta K_2(P)$ are defined

$$K_1(P) = \tfrac{1}{2}[s_1(P_1) + s_2(P_2)] \quad \text{and} \quad \Delta K_1(P) = \tfrac{1}{2}[s_1(P_1) - s_2(P_2)],$$
$$K_2(P) = \tfrac{1}{2}[b_1(P_1) + b_2(P_2)] \quad \text{and} \quad \Delta K_2(P) = \tfrac{1}{2}[b_1(P_1) - b_2(P_2)]. \quad (6.13)$$

By definition $L(P)$ can be written as $P_1 B - P_2 B = L(P)$ and $SP_2 - SP_1 = L(P)$. This leads to the relation

$$\delta P = [K_1(P) + \Delta K_1(P)] \cdot SP_1 + [K_2(P) + \Delta K_2(P)] \cdot P_1 B$$
$$- [K_1(P) - \Delta K_1(P)] \cdot SP_2 - [K_2(P) - \Delta K_2(P)] \cdot P_2 B$$
$$= K_2(P) \cdot L(P) - K_1(P) \cdot L(P)$$
$$+ \Delta K_2(P) \cdot [P_1 B + P_2 B] + \Delta K_1(P) \cdot [SP_1 + SP_2]. \quad (6.14)$$

The displacements are far smaller compared with the dimensions of the arrangement geometry – $|L|$ is in the micrometer range, $\overline{SP_i}$ and $\overline{P_i B}$ are in the meter range –, the same relation holds between the lengths of ΔK_1 and ΔK_2 compared to the lengths of the unit vectors K_1 and K_2. Furthermore the vector ΔK_1 is nearly orthogonal to $SP_1 + SP_2$, and ΔK_2 nearly orthogonal to $P_1B + P_2B$. Therefore the scalar products comprising ΔK_1 and ΔK_2 can be neglected, it remains

$$\delta(P) = L(P) \cdot [K_2(P) - K_1(P)]. \tag{6.15}$$

Introducing the so-called sensitivity vector

$$K(P) = \frac{2\pi}{\lambda}[K_2(P) - K_1(P)], \tag{6.16}$$

we get the relation

$$\Delta\phi(P) = L(P) \cdot K(P). \tag{6.17}$$

As explicited in earlier chapters, (6.17) shows that for deformation measurements the interference phase at each point is given by the scalar product of the displacement vector and the sensitivity vector; the sensitivity vector is defined by the geometry of the holographic arrangement.

For divergent illumination and observation the unit vectors $K_1(P)$ and $K_2(P)$ are determined by

$$K_1(P) = \begin{pmatrix} K_{1x}(P) \\ K_{1y}(P) \\ K_{1z}(P) \end{pmatrix} = \frac{1}{\sqrt{(x_P - x_S)^2 + (y_P - y_S)^2 + (z_P - z_S)^2}} \begin{pmatrix} x_P - x_S \\ y_P - y_S \\ z_P - z_S \end{pmatrix} \tag{6.18}$$

and

$$K_2(P) = \begin{pmatrix} K_{2x}(P) \\ K_{2y}(P) \\ K_{2z}(P) \end{pmatrix} = \frac{1}{\sqrt{(x_B - x_P)^2 + (y_B - y_P)^2 + (z_B - z_P)^2}} \begin{pmatrix} x_B - x_P \\ y_B - y_P \\ z_B - z_P \end{pmatrix}. \tag{6.19}$$

The sensitivity vectors vary over the object surface if either one of the illumination or observation beams are divergent. For parallel illumination and observation the sensitivity vector is constant over the investigated surface. This is shown in Fig. 6.4.

6.2.4 Interference Phase Variation due to Refractive-Index Variations

Holographic interferometry has widely replaced Mach–Zehnder interferometry in flow visualization, plasma diagnostics and in heat transfer analysis. An optical path difference can be generated by changing the refractive index of the transparent medium through which the light waves were passing initially. Due to the

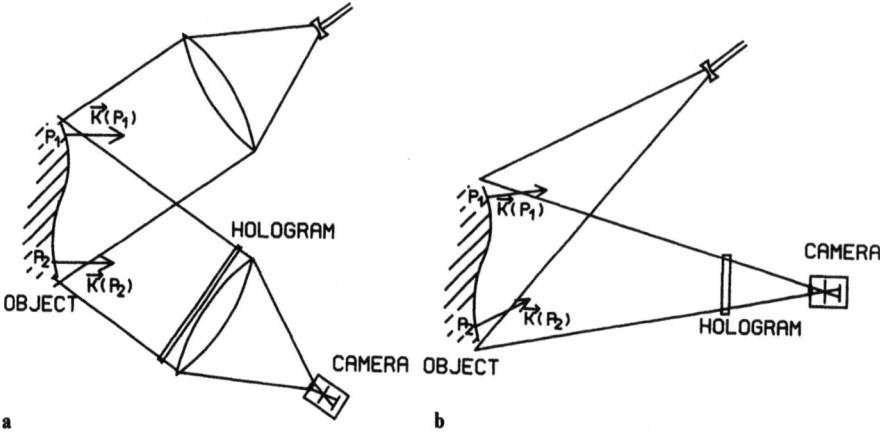

Fig. 6.4. Sensitivity vectors, constant (**a**), not constant (**b**)

common path of the waves in the two states, imperfections of the optical component, such as windows or glass walls of the containments, do not contribute to the generated patterns.

Let the rays propagate along straight lines parallel to the z-axis, and let the change in the refractive index distribution be $\Delta n(x, y, z)$. Then the interference phase distribution $\Delta \phi(x, y)$ in the plane perpendicular to the z-axis is [6.6]

$$\Delta \phi(x, y) = \frac{2\pi}{\lambda} \int_{z_1}^{z_2} \Delta n(x, y, z) \, dz. \tag{6.20}$$

Special cases which are easy to handle are phase objects with constant refractive index in the z-direction $\Delta n(x, y, z) = \Delta n(x, y)$, namely concentrations at boundary layers, and radially symmetric refractive index distributions $\Delta n(x, y, z) = \Delta n(r)$. If in this latter case the axis of symmetry is the y-axis, the interference phase distribution is given by the Abel transform of $\Delta n(r)$

$$\Delta \phi(x, y) = \frac{4\pi}{\lambda} \int_{x}^{R} \frac{\Delta n(r)}{\sqrt{r^2 - x^2}} r \, dr. \tag{6.21}$$

6.2.5 Ambiguity of Fringe Patterns

As was pointed out in (6.4 and 7) in two-wave interferometry, and thus for double-exposure, real-time, or stroboscopic methods of holographic interferometry, the intensity is given by the cosine of the interference phase [6.24]. Unfortunately the cosine is not a one-to-one function, but is even and periodic.

$$\cos \Delta \phi = \cos(s \Delta \phi + 2\pi m), \quad s \in \{-1, 1\}, \quad m \in \mathcal{Z}. \tag{6.22}$$

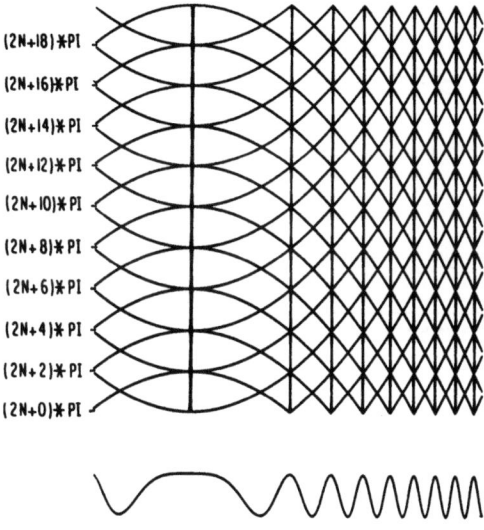

Fig. 6.5. Ambiguity of interference fringes

Each interference phase distribution determined from one intensity distribution is indefinite to an additive integer multiple of 2π and to the sign s. Figure 6.5 displays a part of a graph, which extends to infinity upward and downward. Each path from the left to the right through this graph represents a one-dimensional interference phase distribution belonging to the intensity distribution given at the bottom of the figure.

In most practical applications one can assume that the measured interference phase distribution is not only continuous but as well differentiable, meaning a smoothly varying function. Depending upon the known direction in which the load acts on the object, say, compressive or tensile, a proper sign distribution assignment norm can be fixed based on the increase or decrease of the interference phase. Practical ways to eliminate the sign problem consist of either recording multiple phase stepped interferograms or introducing experimentally a linear phase carrier with a positive slope higher than the steepest descent of the interference phase, thus producing only increasing phase [6.25–6.27].

The problem of the 2π-multiples is solved by the processing step called *demodulation*, *continuation*, or *phase unwrapping*. By addition or subtraction of integer multiples of 2π the phase jumps are eliminated. The absolute additive term in some applications can be determined if there is a point P on the object surface where the value of the displacement is known. Preferably this value is $L(P) = 0$. Sometimes an elastic ribbon is tied from one point on the tested surface to a point which has undergone no displacement. Then the fringes along this ribbon are counted starting from zero.

Defining the interference phase at an arbitrary point of the investigated surface as zero, if one is only interested in the deformation relative to this point but not in the additional rigid body translation, is only admissible for constant

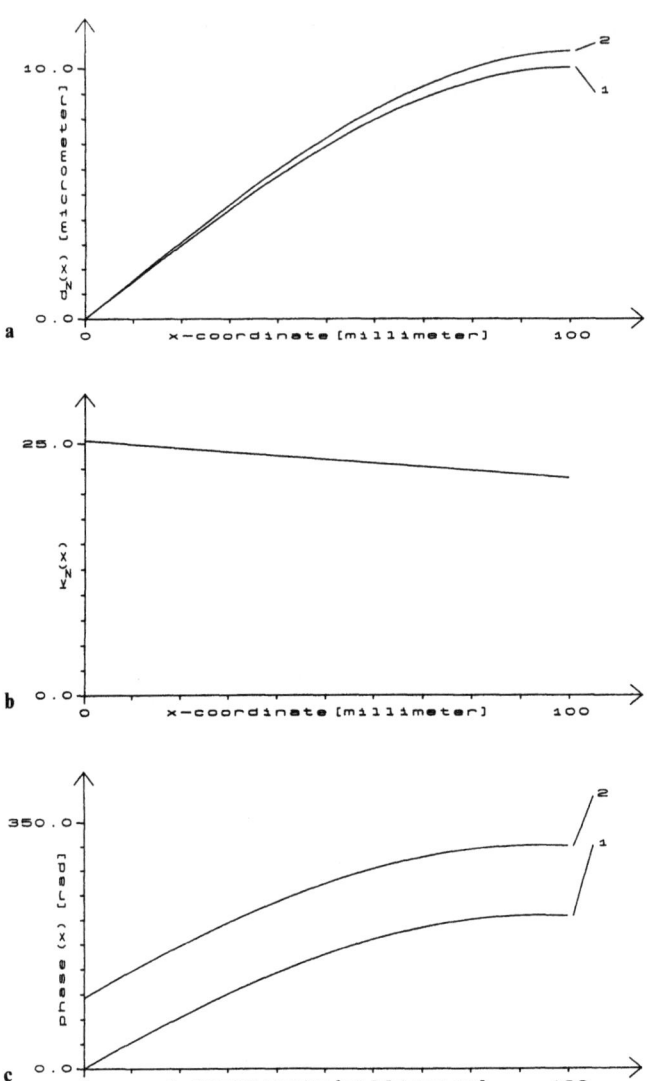

Fig. 6.6. Effect of additional interference phase

sensitivity vectors. If the sensitivity vector varies over the surface, the constant additive term has to be taken into account.

This is demonstrated in Fig. 6.6 for the example of a bended cantilever beam clamped at one end and with a point load at the other end. Curve 1 in Fig. 6.6a gives the z-displacement along the x-axis of the cantilever beam of length 100 mm. The z-component of the sensitivity vector is shown in Fig. 6.6b, where the coordinates of the illumination point are $S = (-100, 0, 300)$ mm, and of the observation point are $B = (50, 0, 300)$ mm. The surface point P varies as

$P = (x, 0, 0)$ mm, with x running from $x = 0$ mm to $x = 100$ mm. The wavelength is 514.5 nm. The one-dimensional interference phase distribution corresponding to this deflection and arrangement is shown in curve 1 of Fig. 6.6c. If this interference phase is now changed to curve 2 by adding a constant offset of 100 rads, the evaluated $L_z(x)$ is shifted non-uniformly. Even if the evaluated z-displacement at the loading point is brought to coincidence with the original one by subtraction of the displacement offset, curve 2 in Fig. 6.6a, there still remains a difference at the other end of the cantilever which is of the order of one fringe period.

6.2.6 Distortions of Holographic Interferograms

In practical applications a computer-aided evaluation not only suffers from ambiguities but also from a number of distortions degrading the interference pattern [6.28]. Thus a typical intensity distribution is not as in (6.4 or 7) but is of the form

$$I(P) = I_0(P)[1 + V(P)\cos \Delta \phi(P)] R_S(P) + R_E(P) + R_D(P). \qquad (6.23)$$

Here $I_0(P)$ denotes the low frequency background intensity caused by a varying illumination, e.g., a Gaussian profile of the enlarged laser beam, or fluctuating object reflectivity. $V(P)$ is the fringe visibility influenced mainly by speckle decorrelation and the ratio between the reference and object wave amplitudes. $R_S(P)$ describes the contrast variation caused by the speckles, which act as signal dependent coherent noise. $R_E(P)$ contains the time-dependent electronic noise due to the electronic components of the image processing equipment and $R_D(P)$ describes diffraction patterns of dust particles in the optical paths.

All these degradations can be summarized by the additive noise $a(P)$ and the multiplicative noise $b(P)$, so that the intensity distribution can be written as

$$I(P) = a(P) + b(P)\cos[\Delta \phi(P)]. \qquad (6.24)$$

Equation (6.24) represents the basic expression from which the interference phase has to be determined.

6.2.7 Simulation of Holographic Interference Patterns

Based on the equations given in Sects. 6.2.1-4, one can calculate the expected holographic interference patterns for a given set-up geometry, including laser wavelength and object shape, and the loading or deformation applied to the object. For each pixel of interest the corresponding sensitivity vector has to be first calculated, followed by the displacement vector at that particular surface point. From these, the resulting intensity is determined. Factors such as distortions, e.g., Gaussian background or stochastic speckle noise, can also be taken into account while simulating the interference pattern.

The simulation of holographic interference patterns is helpful in a number of ways:

- It helps in the planning phase to design optimized holographic arrangements. One can check whether an expected deformation produces fringes of sufficient but not too high density on the basis of cheap computer experiments instead of expensive practical experiments.
- If it is impossible to perform a three-dimensional evaluation with multiple interferograms, one can compare the calculated interferogram of an expected three-dimensional deformation field with the experimentally produced single interference pattern.
- In developing advanced evaluation methods, e.g. fault detection with neural networks, test data or samples for network training can be generated in high numbers and with high diversification. This would be impossible experimentally.

Figure 6.7 illustrates a simulated holographic interference pattern, where a rectangular plate was subjected to strain and torsion; the inhomogeneity in the fringe pattern results from the presence of a subsurface void in the plate.

6.3 Evaluation of Holographic Interferograms

6.3.1 Qualitative Evaluation

Most applications of holographic interferometry are still in the field of holographic nondestructive testing. The inspected sample is loaded and the resulting deformations are made visible as fringe patterns. Local faults, like voids, cracks or other defects, lead to typical local deformations which deviate from the global deformation. Thus the faults are detected by their characteristic local interference patterns.

Fig. 6.7. Simulated holographic interference pattern

While a lot has been done in computerized quantitative evaluation of holographic interferograms, little is reported on automatic computer aided detection of local fringe patterns characterizing material defects. The main reasons are the manifold possible global fringe patterns, the variety of local patterns typical of the defects, and the difficulty to translate into computer software the knowledge of the experienced personnel judging the interferograms.

A few approaches currently in use in automatic qualitative evaluation are based on dividing the whole pattern into a number of sections and comparing the fringe densities in these sections. In [6.29], the interference patterns of pressure vessels are partitioned into squares and the number of fringes in each square is counted. Based on holographic interferograms of proven intact specimens, each square gets a minimum and maximum acceptable fringe count. The fringes are counted automatically in each square, regardless of their orientation. If all fringe counts fall between the predetermined thresholds, the pressure vessel is accepted, otherwise it is rejected as a defective one.

The sections, the patterns are divided in, can even be degenerated rectangles, like rows or columns [6.31, 30]. In [6.31], fringe peaks are counted along horizontal lines. If the number of peaks exceeds a threshold in one line, fringes are counted along short vertical columns centring at this horizontal line until again a threshold is reached. To ensure that a closed ring pattern is detected, the fringes are equally counted along inclined vectors as well. A fault is considered to be detected if the fringe count in each direction is above the threshold value. The process can be repeated to detect multiple defects. The procedure is applied to the detection of debrazes in brazed cooling panels.

In [6.30], Fourier amplitude spectra are calculated one-dimensionally along lines and columns or two-dimensionally in small rectangles. From these spectra an average amplitude spectrum together with an acceptance band broader than its band-width, is determined. If the fringe density along one line, column, or rectangle differs significantly from the average density, then a locally higher or lower fringe density must have occurred, indicating a defect.

Actual approaches to computerized qualitative evaluation use modern software concepts, like knowledge based expert systems and neural networks [6.32]. Main problems are the effective data reduction, extraction of relevant features from the fringe patterns, and the generation of manifold training samples by computer simulation of holographic interference patterns.

6.3.2 Quantitative Evaluation

The task of quantitative evaluation is the determination of the quantity to be measured, in most applications the displacement vector field, from the intensity pattern. This requires two steps: first the geometry of the holographic set-up and the object surface is determined to calculate the sensitivity vectors $K(P)$, and second the interference phase distribution $\Delta\phi(P)$ is calculated from the recorded interferogram. This is followed by a pointwise inversion of (6.17) to obtain the displacements.

In the static method of quantitative evaluation, the virtual image as seen through the holographic plate is recorded and stored. In the general case of a three-dimensional displacement vector L where all the three Cartesian displacement components have to be determined, we require three equations of the type of (6.17) with different sensitivity vector each, but under the assumption of the same surface and loading. Thus, one records three interference patterns; preferably with the same observation but different illuminations, to avoid perspective distortion and difficult identification of corresponding pixels in each pattern.

A threefold interference phase evaluation yields the three interference phase distributions $\Delta\phi_1(P)$, $\Delta\phi_2(P)$, and $\Delta\phi_3(P)$ corresponding to the sensitivity vectors $K^1(P)$, $K^2(P)$ and $K^3(P)$ respectively. Thus at each point P we have to solve a system of linear equations

$$\begin{pmatrix} \Delta\phi_1(P) \\ \Delta\phi_2(P) \\ \Delta\phi_3(P) \end{pmatrix} = \begin{pmatrix} L_x(P) \\ L_y(P) \\ L_z(P) \end{pmatrix} \begin{pmatrix} K_x^1(P) & K_y^1(P) & K_z^1(P) \\ K_x^2(P) & K_y^2(P) & K_z^2(P) \\ K_x^3(P) & K_y^3(P) & K_z^3(P) \end{pmatrix} \tag{6.25}$$

to obtain $L(P)$ [6.33].

This procedure is valid if at each point of the evaluated surface only the interference phase is known. On the other hand, the knowledge of fringe density and fringe orientation at each point permits one to define the two-dimensional fringe vector [6.34]. This leads to an over-determined system of equations which can be solved by the Gaussian least-squares method. Two holographic recordings with different sensitivity vectors thus suffice with this procedure.

In many applications only the normal displacements are of interest and not the whole three-dimensional deformation field. In this case an optimized holographic set-up with parallel illumination and observation waves, whose bisectrix is orthogonal to the surface, is implemented [6.35]. If the illumination and observation centres are far enough from the surface, a sufficiently constant sensitivity vector can be assumed even with divergent waves. Let z be the coordinate normal to a plane surface under test, then in this case $K_x(P)$ and $K_y(P)$ can be neglected and $K_z(P)$ is nearly constant

$$K_z(P) = K_z = \frac{4\pi}{\lambda} \cos\frac{\alpha}{2}, \tag{6.26}$$

where $\alpha/2$ is the angle between the surface normal and the illumination direction, respectively the observation direction. Now the normal displacement field $L_z(P)$ is proportional to the interference phase

$$L_z(P) = \frac{\lambda}{4\pi \cos(\alpha/2)} \Delta\phi(P). \tag{6.27}$$

In the dynamic method of fringe evaluation the observation direction is changed for each fixed point in the real image of the object surface. Due to the localization of the interference fringes the interference order varies with the change of the observation direction [6.36, 37]. In practice, the real image is formed by illuminating a double-exposure hologram via a conjugate reference beam, which means the wave fronts are exactly reversed with respect to the original reference wave. Such a beam is made to illuminate only a small region of the hologram. The real image exhibits a fringe system corresponding to the viewing direction, defined by the observed object point and by the illuminated spot on the hologram. While the spot moves across the hologram, the fringe system changes, which is recorded by a suitable detector placed at the observed point of the real image. The scanning of the reference beam is performed by an oscillating or rotating mirror or a lens. The detected signal is analysed by an oscilloscope and computer.

By altering the observation unit vector from $K_2^1(P)$ to $K_2^2(P)$ and thus the sensitivity vector from $K^1(P)$ to $K^2(P)$, one changes the interference phase from $\Delta\phi_1(P)$ to $\Delta\phi_2(P)$. This leads to a variation of the detected intensity. The measurement of the phase differences leads to

$$\Delta\phi_2(P) - \Delta\phi_1(P) = L(P)[K^2(P) - K^1(P)]$$

$$= L(P)\frac{2\pi}{\lambda}[K_2^2(P) - K_2^1(P)]. \qquad (6.28)$$

Since $L(P)$ consists of three components, three such phase differences have to be gathered with respect to the same point P but each for different observation directions. This gives rise to the following system of equations

$$\begin{pmatrix} \Delta\phi_2(P) - \Delta\phi_1(P) \\ \Delta\phi_4(P) - \Delta\phi_3(P) \\ \Delta\phi_6(P) - \Delta\phi_5(P) \end{pmatrix}$$
$$= \frac{2\pi}{\lambda} \begin{pmatrix} L_x(P) \\ L_y(P) \\ L_z(P) \end{pmatrix} \begin{pmatrix} K_{2x}^2(P) - K_{2x}^1(P) & K_{2y}^2(P) - K_{2y}^1(P) & K_{2z}^2(P) - K_{2z}^1(P) \\ K_{2x}^4(P) - K_{2x}^3(P) & K_{2y}^4(P) - K_{2y}^3(P) & K_{2z}^4(P) - K_{2z}^3(P) \\ K_{2x}^6(P) - K_{2x}^5(P) & K_{2y}^6(P) - K_{2y}^5(P) & K_{2z}^6(P) - K_{2z}^5(P) \end{pmatrix}$$
$$(6.29)$$

which after solution yields the three components of displacement undergone by the object point.

The scanning of the reference spot along a closed curve on the hologram provides the possibility to generate multiple, three or more, fringe patterns. An overdetermined system of many equations of the form (6.28) is then solved by Gaussian least squares.

Equation (6.28) is independent of the illumination direction, as long as this direction is not changed, and is also independent of the additive phase offset.

The sensitivity vector for dynamic evaluation $K_2^2 - K_2^1$ has a direction nearly orthogonal to that obtained in the static methods. Despite these benefits, the dynamic method has not attracted widespread applications. Hence this method will not be considered further in the following discussions.

6.3.3 Condition of the Evaluation Matrix

The determination of three-dimensional displacement fields requires the solution of linear systems of equations given in (6.25 or 29). It is well known that if the equations are linearly dependent, leading to a singular matrix whose determinant is zero, the system is not solvable. However, even if the matrices on the right-hand side of (6.25 or 29) are not singular, the solutions may show errors of some 100% in magnitude and direction if we have an ill-conditioned matrix.

A figure to measure the condition of a system of linear equations, or equivalently that of the matrix A of this system, is the Hadamard condition number $K_H(A)$ defined as

$$K_H(A) = \frac{|\det A|}{\prod_{i=1}^{n} \alpha_i} \tag{6.30}$$

with

$$\alpha_i = \sqrt{\sum_{k=1}^{n} a_{ik}^2}, \tag{6.31}$$

where n is the rank of matrix A, whose elements are given by a_{ik}; det denotes the determinant. The system is ill-conditioned if $K_H(A) \ll 1.0$, and is optimum for $K_H(A) = 1.0$. A condition number of $K_H(A) > 0.1$ is considered to be desirable.

The elements of the matrix are the components of the sensitivity vectors; these are derived from measurements of the geometry of the holographic arrangement. A good condition means that the directions of the sensitivity vectors are as different as possible and linearly independent. The best condition would be if these vectors were mutually orthogonal, which is impossible in practice.

Of course, the geometry has to be measured precisely. However, even more important than a precise measurement of the geometry is a proper separation of the sensitivity vectors. With an ill-conditioned system even the smallest errors in the geometry values would cause severe measurement errors. The use of an overdetermined system of equations, which is solved by Gaussian least squares, will cause the same errors if the sensitivity vectors are not separated far enough.

6.3.4 Determination of Sensitivity Vectors

In setting up the equations to solve for the displacement vector, the sensitivity vectors have to be determined from the geometry of the holographic arrangement. If only one sensitivity vector and a plane surface are present, a simple yardstick measurement would suffice. Difficulties arise if the object surface has a complicated shape. The surface contours are only sometimes defined in the design data, and in general they have to be measured. Laser triangulation or moiré-based methods can then be applied to determine the surface contour. Since laser, optics, and computer evaluation are all needed, these methods can be integrated into an optical arrangement combining holography and triangulation or holography and moiré techniques.

Moreover, things get more complicated if we require to generate multiple sensitivity vectors, by means of multiple observation or multiple illumination directions. Sometimes, if the objects have a complicated shape, to achieve a well-conditioned system of equations, mirrors are used for back or side views. In these cases the sensitivity vectors can be determined by holographic interferometry using known rotations $\theta(P)$ or displacements $L(P)$ and solving a system of equations for the now unknown sensitivity vectors $K(P)$ [6.38].

One practical way is to record three separate double exposure holograms of the object undergoing different rigid-body rotations. The rotation angles $\theta_1(P)$, $\theta_2(P)$ and $\theta_3(P)$ and the directions of their respective rotation axes are considered to be known. These axes must not be coplanar. The recording and observation geometry of the three holograms, with respect to the object, are assumed to be the same. This insures an invariant sensitivity vector from one recording to the other.

For all points P of interest the fringe vector $K_f(P)$ is determined and the fringe locus function Ω computed for each hologram corresponding to different rotations [6.34, 39]. The equivalent of (6.17) for rotations is

$$K_f(P) = -\theta \times K(P), \tag{6.32}$$

where $\times$ denotes the vectorial product. The threefold evaluation enables one to set up the system of equations,

$$\begin{pmatrix} K_{fx}^1(P) & K_{fy}^1(P) & K_{fz}^1(P) \\ K_{fx}^2(P) & K_{fy}^2(P) & K_{fz}^2(P) \\ K_{fx}^3(P) & K_{fy}^3(P) & K_{fz}^3(P) \end{pmatrix} =$$

$$\begin{pmatrix} \theta_{1x} & \theta_{1y} & \theta_{1z} \\ \theta_{2x} & \theta_{2y} & \theta_{2z} \\ \theta_{3x} & \theta_{3y} & \theta_{3z} \end{pmatrix} \begin{pmatrix} 0 & K_z(P) & -K_y(P) \\ -K_z(P) & 0 & K_x(P) \\ K_y(P) & -K_x(P) & 0 \end{pmatrix}, \tag{6.33}$$

which is then solved to obtain $K(P)$ for each P of interest.

170   6. Computer-Aided Evaluation of Holographic Interferograms

Nearly the same procedure can be used directly with (6.17). Three different, but known, rigid body displacements L^1, L^2, and L^3 are effected in non-coplanar directions; $\Delta\phi_1(P)$, $\Delta\phi_2(P)$ and $\Delta\phi_3(P)$ are determined for the three displacements, assuming equal sensitivity vectors. The sensitivity vectors are obtained as the solutions of

$$\begin{pmatrix} \Delta\phi_1(P) \\ \Delta\phi_2(P) \\ \Delta\phi_3(P) \end{pmatrix} = \begin{pmatrix} K_x(P) \\ K_y(P) \\ K_z(P) \end{pmatrix} \begin{pmatrix} L_x^1 & L_y^1 & L_z^1 \\ L_x^2 & L_y^2 & L_z^2 \\ L_x^3 & L_y^3 & L_z^3 \end{pmatrix}. \tag{6.34}$$

6.3.5 Correction of Perspective Distortion

The determination of more than one component of the displacement vector necessitates the generation of several interference patterns which, in turn, are produced with different configuration geometries. If the observation direction varies between the recordings, the fringe patterns suffer from different perspective distortion. For an oblique sight of the inspected surface, this surface appears deformed in the recorded image and may not fill the full frame. A spatial transform maps the pixels of the recorded interferograms to new pixels in the output interferogram, such that identical points of the object surface, which are imaged to different pixels in the recorded images, are mapped to identical pixels in the output images.

In the following, an algorithm is given for the spatial transform to correct perspective distortion in the case of plane surfaces. Without loss of generality, we assume a rectangular surface area to be evaluated, which by perspective distortion is deformed to an arbitrary convex quadrangle. The spatial transform is performed by a bilinear interpolation, which on the one hand is fast and computationally simple, and on the other hand produces a smooth mapping that preserves continuity and connectivity.

Let $I(x, y)$ be the recorded and stored perspectively distorted interferogram with pixel coordinates (x, y). Then the corrected interference pattern is $I'(x, y)$:

$$I'(x, y) = I(x', y') = I(ax + by + cxy + d, ex + fy + gxy + h). \tag{6.35}$$

This bilinear transformation is defined by the values of the eight coefficients a through h. By specifying the mapping of the four vertices (x_a, y_a), (x_b, y_b), (x_c, y_c), (x_d, y_d) of the quadrangle (Fig. 6.8) to the four vertices (x_0, y_0), (x_1, y_0), (x_1, y_1), (x_0, y_1) of the output rectangle, we create a system of four equations

$$\begin{pmatrix} x_a \\ x_b \\ x_c \\ x_d \end{pmatrix} = \begin{pmatrix} x_0 & y_0 & x_0 y_0 & 1 \\ x_1 & y_0 & x_1 y_0 & 1 \\ x_1 & y_1 & x_1 y_1 & 1 \\ x_0 & y_1 & x_0 y_1 & 1 \end{pmatrix} \begin{pmatrix} a \\ b \\ c \\ d \end{pmatrix}. \tag{6.36}$$

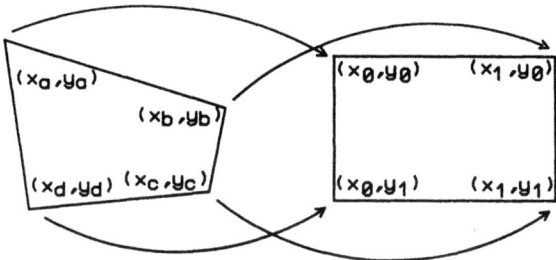

Fig. 6.8. Bilinear mapping

After inversion of the matrix we get the four coefficients a, b, c, and d as

$$\begin{pmatrix} a \\ b \\ c \\ d \end{pmatrix} = \frac{1}{(x_0 - x_1)(y_1 - y_0)} \begin{pmatrix} y_1 & -y_1 & y_0 & -y_0 \\ x_1 & -x_0 & x_0 & -x_1 \\ -1 & 1 & -1 & 1 \\ -x_1 y_1 & x_0 y_1 & -x_0 y_0 & x_1 y_0 \end{pmatrix} \begin{pmatrix} x_a \\ x_b \\ x_c \\ x_d \end{pmatrix}. \quad (6.37)$$

The y_a through y_d are given by a system of equations similar to (6.36).

$$\begin{pmatrix} y_a \\ y_b \\ y_c \\ y_d \end{pmatrix} = \begin{pmatrix} x_0 & y_0 & x_0 y_0 & 1 \\ x_1 & y_0 & x_1 y_0 & 1 \\ x_1 & y_1 & x_1 y_1 & 1 \\ x_0 & y_1 & x_0 y_1 & 1 \end{pmatrix} \begin{pmatrix} e \\ f \\ g \\ h \end{pmatrix}, \quad (6.38)$$

where the four coefficients e, f, g, and h are given by

$$\begin{pmatrix} e \\ f \\ g \\ h \end{pmatrix} = \frac{1}{(x_0 - x_1)(y_1 - y_0)} \begin{pmatrix} y_1 & -y_1 & y_0 & -y_0 \\ x_1 & -x_0 & x_0 & -x_1 \\ -1 & 1 & -1 & 1 \\ -x_1 y_1 & x_0 y_1 & -x_0 y_0 & x_1 y_0 \end{pmatrix} \begin{pmatrix} y_a \\ y_b \\ y_c \\ y_d \end{pmatrix}. \quad (6.39)$$

Thus the coefficients of the bilinear transform are defined. The transform can be implemented by (6.35), but a computationally more efficient algorithm is based on a line by line processing of the output image. Each new pixel coordinate is calculated by an increment from the foregoing one. The increments are constant along one line and are raised by a constant increment from line to line.

The spatial transform for each pixel calculates the origin of this pixel in the input image. In most cases they stem from fractional positions in the input pattern, meaning that their origin is between four adjacent pixels, so an interpolation is necessary to determine the gray level of the output pixel. The simplest interpolation is the nearest neighbour interpolation. A better solution, especially

172   6. Computer-Aided Evaluation of Holographic Interferograms

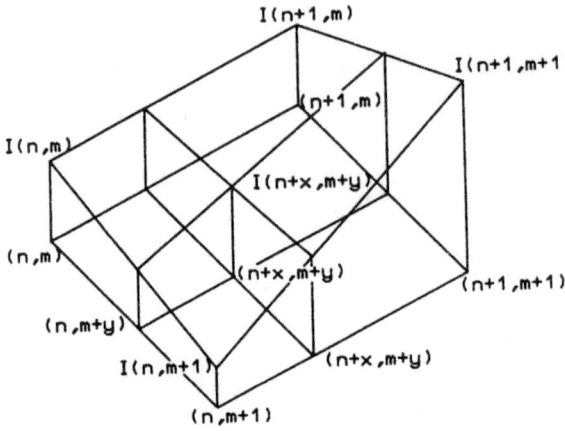

Fig. 6.9. Bilinear mapping

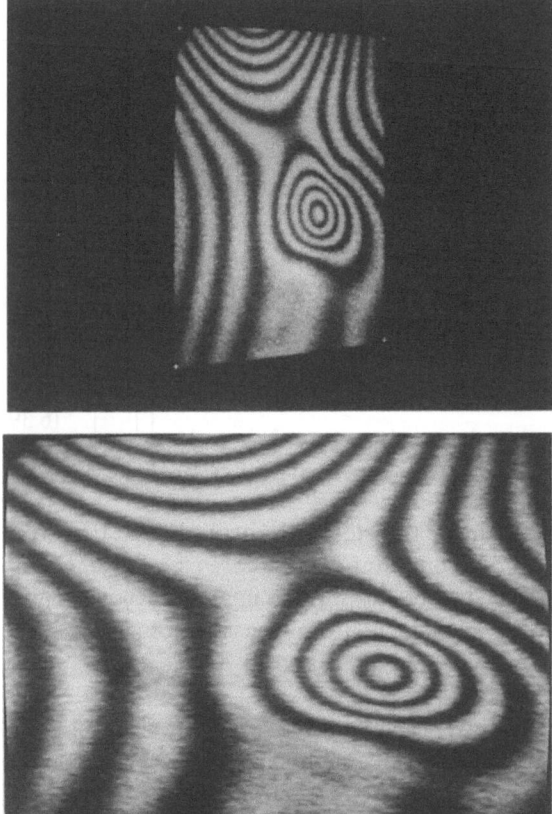

Fig. 6.10. Correction of perspective distortion

if significant gray level changes occur over one unit of pixel spacing, consists of applying the bilinear interpolation. Let us look for the intensity $I(n + x, m + y)$ with integers n and m and $x, y \in (0., 1.)$ (Fig. 6.9). The bilinear interpolation then gives

$$I(n + x, m + y) = I(n, m) + [I(n + 1, m) - I(n, m)]x$$
$$+ [I(n, m + 1) - I(n, m)]y + [I(n + 1, m + 1) + I(n, m)$$
$$- I(n, m + 1) - I(n + 1, m)]xy. \qquad (6.40)$$

An example for a correction of perspective distortion of a plane rectangular plate is presented in Fig. 6.10.

6.4 Interference Phase Determination

As was pointed out in Sect. 6.3, the main goal of the computer-aided fringe analysis task resides in the determination of the interference phase distribution. After having recorded the single or multiple holographic interferograms, the intensity distributions are digitized and quantized before being stored in the computer memory. The stored intensity is of the form of (6.24)

$$I(x, y) = a(x, y) + b(x, y) \cos[\Delta\phi(x, y)] \qquad (6.41)$$

with (x, y) denoting the pixel coordinates. Then the interference phase distribution has to be determined by one of the methods to be described in the sequel [6.40–44].

Since the interference phase is contained in the cosine function, an inverse trigonometric function in principle has to be applied to determine the phase. The arctangent gives values between $-\pi$ and $+\pi$, termed the principal values, if, in applying the arctangent of a fraction, the signs of the numerator and the denominator are taken into account separately, as is done by FORTRAN-function ATAN2(x, y).

6.4.1 Fringe Skeletonizing

The fringe-skeletonizing methods are computerized forms of the former manual fringe counting methods. It is assumed that the local extrema of the intensity distribution correspond to the maxima and minima of the cosine-function, (6.5 and 6). In this case the interference phase at pixels, where an intensity maximum or minimum is detected, is an even or odd integer multiple of π.

The methods for fringe skeletonizing can be divided into those based on fringe tracking and into those related to segmentation. Both techniques require a careful preprocessing to minimize speckle noise by lowpass filtering and to correct an uneven brightness distribution in the background illumination.

In methods based on fringe tracking, the algorithm looks for neighbouring pixels which correspond to local maxima or minima in the gray-values [6.45, 46]. The direction for search is predetermined by the precidingly found pixels. Starting points for fringe tracking are defined interactively by the user. Interaction by the user may likewise be necessary if obstacles are met, the fringe position is wrongly determined, or the algorithm is captured in small circles due to speckle noise. The operator can link data points belonging together or correct wrong decisions.

The segmentation techniques divide the pixels of the interference pattern into different regions representing ridges, valleys, and slopes in the gray-value landscape [6.47–50]. A general processing scheme for skeletonizing by segmentation consists of the following steps [6.28, 51]:

- Improvement of the signal-to-noise-ratio in the interference pattern by spatial and temporal filtering.
- Segmentation of intensity into maxima, minima and slopes by using adaptive thresholds, gradient operators, or piecewise approximation by elementary functions.
- Enhancement of the regions; elimination of the isolated points.
- Production of the fringe skeleton by thinning of the regions to line-structures.
- Enhancement of the skeleton by linking together interrupted lines, by adding missing points, and by removal of artifacts, line crossings or interconnections. This step may be performed interactively.
- Numbering of the fringes by attaching interference order numbers to them. Neighbouring lines indicating adjacent fringe maxima can only differ by -1, 0, or $+1$, if a continuous interference phase distribution is assumed.
- Interpolation of the interference phase distribution between the skeleton lines. This may be done by linear or bilinear functions, by spline-functions, by Bezier polynomials, or others.
- Calculation of the values to be measured from the interference phase distribution.

Some of these steps are demonstrated in Fig. 6.11 by means of an example of a valve loaded by internal pressure. Figure 6.11a depicts the holographic interferogram together with an intensity profile along the dotted horizontal line. The enhanced intensity distribution, after averaging and shading correction, is given in a gray-scale display in Fig. 6.11b and as a pseudo-3D-display in Fig. 6.11c. The enhanced intensity is segmented (Fig. 6.11d) where the ridges are white, the valleys are gray, and unidentified pixels are black. This pattern is enhanced by region-growing and binary filtering; the result is given in Fig. 6.11e. These regions are then thinned to a skeleton form, as shown in Fig. 6.11f. Numbering and interpolation lead to a continuous phase distribution as depicted in Fig. 6.11g.

A special fringe-contour detection scheme was proposed in [6.52]. It only works with interferograms fulfilling the prerequisites: (i) the presence of a dominant spatial frequency associated with the fringe pattern, (ii) the near invariance of this frequency with position. Then, along lines normal to the fringes, the one-dimensional Fourier-spectrum is calculated by the FFT algorithm. The phase of the dominant spatial frequency computed for each image line is a quantitative measure of fringe displacement at each line.

6.4.2 Evaluation by Temporal Heterodyning

In temporal heterodyning a small frequency difference $\Delta\omega/2\pi$ – in the range of 100 kHz – between the two interfering wave fields is introduced [6.53, 54]. In holographic interferometry this is realized in a double-reference-wave set-up, where the two states of the object are recorded by double exposure, each state with one of the two reference waves. In one of the reference waves two acousto-optic modulators in cascade introduce frequency shifts with opposite sign. During recording, both modulators are driven with 40 MHz, so the net shift is zero. The reconstruction of the holographic interference pattern is performed with both reference waves together, while one modulator is driven with 40 MHz and the other one with 40.1 MHz. Thus one gets an interference pattern oscillating with the beat frequency of 100 kHz. Chapter 4 considers these aspects in detail. Other means for introducing frequency shifts are rotating radial gratings, rotating birefringent plates, or two-frequency lasers, but the acousto-optic modulators are most common.

Because of the 100 kHz-oscillation, two-dimensional detectors like TV-cameras cannot record the interferograms; only point-detectors can be used. Photodetectors have to be scanned mechanically over the reconstructed real image. The phase difference of the oscillating signals at two points in the real image equals the interference phase difference between these two points.

One way to evaluate the interference phase is to keep one detector fixed at a reference point whilst the other is scanned. This way one measures the phase difference with respect to the phase obtained at the reference point. The phase differences are stored and can be displayed arranged in accordance to the points where they have been gathered.

Another evaluation technique scans an array of three, four, or five photo-detectors over the real image [6.55]. In this way the phase differences $\Delta\phi_x$ and $\Delta\phi_y$ between adjacent points in x- and y-directions, with known separations, are measured. Numerical integration of the recorded and stored phase values again gives the interference phase distribution. The detector array is realized by the ends of optical fibers, which transmit the signals to photomultipliers or photo-transistors, electronic filters and amplifiers, and the phase meter.

Temporal heterodyning provides the phase values independently of background intensities. Fractional phase values can be measured, so the problem of interpolation does not arise. Moreover the sign ambiguity problem is solved, since it is uniquely the increase or decrease of the phase which is detected, provided the scanning steps are smaller than half a fringe period as required by the sampling theorem.

176 6. Computer-Aided Evaluation of Holographic Interferograms

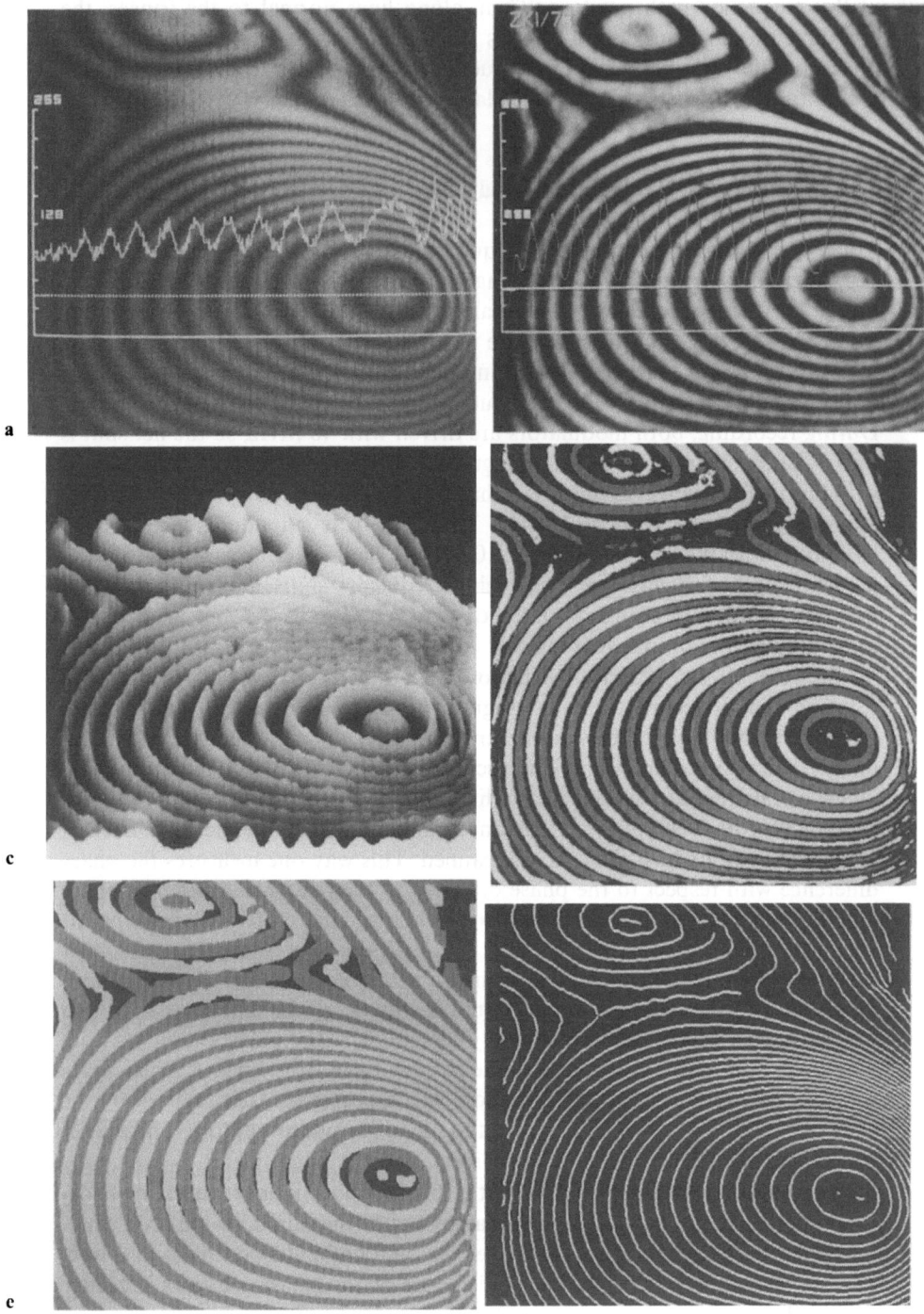

a

c

e

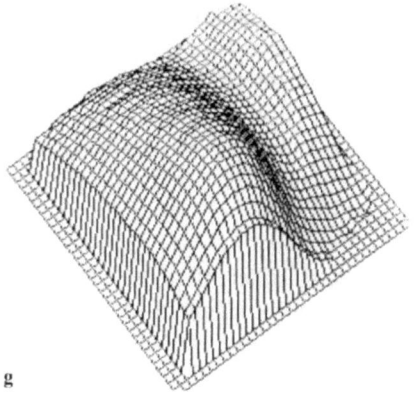

Fig. 6.11a–g. Interferogram evaluation by skeletonizing

g

6.4.3 Phase-Sampling Evaluation

In temporal heterodyning, point detectors have to be used due to the required detector bandwidth, which must be higher than the beat frequency. This high bandwidth is not achievable with the two-dimensional image-recording detectors like TV tubes or CCD arrays. However, if the reference phase is varied in discrete steps instead of in a continuous manner as in temporal heterodyning, the requirements on the bandwidth are considerably lowered, permitting the use of conventional TV detectors [6.56].

If the reference phase ϕ_R is varied in steps of ϕ_{Rn}, $n = 1,\ldots,m$ (6.24) accordingly provides m intensity distributions

$$I_n(x,y) = a(x,y) + b(x,y)\cos[\Delta\phi(x,y) + \phi_{Rn}], \quad n = 1,\ldots,m, \quad m \geq 3.$$
(6.42)

If the reference phases are known, at least three recordings of intensities are necessary to solve the nonlinear system of equations (6.42) for the three unknowns $a(x,y)$, $b(x,y)$, and $\Delta\phi(x,y)$ [6.57, 58]. Chapter 5 provides a more thorough analysis of the fringe patterns using phase shifting techniques.

Modulation of the reference phase in the holographic arrangement might be induced by a tilted glass plate, a moving grating, a rotating half-wave plate, acousto-optic or electro-optic modulators, or a Zeeman-laser. However, the most often used technique to-date consists of moving the mirror reflecting the reference wave to the holographic plate. The mirror is fixed on a piezo-electric transducer, which allows for a precise control of the mirror-shift in fractions of the wavelength of the used laser-light [6.59].

Phase sampling can also be performed with a dual reference-beam arrangement as described in Sect. 6.4.2. The two recordings of a double exposure hologram are obtained with distinct reference beams. During reconstruction with both references together, one of the reference waves is shifted in phase

6.4.3.1 Phase Shifting and Phase Stepping

Instead of recording the stationary intensities with different phase steps, the reference phase can also be shifted linearly during the recording of each interference pattern. This results in an integration of the varying intensity over the phase interval. Let the reference phase be shifted from $\phi_{Rn} - \Delta\phi_R/2$ to $\phi_{Rn} + \Delta\phi_R/2$ during the n-th recording step. The recorded intensity will then be given by

$$I_n(x,y) = \frac{1}{\Delta\phi_R} \int_{\phi_{Rn} - \Delta\phi_R/2}^{\phi_{Rn} + \Delta\phi_R/2} \{a(x,y) + b(x,y)\cos[\Delta\phi(x,y) + \phi_R(t)]\} d\phi_R(t). \quad (6.43)$$

The evaluation of the integral gives

$$I_n(x,y) = a(x,y) + \mathrm{sinc}\left(\frac{\Delta\phi_R}{2}\right) b(x,y) \cos[\Delta\phi(x,y) + \phi_{Rn}]. \quad (6.44)$$

This expression is equivalent to (6.42); only the contrast term $b(x,y)$ is modified by the constant factor $\mathrm{sinc}(\Delta\phi_R/2)$. In this sense phase shifting is equivalent to phase stepping, therefore the notations are used synonymously.

For solution of the nonlinear system of equations, (6.42), one uses a Gaussian least squares approach [6.60] by introducing

$$u(x,y) = b(x,y)\cos[\Delta\phi(x,y)] \quad \text{and} \quad v(x,y) = -b(x,y)\sin[\Delta\phi(x,y)] \quad (6.45)$$

and rewriting (6.42)

$$I_n = a + b\cos[\Delta\phi + \phi_{Rn}] = a + u\cos\phi_{Rn} + v\sin\phi_{Rn}, \quad (6.46)$$

where the pixel coordinates (x,y) are omitted for convenience. The sum of the quadratic errors

$$\sum_{n=1}^{m} (a + u\cos\phi_{Rn} + v\sin\phi_{Rn} - I_n)^2 \quad (6.47)$$

has to be minimized. Partial differentiation of this function with respect to a, u, and v and equating the derivatives to zero gives a linear system of three equations:

$$\begin{vmatrix} m & \sum\cos\phi_{Rn} & \sum\sin\phi_{Rn} \\ \sum\cos\phi_{Rn} & \sum\cos^2\phi_{Rn} & \sum\sin\phi_{Rn}\cos\phi_{Rn} \\ \sum\sin\phi_{Rn} & \sum\sin\phi_{Rn}\cos\phi_{Rn} & \sum\sin^2\phi_{Rn} \end{vmatrix} \begin{pmatrix} a \\ u \\ v \end{pmatrix} = \begin{pmatrix} \sum I_n \\ \sum I_n \cos\phi_{Rn} \\ \sum I_n \sin\phi_{Rn} \end{pmatrix}. \quad (6.48)$$

This system has to be solved pointwise for u and v, while the solution for a may be omitted. Then the interference phase is determined modulo 2π by

$$\Delta\phi(x, y) = \arctan \frac{-v(x, y)}{u(x, y)}. \tag{6.49}$$

6.4.3.2 Solution of Phase Sampling Equation

The pointwise solution of system (6.48) requires only a single inversion of the matrix to the left, provided the phase steps ϕ_{Rn} are the same for all pixels. Simple solutions of (6.48) are given for phase steps which are constant and have special values. Without loss of generality ϕ_{R1} is assumed to be zero. If the phase-step is known to be $\Delta\phi_R = 60°$ as we have $m = 3$, implying $\phi_{R1} = 0°$, $\phi_{R2} = 60°$, $\phi_{R3} = 120°$, the interference phase is calculated pointwise from

$$\Delta\phi(x, y) = \arctan \frac{2I_1(x, y) - 3I_2(x, y) + I_3(x, y)}{\sqrt{3}[I_2(x, y) - I_3(x, y)]}. \tag{6.50}$$

For $\Delta\phi_R = 90°$ and $m = 3$ we get

$$\Delta\phi(x, y) = \arctan \frac{I_1(x, y) - 2I_2(x, y) + I_3(x, y)}{I_1(x, y) - I_3(x, y)}. \tag{6.51}$$

With $\Delta\phi_R = 120°$ and $m = 3$ the interference phase is determined by

$$\Delta\phi(x, y) = \arctan \frac{\sqrt{3}[I_3(x, y) - I_2(x, y)]}{2I_1(x, y) - I_2(x, y) - I_3(x, y)}; \tag{6.52}$$

$m = 4$ reconstructions with a mutual phase step of $\Delta\phi_R = 90°$ furnish a relatively simple looking formula [6.61, 62]

$$\Delta\phi(x, y) = \arctan \frac{I_4(x, y) - I_2(x, y)}{I_1(x, y) - I_3(x, y)}. \tag{6.53}$$

For $\Delta\phi_R = 90°$ and $m = 5$ we have

$$\Delta\phi(x, y) = \arctan \frac{7[I_4(x, y) - I_2(x, y)]}{4I_1(x, y) - I_2(x, y) - 6I_3(x, y) - I_4(x, y) + 4I_5(x, y)}. \tag{6.54}$$

and for $\Delta\phi_R = 60°$ and $m = 4$ we get

$$\Delta\phi(x, y) = \arctan \frac{5[I_1(x, y) - I_2(x, y) - I_3(x, y) + I_4(x, y)]}{\sqrt{3}[2I_1(x, y) + I_2(x, y) - I_3(x, y) - 2I_4(x, y)]}. \tag{6.55}$$

This progression can be continued arbitrarily.

6.4.3.3 Phase Sampling with Unknown Phase Steps

A procedure which is different in principle from the aforementioned is the one with constant but unknown phase steps [6.63–66]. This gives a nonlinear system of four equations

$$I_1(x,y) = a(x,y) + b(x,y)\cos[\Delta\phi(x,y)],$$
$$I_2(x,y) = a(x,y) + b(x,y)\cos[\Delta\phi(x,y) + \Delta\phi_R],$$
$$I_3(x,y) = a(x,y) + b(x,y)\cos[\Delta\phi(x,y) + 2\Delta\phi_R],$$
$$I_4(x,y) = a(x,y) + b(x,y)\cos[\Delta\phi(x,y) + 3\Delta\phi_R] \qquad (6.56)$$

with the four unknowns a, b, $\Delta\phi$, and $\Delta\phi_R$. The phase step $\Delta\phi_R$ can now be determined as

$$\Delta\phi_R(x,y) = \arccos\frac{I_1(x,y) - I_2(x,y) + I_3(x,y) - I_4(x,y)}{2[I_2(x,y) - I_3(x,y)]}. \qquad (6.57)$$

By assumption $\Delta\phi_R$ has to be constant over the whole image, so we can take the average $\overline{\Delta\phi_R}$ over all pixels remaining after discarding the outliers. These outliers preferably occur where the denominator of (6.57) is zero or near to zero. Based on this $\overline{\Delta\phi_R}$, the interference phase at each point is given by

$$\Delta\phi(x,y) = \arctan\frac{(I_3 - I_2) + (I_1 - I_3)\cos\overline{\Delta\phi_R} + (I_2 - I_1)\cos 2\overline{\Delta\phi_R}}{(I_1 - I_3)\sin\overline{\Delta\phi_R} + (I_2 - I_1)\sin 2\overline{\Delta\phi_R}} \qquad (6.58)$$

and

$$\Delta\phi(x,y) = \arctan\frac{(I_4 - I_3) + (I_2 - I_4)\cos\overline{\Delta\phi_R} + (I_3 - I_2)\cos 2\overline{\Delta\phi_R}}{(I_2 - I_4)\sin\overline{\Delta\phi_R} + (I_3 - I_2)\sin 2\overline{\Delta\phi_R}} - \overline{\Delta\phi_R}. \qquad (6.59)$$

The (x,y) on the right-hand side have been omitted for convenience. The 2π-steps of the interference phase distribution modulo 2π in (6.58 and 59) occur at different points. This information can be used in the subsequent demodulation by considering the continuous phase variation of the two terms at each pixel.

Phase sampling with unknown phase steps works, in principle, for phase steps $\Delta\phi_R > 0°$ and $\Delta\phi_R < 180°$. For reliable results phase shifts between 30° and 150° are recommended. The presented method has advantages over the Carre method [6.67], which also works with four patterns with an unknown but constant phase step, and which calculates the interference phase according to

$$\Delta\phi(x,y) = \arctan\frac{\sqrt{(-I_1 + 3I_2 - 3I_3 + I_4)(I_1 + I_2 - I_3 - I_4)}}{-I_1 + I_2 + I_3 - I_4}. \qquad (6.60)$$

The described method, (6.57 to 59) enables several signal enhancement procedures. The constancy of the performed phase step can be checked via the calculated phase step $\Delta\phi_R(x, y)$, (6.57). Strong variations would indicate non-constant phase steps. Averaging the calculated phase steps results in an interference phase determination with the most probable phase step value and not with different calculated phase steps at each point, especially at the points with outliers. The two formulas to calculate the interference phase, (6.58 and 59), give redundant information for the demodulation process. All these facts contribute to better accuracy and reliability.

Furthermore, this method has advantages over the procedures with fixed known phase steps because it may be technically difficult to adjust the phase steps to exact values. If the four-phase-stepped recordings for the present method are taken in sequence with very short time gaps, e.g., with video-frequency, spurious vibrations can be assumed to contribute linearly to the phase and are compensated inherently by the evaluation process.

While evaluating by the phase sampling methods we still have the inverse trigonometric functions leading to interference phase modulo 2π. This means, there remains an uncertainty about the additive integer multiple of 2π. On the other hand, the sign ambiguity is uniquely resolved, due to the redundancy achieved by the multiple recorded interference patterns.

An example of phase stepping with unknown but constant phase steps is shown in Figs. 6.12 and 13. Figure 6.12 represents the four-phase-stepped holographic interferograms of a tensile test specimen with an internal crack obtained by the real-time method. The evaluation along one line is shown in several steps in Fig. 6.13. The four recorded intensity distributions along the line are displayed in Figs. 6.13a to d. In Fig. 6.13e the phase step $\Delta\phi_R(x, y)$, calculated by (6.57), is given together with the straight line which corresponds to the averaged $\overline{\Delta\phi_R}$. Fig. 6.13f and g display the interference phase distributions

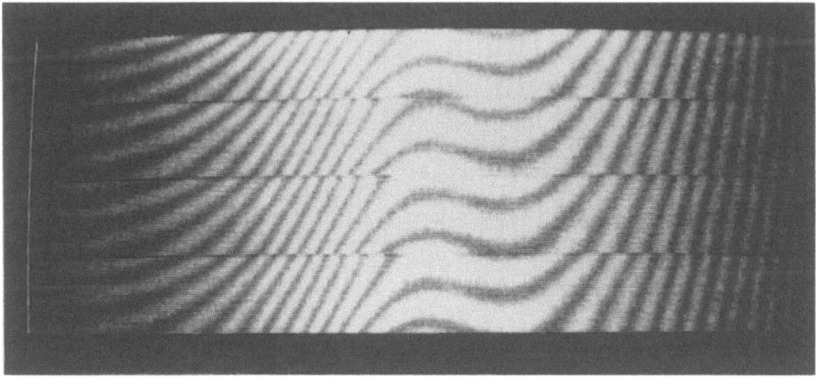

Fig. 6.12. Phase stepped holographic interferograms of tensile test specimen

182   6. Computer-Aided Evaluation of Holographic Interferograms

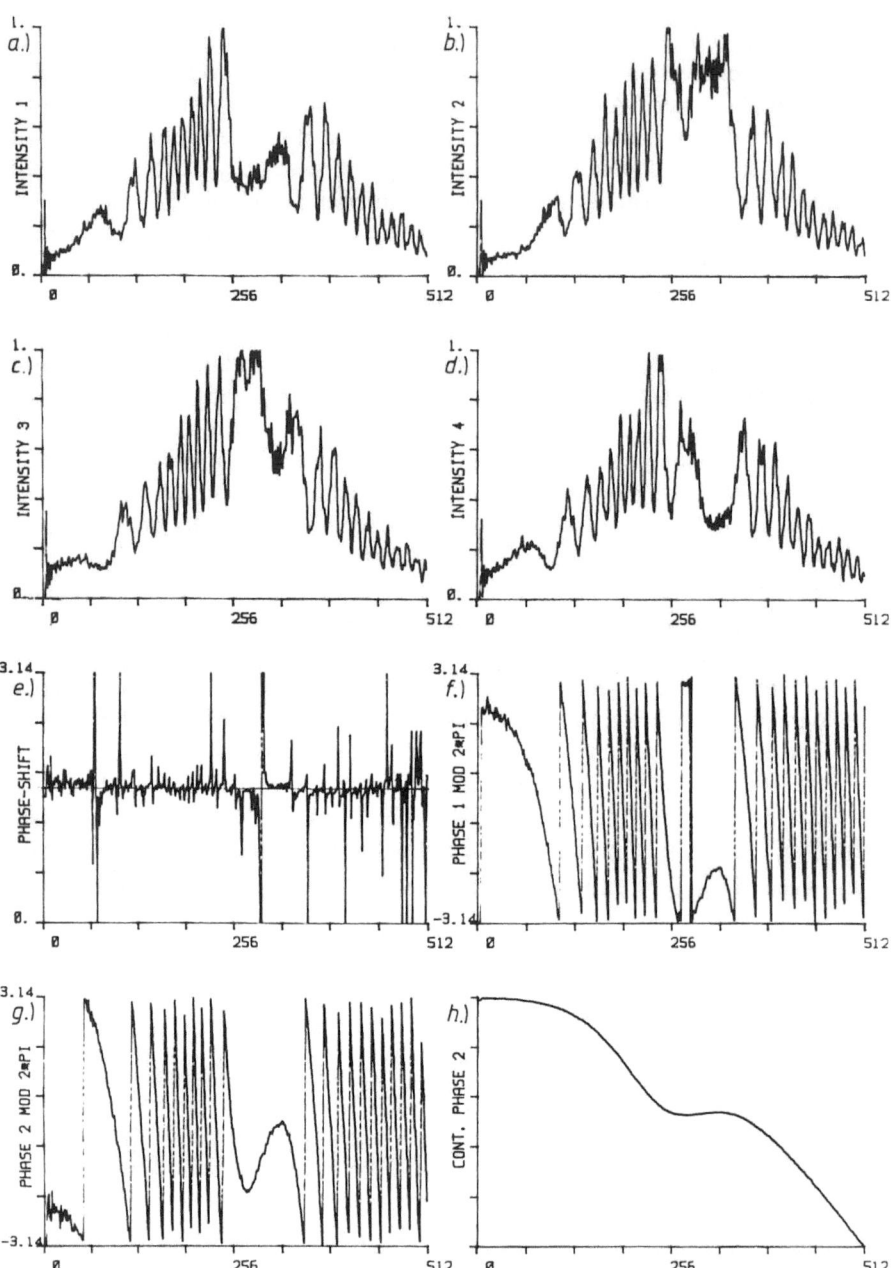

Fig. 6.13a–h. Evaluation by phase stepping

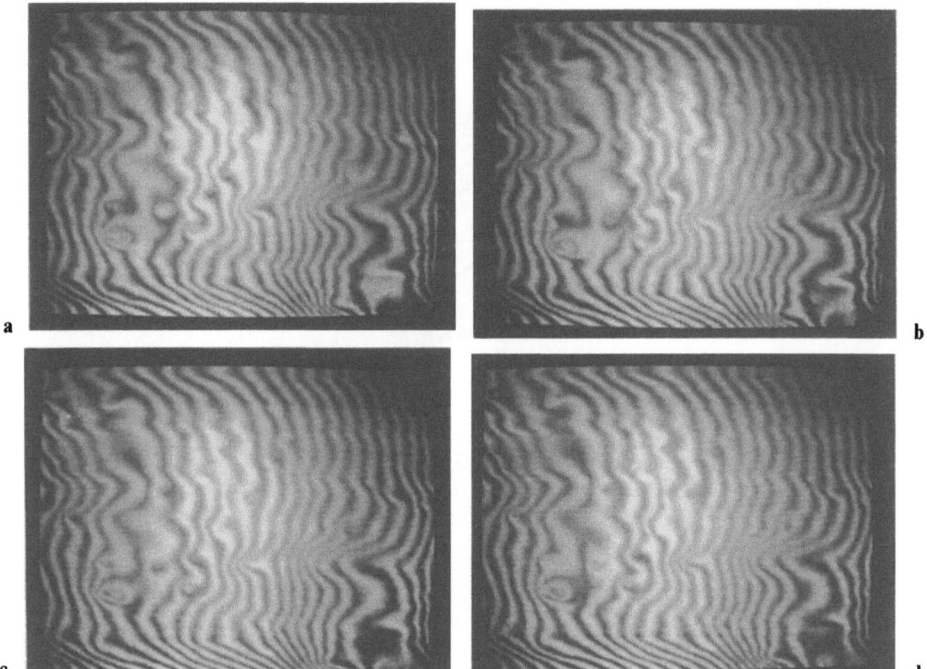

Fig. 6.14a–d. Phase stepped holographic interferogram of thermally loaded panel

modulo 2π determined by (6.58 and 59). The continuous interference phase, after unwrapping the 2π-discontinuities, is displayed in Fig. 6.13h. Although a varying background intensity, varying contrast, and even local saturation in the recorded intensities occur, the calculated interference phase distribution is clean and smooth. The changes in the slope, decreasing to increasing and vice versa, are uniquely detected.

A two-dimensional evaluation by phase sampling is demonstrated by the example of a thermally loaded panel, consisting of an internal aluminum honeycomb structure with surface layers of carbon fiber reinforced plastic. Figure 6.14a to d exhibit the four-phase-stepped interferograms arising from a temperature difference of 2 °C. The size of the panel was about 80×80 cm. The resulting interference phase distribution, which can be interpreted as proportional to the normal displacement field due to the optimized sensitivity vectors, is presented in Fig. 6.15. Several debonds of the surface layer from the internal structure are clearly detectable.

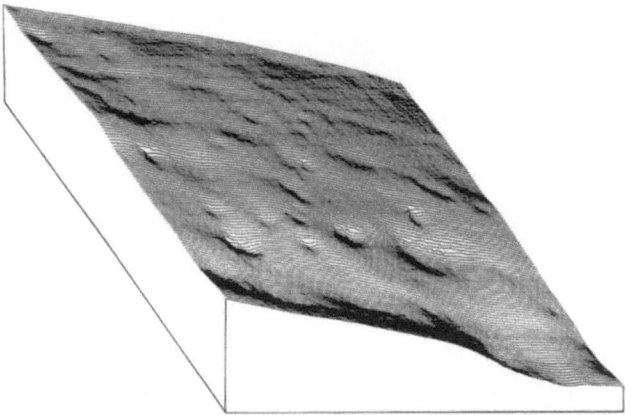

Fig. 6.15. Evaluated interference phase field

6.4.4 Fourier-Transform Evaluation

6.4.4.1 Spatial Heterodyning

In spatial heterodyning an additional carrier frequency is added to the interference pattern [6.68–74]. In Michelson interferometry this is done by tilting a mirror in one arm of the interferometer. This leads to the generation of a spatial carrier frequency f_0 in, say, the x-direction of the interference pattern

$$I(x, y) = a(x, y) + b(x, y) \cos[2\pi f_0 x + \Delta\phi(x, y)]. \tag{6.61}$$

By introducing

$$c(x, y) = \tfrac{1}{2} b(x, y) \exp[i\Delta\phi(x, y)] \tag{6.62}$$

the cosine is expressed as a complex exponential, with i being the imaginary unit and * denoting complex conjugation,

$$I(x, y) = a(x, y) + c(x, y) \exp(2\pi i f_0 x) + c^*(x, y) \exp(-2\pi i f_0 x). \tag{6.63}$$

The Fourier transform of the intensity with respect to x then yields [6.75]

$$\mathcal{I}(u, y) = \mathcal{A}(u, y) + \mathcal{C}(u - f_0, y) + \mathcal{C}^*(u + f_0, y). \tag{6.64}$$

Here the italics denote Fourier spectra and u is the spatial frequency in the x-direction. Since the spatial carrier frequency chosen is higher than the spatial variations $a(x, y)$, $b(x, y)$, and $\Delta\phi(x, y)$, the partial spectra $\mathcal{A}, \mathcal{C}$, and $\mathcal{C}^*$ are well separated. $\mathcal{A}$ is concentrated around the dc term at $u = 0$ and carries the low frequency background illumination. $\mathcal{C}$ and $\mathcal{C}^*$ are placed symmetrically to the dc term and are centred around $u = f_0$ and $u = -f_0$. If by an adequate bandpass-filter, $\mathcal{A}$ and $\mathcal{C}^*$ are eliminated and $\mathcal{C}(u - f_0, y)$ is shifted by f_0 toward the origin, the carrier is removed and we obtain $\mathscr{C}(u,y)$. Taking the inverse 1D-Fourier-transform of $\mathscr{C}(u,y)$ with respect to u yields $c(x,y)$ defined by (6.62). From this $c(x,y)$ the interference phase is calculated as

$$\Delta\phi(x,y) = \arctan\frac{\operatorname{Im} c(x,y)}{\operatorname{Re} c(x,y)} \tag{6.65}$$

with phase values lying between $-\pi$ and $+\pi$.

The combination of spatial and temporal heterodyning has been performed with a Michelson interferometer. This enables simultaneous recording of multiple phase objects on a single space-time interferogram [6.76]. Spatial heterodyning is not recommended for use in most applications of holographic interferometry since, especially for arrangements with varying sensitivity vectors, it is technically not possible to produce a carrier of equidistant straight fringes.

6.4.4.2 Fourier-Transform Evaluation Without Carrier

The Fourier-transform evaluation method presented below shows that it is not necessary to introduce a carrier of equidistant fringes; nevertheless reliable results can be expected [6.24]. Although the method can be applied in one dimension along rows or along columns, we will here describe the more general two-dimensional case. Again by using (6.62), the intensity without additional carrier, see (6.61), is expressed as

$$I(x,y) = a(x,y) + c(x,y) + c^*(x,y). \tag{6.66}$$

The discrete two-dimensional Fourier-transform via the FFT algorithm applied to $I(x,y)$ yields

$$\mathscr{I}(u,v) = \mathscr{A}(u,v) + \mathscr{C}(u,v) + \mathscr{C}^*(u,v). \tag{6.67}$$

with (u,v) being the spatial frequency coordinates. Since $I(x,y)$ is a real distribution in the spatial domain, $\mathscr{I}(u,v)$ is a Hermitean distribution in the spatial frequency domain, which means

$$\mathscr{I}(u,v) = \mathscr{I}^*(-u,-v). \tag{6.68}$$

The real part of $\mathscr{I}(u,v)$ is even and the imaginary part is odd. The amplitude spectrum $|\mathscr{I}(u,v)|$ thus looks point-symmetric with respect to the dc term $\mathscr{I}(0,0)$. $\mathscr{A}(u,v)$ contains the zero-peak $\mathscr{I}(0,0)$ and the low frequency variations of the background. $\mathscr{C}(u,v)$ and $\mathscr{C}^*(u,v)$ carry the same information as evident from (6.68).

By band-pass filtering in the spatial frequency domain, $\mathscr{A}(u,v)$ and one of the terms $\mathscr{C}(u,v)$ or $\mathscr{C}^*(u,v)$ are eliminated. The remaining spectrum, $\mathscr{C}^*(u,v)$ or $\mathscr{C}(u,v)$, is no longer Hermitean, so the inverse Fourier-transform applied to, e.g. $\mathscr{C}(u,v)$, gives a complex $c(x,y)$ with non-vanishing real and imaginary parts. The interference phase is calculated from (6.65). The inverse transform of $\mathscr{C}^*(u,v)$

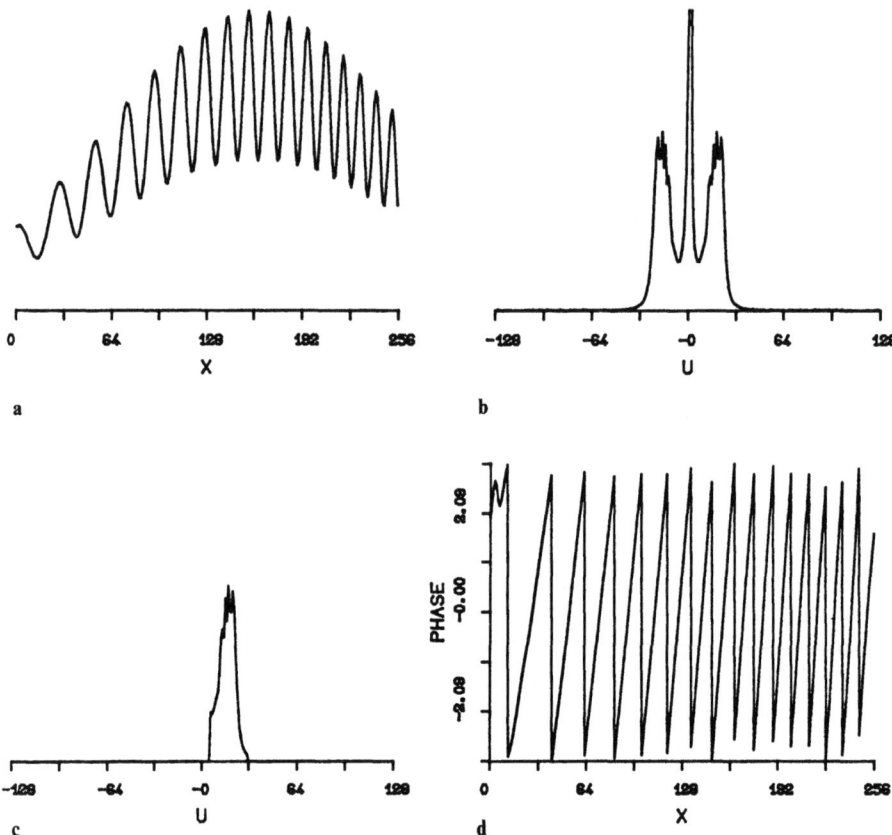

Fig. 6.16a–d. Fourier-transform evaluation

instead of $\mathscr{C}(u,v)$ would result in $-\Delta\phi(x,y)$. The uncertainty about which of the symmetric parts of the spectrum belongs to $\mathscr{C}(u,v)$ and which to $\mathscr{C}^*(u,v)$ is a manifestation of the sign ambiguity, (6.22).

A one-dimensional example for the Fourier-transform evaluation is given in Fig. 6.16. The intensity distribution with varying contrast and varying background of Fig. 6.16a is Fourier transformed; the amplitude spectrum is shown in Fig. 6.16b. After filtering the spectrum of which the amplitude is plotted in Fig. 6.16c, remains. The application of the inverse transform and (6.65) result in the phase modulo 2π shown in Fig. 6.16d.

The band-pass filtering in the spatial frequency domain not only makes the spectrum non-Hermitean, but also enables a reasonable image enhancement [6.77]. Low-frequency background variations, e.g. a Gaussian illumination, lead to spectral components centred around the zero component. Their influence is minimized by a band-pass filter which eliminates all spectral components up to a certain lower cutoff frequency. High-frequency components, like speckle noise are suppressed if the filter stops all frequencies higher than an upper cutoff frequency.

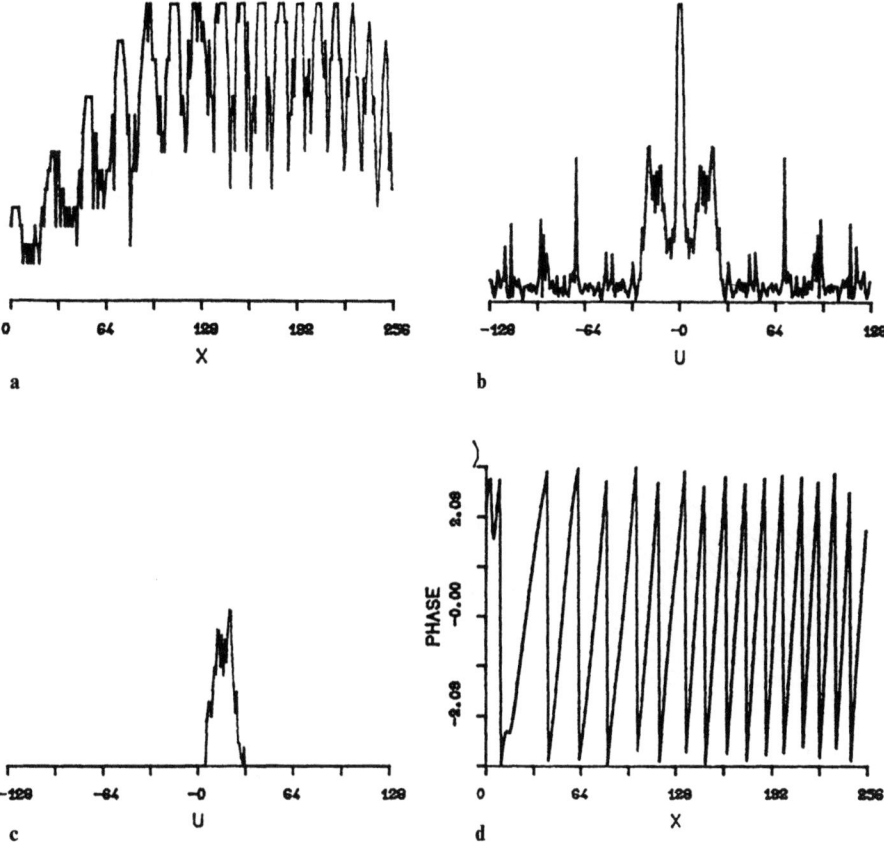

Fig. 6.17a–d. Image enhancement by Fourier-transform evaluation

This capability is depicted in Fig. 6.17. The intensity (Fig. 6.17a), is degraded by a Gaussian background, varying contrast, speckle noise, reduced quantization into 16 gray-levels, and non-linear response with saturation. Nevertheless the finally evaluated interference phase distribution (Fig. 6.17d) is clean and fully modulated.

In eliminating high frequencies, one has to consider the fact that the finite discrete Fourier transform assumes a periodic input signal, whereas, in practice, data consists of one non-periodic stretch of finite length. The discontinuities from the right to the left or from the lower to the upper edge of the image lead to high-frequency components in the spectrum. If these are filtered away, the resulting phase distribution suffers from a wrap-around pollution by a forced smooth continuation at the edges of the frame. Thus the marginal pixels at the edges of the frame get no reliable phase values; the number of these pixels depend on the choice of the upper cutoff frequency. The run-out at the left edge in Fig. 6.17d is caused by this effect.

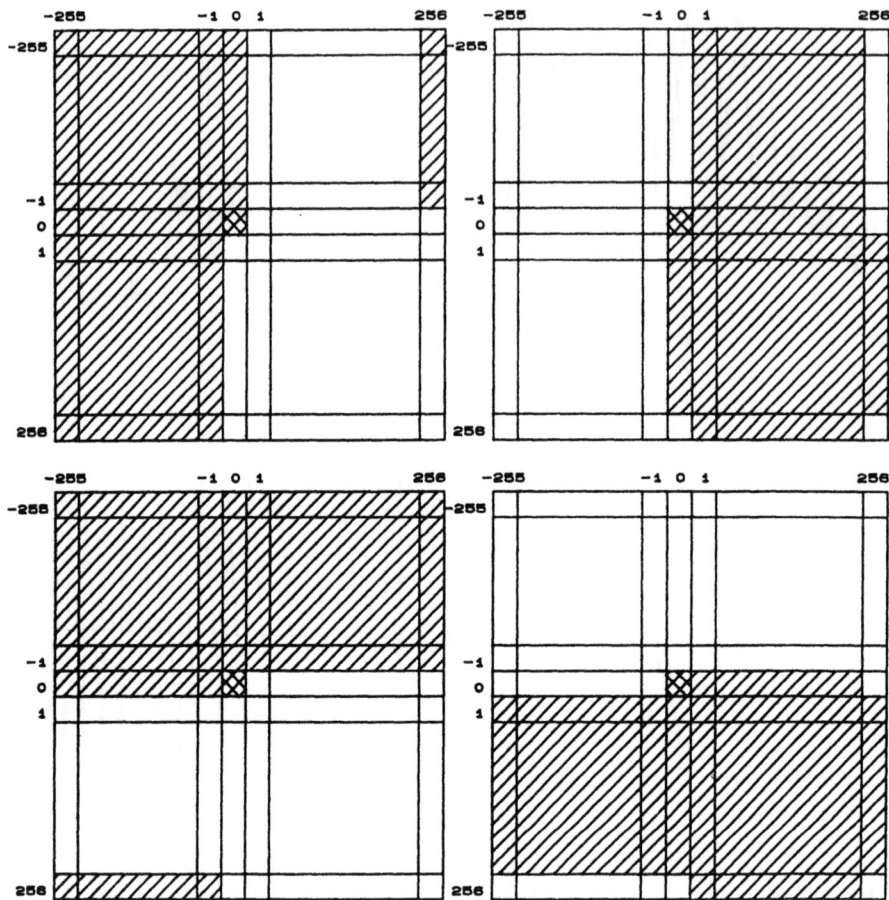

Fig. 6.18. Halfplanes for filtering in the spatial frequency domain, reordered display

Making the spectrum non-Hermitean means setting to zero the spectral value at each spatial frequency or at its symmetric counterpart. Since there is no general way to decide whether $\mathscr{I}(u,v)$ at a certain (u,v) belongs to $\mathscr{C}(u,v)$ or to $\mathscr{C}^*(u,v)$, or is a combination of contributions belonging to both of them, the easiest way is to eliminate one half-plane of the spatial frequency plane. Some of these halfplanes are displayed in Fig. 6.18. In Fig. 6.18a the pass band is the $+u$-halfplane, displayed in white, and the cross-hatched region is the stop band. In Fig. 6.18b the pass band is the $-u$-halfplane, in Fig. 6.18c the pass band is the $+v$-halfplane and in Fig. 6.18d we have the pass band in the $-v$-halfplane. For Fourier-transform evaluation the component at spatial frequency $(0,0)$ is always set to zero. Filters which destroy the Hermitean property and eliminate low-frequency background and high-frequency speckle noise have the form, as shown in Fig. 6.19, with pass bands in halfplanes and with defined lower and upper cutoff frequencies.

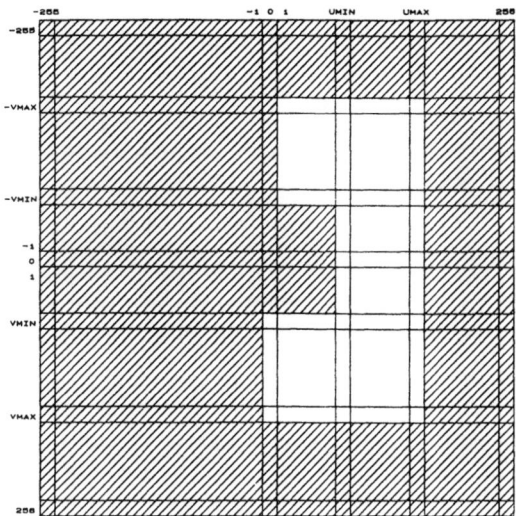

Fig. 6.19. Bandpass filters, reordered display

A two-dimensional Fourier-transform calculated by the FFT-algorithm normally has its zero-frequency in the upper left corner when displayed. A Fraunhofer diffraction pattern of a photographic transparancy produced by a coherent optical system has its zero-frequency term at the centre of the display. For the display of Fig. 6.18 the quadrants have been reordered to bring the spatial frequency (0,0) to the centre. In the examples demonstrated below, we use only the reordered display of amplitude spectra.

The filter of Fig. 6.18a, where only positive spatial frequencies in the horizontal u-direction, and both positive and negative frequencies in the vertical v-direction can pass, gives an interference phase distribution with increasing phases in the horizontal direction, but increasing and decreasing phases in the vertical direction. For a pass band in the $-u$-halfplane the phases are decreasing in the horizontal direction. In the cases of Figs. 6.18c and d, the roles of the directions interchange with respect to Figs. 6.18a and b.

These facts can be used for detection of sign changes: two phase distributions determined from the same pattern, but with orthogonally oriented pass bands are compared to resolve a local sign ambiguity. Only the global sign is left indefinite, and this has to be decided from knowledge of the specific experiment. From a computational view the elimination of a halfplane oriented parallel to the u- or v-axes would look to be the easiest way to render the spectrum non-Hermitean. Other orientations of halfplanes, or even other filter strategies destroying the Hermitean property, are also feasible.

The Fourier-transform evaluation of the holographically measured deformation of a plate thermally loaded in a microwave-oven is shown in Fig. 6.20. Overlayed on the interference pattern we see the metallic grid in the window protecting the environment from microwave radiation (Fig. 6.20a). The bright

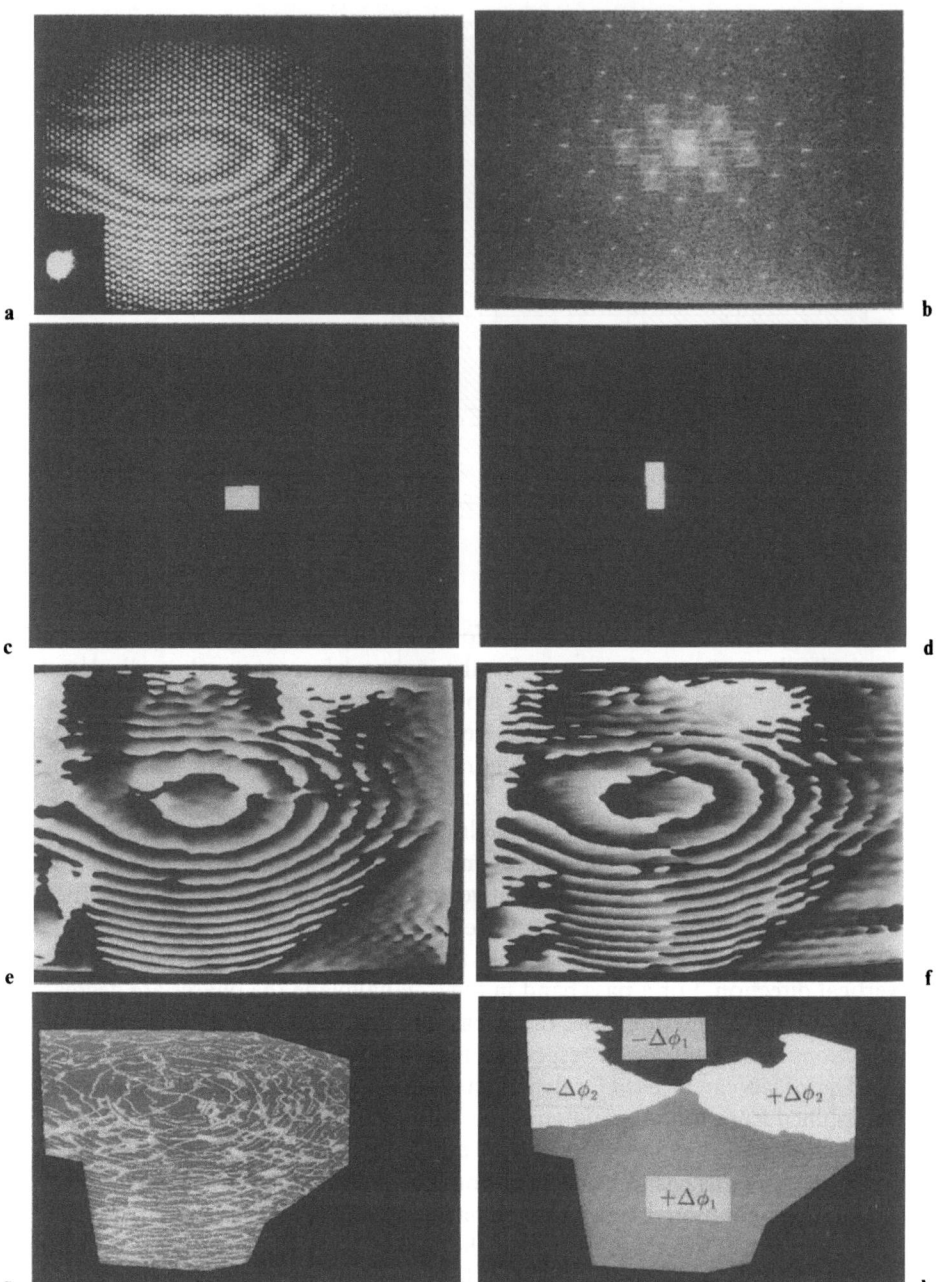

Fig. 6.20a–j. Fourier-transform evaluation of thermally loaded plate

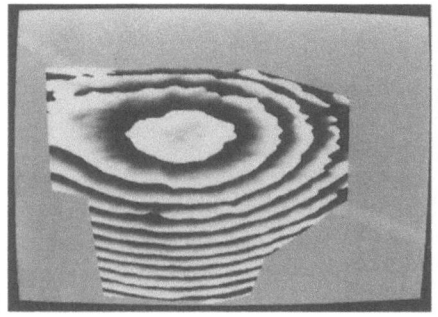
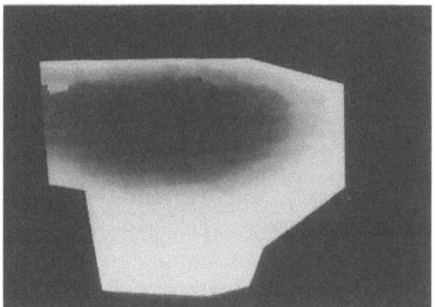

Fig. 6.20. Continued

spot in the lower left is a lens in the oven's door transmitting the illumination wave. The amplitude spectrum of this pattern is displayed in Fig. 6.20b in a logarithmic gray-scale to compress the high dynamic range, especially that of the zero peak. Two mutually orthogonally oriented band-pass filters with lower cutoff frequencies $u_{min} = v_{min} = 2$ and upper cutoff frequencies $u_{max} = 18$, $v_{max} = 34$ in both cases, are shown in Figs. 6.20c and d. The units are cycles per 512 pixels. The upper cutoff frequencies are chosen lower than the least frequency of the grid pattern. Therefore only fringe periods greater than the grid spacing can be evaluated.

The resulting interference phase distributions modulo 2π are given in the gray-scale display in Figs. 6.20e and f. The phase $\Delta\phi_1$ of Fig. 6.20e was produced with the filter in the $+u$-halfplane of Fig. 6.20c. The phase $\Delta\phi_2$ of Fig. 6.20f stems from the filter in the $+v$-halfplane of Fig. 6.20d. Artificial phases are given to pixels where no interference was present and these regions are masked out during further processing.

For the subsequent interactive sign correction, all pixels where the two phase distributions differ less than a threshold, here chosen as 0.1, are marked.

$$\| \Delta\phi_1(x,y)| - |\Delta\phi_2(x,y) \| < 0.1. \tag{6.69}$$

The borders of these regions are defined only along those lines (Fig. 6.20g) in which the phase values take the values $+\Delta\phi_1$, $-\Delta\phi_1$, $+\Delta\phi_2$, and $-\Delta\phi_2$. In this way a sufficiently continuous crossing of the borders is provided. The regions are shown in Fig. 6.20h, and the sign corrected interference phase distribution modulo 2π is displayed in Fig. 6.20i. The demodulation (Sect. 6.4.6), now leads to the continuous phase distribution of Fig. 6.20j, which as a pseudo-3D-display is shown in Fig. 6.21.

In many holographic applications it is possible to record and store the illuminated surface before the interference pattern is produced [6.78]. So in real-time holographic interferometry one may

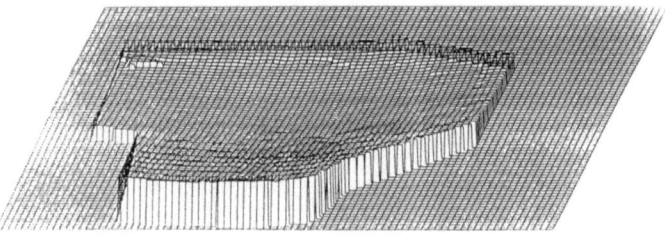

Fig. 6.21. Evaluated interference phase distribution

- record the object surface illuminated only by the object wave with the reference wave blocked. The hologram plate may be present or absent.
- record a holographic reconstruction of the object with the object wave blocked. Only the reference wave illuminates the hologram plate.
- record the zero interference pattern, still without fringes, before starting the loading.

Let this recorded background be $a'(x, y)$ and its Fourier-transform be $\mathscr{A}'(u, v)$. A normalized version of $\mathscr{A}'(u, v)$ is then subtracted from $\mathscr{I}(u, v)$

$$\mathscr{I}'(u, v) = \mathscr{I}(u, v) - \frac{\operatorname{Re}\mathscr{I}(0, 0)}{\mathscr{A}'(0, 0)} \mathscr{A}'(u, v). \tag{6.70}$$

Since the imaginary part of the dc term of a Fourier-transform is always zero, we have zeroed the dc term and eliminated the background. Now from $\mathscr{I}'(u, v)$ one halfplane is eliminated and after the inverse transform has been obtained the interference phase is calculated as described above. This procedure is recommended if the interference pattern or the object does not cover the whole frame or if we have complicated background variations in the frequency range of the interference pattern.

6.4.4.3 Fourier-Transform Evaluation of Phase Shifted Interferograms

Sign ambiguity is always present in the case of the evaluation of a single interferogram. After interactive sign-correction, as demonstrated above, there remains at least a global sign-ambiguity, which means all phases have either the correct or the wrong sign. An exact determination of the sign distribution is achieved if an additional phase stepped interferogram is produced with a mutual phase step ϕ_R. Theoretically, ϕ_R must be in the range $0 < \phi_R < \pi$ but, in practice, values $\pi/3 < \phi_R < 2\pi/3$ are recommended. If this condition is fulfilled, the exact value of ϕ_R does not need to be known [6.24, 79].

The single-phase stepping complicates the experimental procedure, but less than with the phase step methods, where several constant phase steps have to be provided. Let us now write the two intensity distributions

$$I_1(x,y) = a(x,y) + b(x,y)\cos[\Delta\phi(x,y)],$$
$$I_2(x,y) = a(x,y) + b(x,y)\cos[\Delta\phi(x,y) + \phi_R]. \tag{6.71}$$

Fourier-transform processing of each intensity with the same band-pass filter parameters yields

$$c_1(x,y) = \tfrac{1}{2} b(x,y)\exp[i\Delta\phi(x,y)],$$
$$c_2(x,y) = \tfrac{1}{2} b(x,y)\exp[i\Delta\phi(x,y) + i\phi_R(x,y)] \tag{6.72}$$

From (6.72) $\phi_R(x,y)$ is computed pointwise as

$$\phi_R(x,y) = \arctan \frac{\mathrm{Re}\{c_1(x,y)\}\mathrm{Im}\{c_2(x,y)\} - \mathrm{Im}\{c_1(x,y)\}\mathrm{Re}\{c_2(x,y)\}}{\mathrm{Re}\{c_1(x,y)\}\mathrm{Re}\{c_2(x,y)\} + \mathrm{Im}\{c_1(x,y)\}\mathrm{Im}\{c_2(x,y)\}}. \tag{6.73}$$

The knowledge of $\phi_R(x,y)$ is used for determination of the sign-corrected interference phase distribution via

$$\Delta\phi(x,y) = \mathrm{sign}[\phi_R(x,y)]\arctan\frac{\mathrm{Im}\{c_1(x,y)\}}{\mathrm{Re}\{c_1(x,y)\}}. \tag{6.74}$$

Figure 6.22a and b exhibit two phase-stepped holographic interferograms; the sign of $\phi_R(x,y)$ as calculated by (6.73) is shown in Fig. 6.22d; Fig. 6.22c displays the phase distribution modulo 2π, and Fig. 6.22e shows the sign-corrected phase distribution modulo 2π, calculated by (6.74). The demodulated interference phase distribution and its 3D plot are shown in Figs. 6.22e and 6.23, respectively.

The Fourier-transform calculation of the phase step by (6.73) allows a generalization of the phase step method of Sect. 6.4.3. Recording three or more phase-stepped interference patterns with arbitrary phase steps, provided that these are $< \pi$, the phase steps ϕ_{Rn} can be evaluated by the Fourier-transform procedure described above. Taking the average of the absolute values $|\phi_{Rn}|$, we get mutual phase steps which are used to set up a system of equations similar to (6.48). This system is solved and the interference phase calculated by (6.49). This generalized phase shifting interferometry, using additional parallel Fizeau fringes which are evaluated by the Fourier-transform method with spatial carrier, is presented in [6.80].

6.4.5 Further Methods for Interference Phase Determination

In a version of the spatial-carrier method working in the spatial domain, the recorded interference pattern is multiplied pointwise by $\cos(2\pi f_0 x)$ and

194   6. Computer-Aided Evaluation of Holographic Interferograms

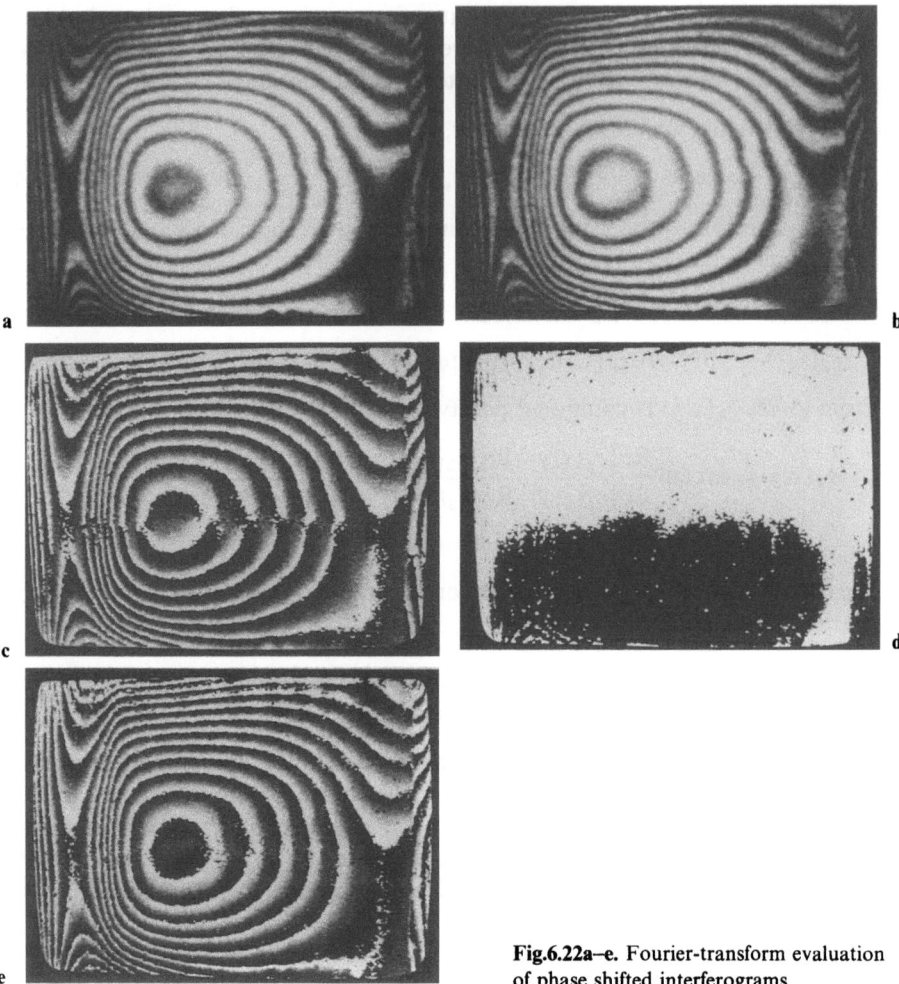

Fig.6.22a–e. Fourier-transform evaluation of phase shifted interferograms

Fig. 6.23. Demodulated interference phase distribution

$\sin(2\pi f_0 x)$ [6.81]. If the difference between $\Delta\phi(x,y)$ and $2\pi f_0 x$ is sufficiently small, one obtains low-frequency components

$$M_1(x,y) = \frac{b}{2}\cos[\Delta\phi(x,y) - 2\pi f_0 x]$$

and

$$M_2(x,y) = -\frac{b}{2}\sin[\Delta\phi(x,y) - 2\pi f_0 x] \tag{6.75}$$

which are isolated by a low-pass filter. The interference phase is then calculated by

$$\Delta\phi(x,y) = 2\pi f_0 x + \arctan\frac{-M_2(x,y)}{M_1(x,y)}. \tag{6.76}$$

This version is only feasible if the interference pattern has fringes of nearly equal inclination and density.

The phase-lock method uses a sinusoidal phase modulation obtained by, e.g., a piezoelectrically excited axially oscillating mirror [6.82]. The intensity can be written as

$$I(x,y,t) = a(x,y) + b(x,y)\cos[\Delta\phi(x,y) + L\sin\omega t]. \tag{6.77}$$

$L < \lambda/2$ is the amplitude and $v = \omega/2\pi$ the frequency of this oscillation. A bandpass filter, centred on $\sin\omega t$, determines the amplitude $U_\omega = 2b(x,y)J_1(L) \times \sin\Delta\phi(x,y)$, which is zero at the points (x,y), where $\Delta\phi(x,y) = N\pi$. These points can thus be detected and give a skeleton whose lines correspond to the interference phase differences of $\pi/2$. The further processing of these skeletons is as with the other skeletonizing methods.

The electro-optic holography method combines the advantages of phase stepping and electronic speckle pattern interferometry to perform an image enhancement in holographic interferometry for deformation and vibration measurements [6.83, 84].

6.4.6 Interference Phase Demodulation

Due to the interferometric process, the evaluated interference phase distributions are ambiguous in having only values between $-\pi$ and $+\pi$. They are said to be wrapped into modulo 2π. The resolution of the 2π-steps in the sawtooth-like phase field is called *continuation, phase unwrapping* or *demodulation*. Demodulation generally requires sign-correct interference phase distributions [6.85].

Demodulation of one-dimensional interference phase distributions $\Delta\phi(x)$ is done by checking phase differences of adjacent pixels $\Delta\phi(x+1) - \Delta\phi(x)$. If this difference is less than $-\pi$, 2π is added to $\Delta\phi$ from $x+1$ onwards; if the difference is greater than $+\pi$, 2π is subtracted from $\Delta\phi$ starting at pixel $x+1$. Several of these 2π terms may cumulate to integer multiples of 2π. The starting point need not necessarily lie at the left-most pixel $x = 1$. If a central starting pixel x_0 is chosen, differences to the right $\Delta\phi(x + 1) - \Delta\phi(x)$ and to the left $\Delta\phi(x - 1) - \Delta\phi(x)$ have to be calculated. If, due to noise, a wrong difference occurs, the resulting phase error spreads up to the outmost pixel (if it is not neutralized by another error in the opposite direction).

This one-dimensional demodulation procedure can be transferred to two dimensions along rows and columns of pixels. Let us start, e.g., with one row. Once this row is demodulated, the pixels of this row act as starting pixels for column demodulation. The 2π-multiples from row demodulation at each of these pixels are added to the 2π multiples from column demodulation. The result is a path-dependent demodulation. Besides the possible expansion of erroneous phase, path-dependent demodulation may fail with images of complex forms, e.g., those containing holes, where no interference phase is defined at all.

To circumvent these difficulties, path-independent demodulation procedures are recommended [6.79, 86]. The following algorithm interprets the interference phase distribution modulo 2π as a graph, where the points are the nodes and the arcs are the connections between neighbouring points. 4-neighbourhoods or 8-neighbourhoods may be used. With each arc a value $d_{2\pi}(\Delta\phi_1, \Delta\phi_2)$ is associated, defined by the phase values $\Delta\phi_1$ and $\Delta\phi_2$ of the two points it connects.

$$d_{2\pi}(\Delta\phi_1, \Delta\phi_2) = \min\{|\Delta\phi_1 - \Delta\phi_2|, |\Delta\phi_1 - \Delta\phi_2 + 2\pi|, |\Delta\phi_1 - \Delta\phi_2 - 2\pi|\}.$$
(6.78)

The values $d_{2\pi}$ may be interpreted as a distance modulo 2π.

The demodulation now proceeds along paths where these distances are least. Along these paths the probability of an erroneous phase is least. Points with wrong phase are surrounded this way and the same is true for regions without interference phase at all. If a point possesses an erroneous phase that cannot be reached correctly along any path, the incorrectly demodulated point still remains isolated in the finally resulting interference phase distribution.

The algorithm proceeds as follows:

1. For a starting point all emanating arcs are recorded in a list together with their values $d_{2\pi}$.
2. The minimal value in the list is searched for. The demodulation term 0, -2π, or $+2\pi$ for this arc is stored in an extra file. The arc, together with its value, is discarded from the list and marked to avoid repeated consideration.
3. The final node of the just considered arc acts as a new starting point in step 1. Only those arcs which have not been stored formerly in the extra file in step 2, are considered as free. If no free arcs emanate from this node, proceed to step 2 directly.
4. If the capacity of the list is exhausted, the list is checked for arcs which may have already entered the extra file along another path. These arcs are deleted.

5. Steps 1–4 are repeated until all points have been an end node of an arc in the extra file.
6. The interference phase distribution is demodulated by using the values in the extra file.

A modification of this algorithm, which shortens the computational effort, allows only arcs with values less than a prescribed threshold to be recorded in the list. Thus the number of comparisons for searching the minimum in step 2 is drastically reduced.

This demodulating algorithm is demonstrated by the numerical example of Fig. 6.24a. Figure 6.24b shows, how the algorithm detects the best track and circumvents the one bad spot. Figure 6.24c displays the resulting unwrapped phase distributions after demodulation along predetermined horizontal paths and after path-independent phase demodulation. The interference phase distributions of Figs. 6.15, 21, and 23 are unwrapped by the presented path-independent algorithm.

An interesting approach to phase unwrapping is offered by a cellular automata method [6.87]. Here a very simple operation for local unwrapping is repeated many times to bring out a consistent phase distribution. Unfortunately, there exist phase distributions which are not consistently unwrapped by this method. In particular, the method is not able to produce a consistent sign correction with the known discontinuity checks. Further research in this field may exhibit more useful operations for phase unwrapping and sign correction.

6.4.7 Comparison of Interference Phase Determination Methods

Given a problem to be solved by interferometric metrology, an appropriate method has to be chosen to address it. In the following the main methods are compared with regard to several criteria. The results of this comparison are summarized in Table 6.1 [6.88].

6.4.7.1 Experimental Requirements

The least requirements from the interferometric arrangement are demanded when using fringe skeletonizing or Fourier-transform evaluation, since both methods evaluate a single interference pattern. These methods even allow for the evaluation of interferograms given in the form of paper photographs, negatives, slides, computerfiles, etc., but there still remains the problem of sign ambiguity which has to be solved with the help of side information. This increases the necessary computational effort. The Fourier-transform method, with one additional phase-stepped pattern to solve for the exact sign, requires a phase step capability in the experimental setup.

198   6. Computer-Aided Evaluation of Holographic Interferograms

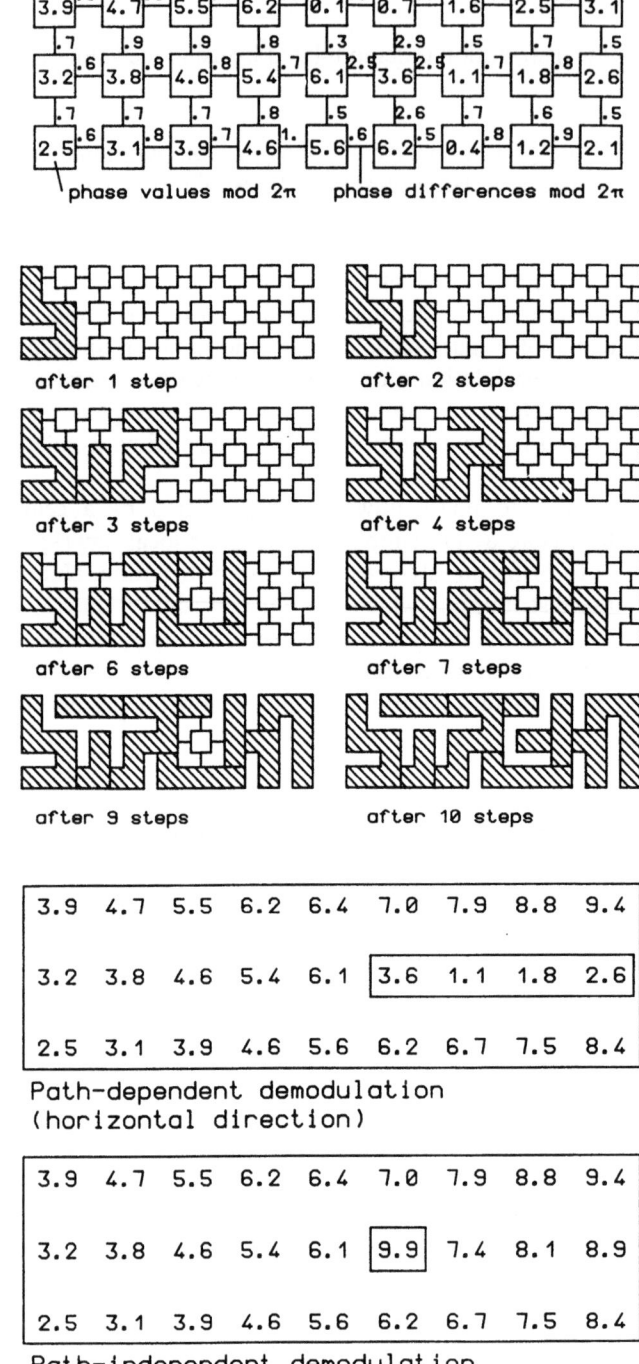

Fig. 6.24a–c. Path independent interference phase demodulation

Table 6.1. Phase determination methods

	Fringe skeletonizing	Phase stepping and shifting	Fourier-transform evaluation	Temporal-heterodyning
Number of interferograms to be reconstructed	1	3 or 4, rarely ≥ 5	1 (2)	One per detection point
Resolution [λ]	1–1/10	1/10–1/100	1/10–1/30	1/100–1/2000
Evaluation between intensity extrema	No	Yes	Yes	Yes
Inherent noise suppression	Partially	Yes	Yes	Partially
Automatic sign detection	No	Yes	No (yes)	Yes
Necessary experimental manipulation	No	Phase shift	No (Phase shift)	Frequency shift
Experimental efforts	Low	High	Low	Extremely high
Sensitivity to external influences	Low	Moderate	Low	Extremely high
Interaction by the operator	Possible	Not possible	Possible	Not possible
Speed of evaluation	Low	High	Low	Extremely low

This capability is also needed by the phase sampling methods, but with the additional requirement of multiple constant phase steps. The procedures with three patterns offer an easy computation; on the other hand, the phase shift must be controlled exactly to a prescribed value. Dealing either with a double reference set-up or using the real-time method, the mechanical stability of the experimental arrangement must be high.

Temporal heterodyning, which also needs two reference waves for performing the mutual frequency shift, demands the highest experimental effort. The mechanical scanning is relatively slow, so the long-term stability of the arrangement must be within a small fraction of the wavelength. The effects of mechanical vibrations, temporal fluctuations, and the variations of the refractive index of surrounding air have to be minimized to meet the stability requirements.

6.4.7.2 Resolution and Precision

The spatial resolution is defined as the distance between the detection points. For fringe skeletonizing this is given by the number of fringes in the interference pattern. Phase shifting and Fourier-transform evaluation offer uniform sampling with a spatial resolution only dependent on the detector array, but independent of the fringe pattern. The resolution can be changed with regard to the pattern by varying the magnification of the recording optics. The spatial resolution of the temporal heterodyne method depends on the distance between the detectors and the stepwidth of the mechanical scanning. Changing the size of the real image optically may also alter the spatial resolution.

The precision of the measured value is normally given in fractions of the wavelength λ, which corresponds to an interference phase difference of 2π. The best precision to be reached by fringe skeletonizing is $\lambda/10$, provided the pattern is of high quality with a reasonable number of fringes and assuming extensive interpolation [6.89]. The Fourier-transform evaluation reaches an accuracy and resolution of $\lambda/20$ at all but the utmost pixel in the marginal five to eight columns and rows of the digitized pattern [6.90]. For well fitting cutoff frequencies of the bandpass filter, even better resolutions are achievable. Phase shifting yields resolutions in the range of $\lambda/100$ [6.91].

The electronic phase-difference measurement of temporal heterodyning enables a resolution of $\lambda/500$ and better. However, these figures are only achievable if all parameters of the process, especially the stepwidth of the detectors and the longterm stability, are kept to the same precision, which is only possible in an ideal laboratory environment.

6.4.7.3 Errors and Distortions

Fringe skeletonizing and Fourier-transform evaluation procedures process one single interferogram and for that reason they are relatively insensitive to external distortions. On the other hand, using a single interferogram does not allow to detect whether the interference phase is increasing or decreasing. A change in the direction of the interference phase is recognized only in special cases with closed ring structures. This change, as well as the increase or decrease in the phase, can best be detected by phase stepping and temporal heterodyning. Moreover, these two methods do not exhibit systematic sign errors.

The sign error is accompanied by a general inaccessibility of the additive multiple of 2π. All methods evaluate only the interference phase modulo 2π; this naturally calls for a demodulation to correct for the 2π ambiguities. Thus, normally, only phase differences between different points are measured and not the absolute interference phase at a single point.

Phase shifting is sensitive to non-constant phase steps and therefore to all distortions which influence the phase steps. In this regard the procedure with four recordings and an unknown, but constant, phase step is advantageous because it inherently corrects for additional linear phase shifts. Finally, temporal heterodyning is extremely sensitive to environmental distortions.

6.4.7.4 Comparison Results

At a first glance phase sampling seems to yield a lower precision than temporal heterodyning. However, the precision of phase sampling is valid even for points far apart in the interference pattern. This is due to the simultaneous recording of the intensity at all points. In temporal heterodyning, small drifts during the evaluation may cumulate and reduce the precision of phase differences of distant points. Phase sampling should be chosen when the most important criteria in the measurement problem one is looking at are accuracy and precision. On the other hand, the Fourier-transform evaluation procedure is to be recommended in the event of the lack of possibility to record phase step or multiple interferograms, as is the case when using a pulsed laser or using a photorefractive crystal for recording, which unfortunately destroys the information during the readout.

The main difference in the concepts of phase sampling and Fourier-transform evaluation resides in the automatic evaluation by phase sampling without interference by the operator. This is in contrast to the Fourier-transform method which though allowing manifold manipulations with the interference pattern requires intervention of the user.

6.5 Processing of Evaluated Data

6.5.1 Curve Fitting by Gaussian Least Squares

For simple objects and loading cases the characteristic deformation, e.g., caused by a defect, can be isolated from rigid body motions and other homogeneous deformations by the method of Gaussian least squares [6.65, 66, 9]. This is demonstrated by the example of a tensile test specimen with an internal crack. Experimental observations and theoretical considerations show that the deformation is a combination of

- a constant translation t_x in longitudinal direction,
- a linearly increasing translation $\varepsilon(x - x_0)$, caused by the strain ε,
- a constant transversal translation t_z,
- a linearly increasing translation in transversal direction $\tan \gamma (x - x_0)$ due to a tilt γ,
- the displacement $u_z(x)$ created by the defect.

According to (6.17) the interference phase outside the defect range if an arrangement with no sensitivity in y-direction is used, can be written as

$$\Delta \phi(x) = L_x(x) K_x(x) + L_z(x) K_z(x)$$

$$= t_x K_x(x) + \varepsilon(x - x_0) K_x(x) + t_z K_z(x) + \tan \gamma (x - x_0) K_z(x). \quad (6.79)$$

The parameters t_x, t_z, ε, and $\tan \gamma$ are now determined by the least-squares method from the measured interference phase values $\Delta \phi(x_i)$ outside the defect area. The system of equations to be solved is

$$\begin{pmatrix} \sum K_{xi}^2 & \sum (x_i-x_0)K_{xi}^2 & \sum K_{xi}K_{zi} & \sum (x_i-x_0)K_{xi}K_{zi} \\ \sum (x_i-x_0)K_{xi}^2 & \sum (x_i-x_0)^2 K_{xi}^2 & \sum (x_i-x_0)K_{xi}K_{zi} & \sum (x_i-x_0)^2 K_{xi}K_{zi} \\ \sum K_{xi}K_{zi} & \sum (x_i-x_0)K_{xi}K_{zi} & \sum K_{zi}^2 & \sum (x_i-x_0)K_{zi}^2 \\ \sum (x_i-x_0)K_{xi}K_{zi} & \sum (x_i-x_0)^2 K_{xi}K_{zi} & \sum (x_i-x_0)K_{zi}^2 & \sum (x_i-x_0)^2 K_{zi}^2 \end{pmatrix}$$

$$= \begin{pmatrix} t_x \\ \varepsilon \\ t_z \\ \tan\gamma \end{pmatrix} \begin{pmatrix} \sum \Delta\phi_i K_{xi} \\ \sum \Delta\phi_i (x_i-x_0)K_{xi} \\ \sum \Delta\phi_i K_{zi} \\ \sum \Delta\phi_i (x_i-x_0)K_{zi} \end{pmatrix}, \qquad (6.80)$$

where $K_{xi} = K_x(x_i)$, $K_{zi} = K_z(x_i)$, $\Delta\phi_i = \Delta\phi(x_i)$; $K_x(x_i)$, $K_z(x_i)$ and the coordinate of the clamping x_0 are determined from the holographic arrangement. After solving (6.80), the normal displacement $u_z(x)$ in the defect range can be expressed as

$$u_z(x) = \frac{\Delta\phi(x)}{K_z(x)} - [t_x + \varepsilon(x-x_0)]\frac{K_x(x)}{K_z(x)} - t_z - \tan\gamma(x-x_0). \qquad (6.81)$$

Figure 6.25a exhibits an evaluated interference phase distribution, and Fig. 6.25b displays the deformation induced by the defect alone.

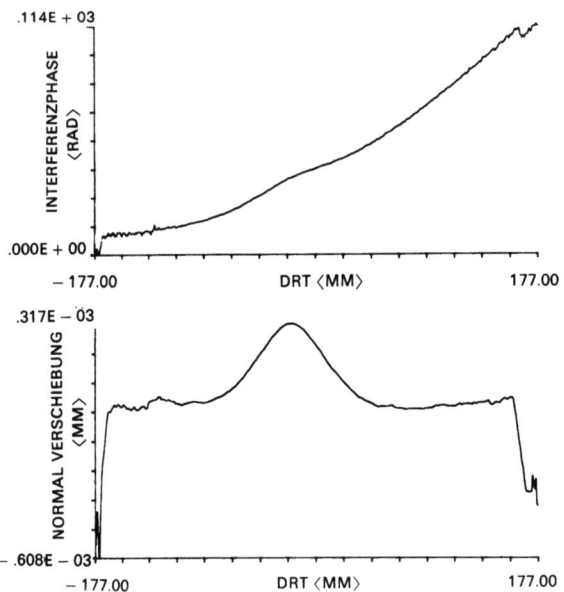

Fig. 6.25. Defect induced deformation

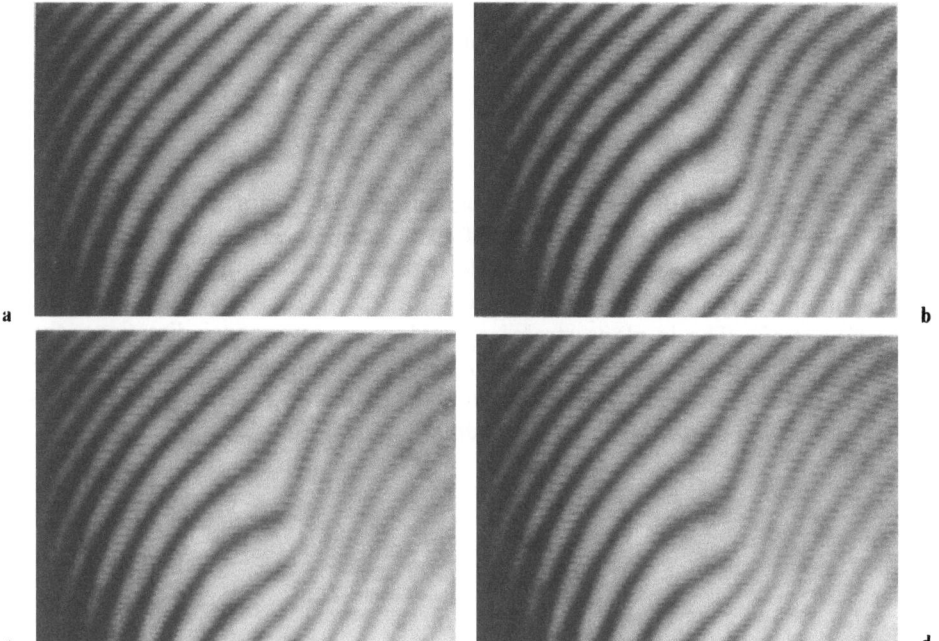

Fig. 6.26a–d. Phase shifted holographic interference patterns

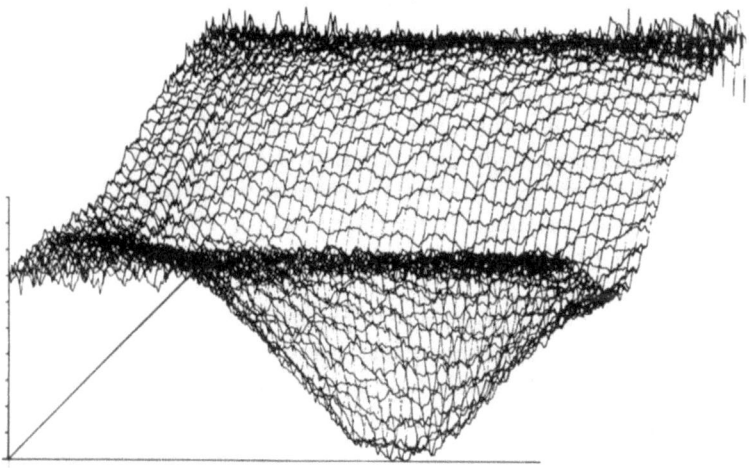

Fig. 6.27. Defect induced deformation

The same method applied in two dimensions is presented in Figs. 6.26 and 27. Figure 6.26 depicts four-phase-stepped holographic interferograms of a pressurized vessel. The interference phase is determined by phase sampling evaluation. The surface deformation caused by an internal defect is given in

Fig. 6.27. The displayed distribution is proportional to the normal displacement with a peak to valley difference of about 700 nm.

6.5.2 Strain and Stress Analysis

In elastomechanics, the interest in the measurement of strains and stresses is far superior than in the measurement of displacements of a deformed object. Therefore, the surface displacements determined with high accuracy by holographic interferometry are often subject to further processing to calculate the surface strain. The in-plane strains $\varepsilon_{\xi\xi}$ and $\varepsilon_{\eta\eta}$ in orthogonal directions, the shearing strains $\varepsilon_{\xi\eta} = \varepsilon_{\eta\xi}$, and the in-plane rotations ω_ζ for surface points are determined by numerical differentiation of the measured in-plane displacements u_ξ, u_η with respect to the surface coordinates (ξ, η, ζ) [6.54]:

$$\varepsilon_{\xi\xi} = \frac{\partial u_\xi}{\partial \xi}, \qquad \varepsilon_{\eta\eta} = \frac{\partial u_\eta}{\partial \eta},$$

$$\varepsilon_{\xi\eta} = \varepsilon_{\eta\xi} = \frac{1}{2}\left(\frac{\partial u_\xi}{\partial \eta} + \frac{\partial u_\eta}{\partial \xi}\right), \qquad \omega_\zeta = \frac{1}{2}\left(\frac{\partial u_\eta}{\partial \xi} - \frac{\partial u_\xi}{\partial \eta}\right). \tag{6.82}$$

The origin of the (ξ, η, ζ) coordinate system is taken to coincide with the surface point under consideration. The remaining components of strain and rotation, which are related to the out-of-plane derivatives of the displacement, are calculated with the aid of the mechanical boundary conditions for an object surface free from external forces.

Optical strain measurements would proceed quite simply, if it were possible to measure displacements perpendicular to the optical axis by observing the object surfaces along their normals. Unfortunately this is not the case for most applications of holographic interferometry; spherical perspective observation implies sensitivity vectors varying over the object surface. This may be the case for arbitrarily shaped objects or for procedures based on multiple oblique observations.

A solution to this problem is given in [6.92] by the mathematical connection between the derivatives of observed displacements in the object plane, as recorded and evaluated by the digital image processing system, and the surface strains and rotations. This connection is given by coordinate transformations for which the matrix formulation [6.93, 94] is used. From this it is generally concluded that derivatives of observed displacement equal the surface strain if and only if the observed displacement lies parallel to the surface.

The numerical differentiation is easily performed by a two-dimensional least-squares fitting of the displacement vector components through the surface point under consideration and its nearest neighbours [6.54]. These four or eight neighbouring points are the projections of the evaluated image plane points onto the plane tangential to the surface at the chosen point. The projected points normally form an irregular grid in the tangential plane due to observation perspective and non-planar surface geometry.

A subsequent calculation of the stresses from the determined strains may be performed by using Hooke's laws for isotropic bodies.

6.5.3 Finite-Element Methods

The finite-element method is a structural analysis method, where a technical structure is divided into a number of discrete elements which have a simple geometry. All these elements are connected via the nodes by which they are defined. The mechanical behaviour in each element can be calculated due to its simple geometry, then the program produces an overall solution that is compatible at all the nodes. Thus displacements, strains and stresses or the thermal behaviour of a structure under mechanical and/or thermal load can be approximated with good accuracy.

Holographic deformation analysis and finite-element calculations can effectively be combined to reach a number of goals [6.10]. One of the most important is the holographic verification of the finite element model. Especially for complex shaped or composite structures it is not an easy task to find a proper discretization and to choose the right material parameters. A comparison of the measured displacements with the calculated displacements can confirm the finite element model, which can then be used for, say, strain and stress calculations.

This strategy has been successfully followed in the investigation of the stress distribution in adhesive bondings [6.95]. It has been determined to what proportions the metal layers and the adhesive layers contribute to the deformation of the specimen. A complete representation of the strain and stress conditions within the specimen has been obtained.

The temperature distribution and the deformation of a thermally loaded overlap adhesive bond with a local void in the adhesive layer was calculated by the finite-element method and measured by holographic interferometry [6.96]. The combined evaluations have shown that the characteristic surface deformation above the defect in the internal adhesive layer is not caused by thermal expansion of the enclosed gas. The inhomogeneities in the surface deformation arise from the disturbed heat transfer in the defect area. This results in locally higher temperature differences, which in consequence leads to locally different thermal deformations. These results explain the aptitude of thermal load for defect detection in holographic non-destructive testing.

The combination of holographic interferometry and finite-element methods not only enables a defect detection but also a defect validation. In systematic calculations a catalogue of surface deformations in the region of internal cracks and voids in, say, steel with variations of defect type, length, orientation, volume or position, is compiled. For a given holographically measured displacement field above a defect, one starts with the best fitting displacement field from this catalogue. In an iterative process in a finite element model, the parameters of the simulated defect are varied. The displacement field is calculated over and over again with varied defects until the agreement with the measured displacement field is sufficiently good [6.9].

Of course, the calculation of a displacement field for a given discrete structure and loading by the finite element method is not a one-to-one mapping. Thus the inverse process of determining the discrete structure from the loading and the measured displacement field is not possible. Nevertheless, with the iterative method we get a defect which is representative of the equivalent class of all defects producing the same deformation under the specific applied load.

6.5.4 Boundary-Element Methods

In the finite-element method, the whole body to be tested is discretized into finite volumes connected at the nodes. Continuity of parameters which are not explicit variables is only warranted at the nodes and not at the borders between the elements. If we have other functions which fulfil exactly the differential equation in the whole region, we have no discretization errors in the interior of the body. Since only the boundary conditions have to be satisfied, the requirement that the boundary is discretized is in itself sufficient. The relating methods are called *boundary-element methods*.

The boundary-element methods can be combined advantageously with experimental methods like holography. Especially, if in practical applications the boundary conditions are too complicated to be described theoretically, they have to be measured. Having measured the displacements by, e.g., holographic interferometry, the boundary of two- or three-dimensional regions is then divided into segments on which the displacements and strains are approximated by polynomials of the first degree. The stress components at prescribed internal points of the region are then calculated by means of the boundary-element method.

In [6.11], this method was applied to transparent models manufactured from PMMA with roughened faces. The inplane components of the displacement vectors are measured by double exposure double aperture speckle interferometry. The objects considered are a three-point loaded beam with an edge crack and a model of a large slab wall stiffened by a frame. Based on the measurements, exact values of stresses σ_x, σ_y, and σ_{xy} in the neighbourhood of the crack tip have been determined. In another application, the friction between the wall and its base was determined by applying the hybrid evaluation method of coherent optical measurement combined with the boundary-element method.

6.5.5 Fracture Mechanics

In linear elastic fracture mechanics the influence of a crack or another defect on the damage of a technical structure is estimated. An important figure is the stress intensity factor K_I and its critical value K_{Ic}, the fracture toughness, which is a material property. The K_I value can be determined holographically by first measuring the deformation field of the structure exhibiting a crack followed by the determination of the boundary of the plastic zone arising above the crack during tensile loading. This is functionally related to the K_I value [6.97]. Another method is based on the integration of the strain equations of Sneddon or of Williams-Irwin. This method only requires a displacement measurement along a line perpendicular to the crack propagation direction [6.98]. In conjunction with these methods holographic interferometry has been used to determine the K_I value for a CT 500 specimen with high accuracy and without any previous knowledge of the specific defect properties like its size and location.

Another criterion for crack propagation is the so-called J-integral, which is a figure independent of the path of integration. Crack propagation occurs if the J-integral exceeds a critical material parameter. In [6.99], the determination of the J-integral is based on measurements of the displacement field by holographic interferometry. Power series estimations up to quadratic terms fulfilling the Lame–Navier equations are set up for the three displacement components. From the coefficients of the power series the integration can then be performed along a rectangular path to yield the J-integral.

6.5.6 Computer Tomography

Holographic interferometric measurements of transparent media are relevant in stress analysis, flow visualization, aerodynamics, heat transfer, concentration analysis, or plasma diagnostics, to name only a few [6.100–103]. The fringe patterns produced by the double-exposure or the real-time method are evaluated, as described in Sect. 6.4 to determine the interference phase field $\Delta\phi(x,y)$. Equation (6.20) is then inverted to yield the change in the refractive index distribution $\Delta n(x,y,z)$.

If Δn is independent of z and the length of the test region is l, (6.20) could be evaluated by substituting

$$\Delta n(x,y,z) = \Delta n(x,y) = \frac{\lambda}{2\pi l} \Delta\phi(x,y). \tag{6.83}$$

In the case of radially symmetric refractive index fields, the Abel transform (6.21) is evaluated by the inverse Abel transform

$$\Delta n(r) = -\frac{\lambda}{2\pi} \int_r^R \frac{(\partial \Delta\phi/\partial x)}{\sqrt{x^2 - r^2}} dx + \frac{\lambda \Delta n(R)}{\pi \sqrt{R^2 - r^2}}. \tag{6.84}$$

For arbitrary three-dimensional refractive-index fields, the optical path length to each point of the hologram is the corresponding line integral through the field [6.104]. The reconstruction of two-dimensional or three-dimensional fields from measurements of line integrals is obtained by methods better known as computer tomography, which is based on the mathematical Radon transform. To give point resolution, computer tomography of refractive-index fields requires multiple views from different angles through the field. Since a holographic interferogram can be viewed from many different directions, a small number of holograms would be sufficient to reconstruct the three-dimensional refractive index field.

In practical measurements the inverse Radon transform is replaced by discrete approximations of one of the following types: The so-called implicit methods divide the test section into a grid of rectangular elements and within each element the refractive index is assumed constant. Other implicit methods represent the refractive-index field by a series expansion. In both cases, a system of linear algebraic equations has to be solved to obtain reconstruction of the refractive-index fields.

The most common explicit methods to achieve reconstruction are the convolution and the filtered back-projection methods. In these methods, the evaluated interference phase distribution or its derivative are described in the form of series expansions. The Fourier-transform methods use the similarity property of the two-dimensional Fourier transform: The Fourier transform of the density function projected in a direction θ is a radial slice of the two-dimensional Fourier spectrum in the same direction and passing through the origin of the spatial-frequency plane. Thus with a number of different directions θ, a good approximation of the two-dimensional Fourier transform of the refractive-index field is generated, which in turn is calculated by the inverse Fourier transform.

Contrary to X-ray or ultrasonic tomography, here we have the additional problem of significant bending of the probing light rays by refraction [6.105]. The above evaluations only apply to the refractionless limit. An iterative procedure has been suggested to consider the ray bending in strongly refracting fields [6.106, 107]. The procedure is based on successive estimation of the deviation of the ray, assumed ray bending and that produced by the calculated refractive index field. After each evaluation, as described for the refractionless limit, the path-length transform is calculated by computational ray tracing from the actually determined refractive index field. From this path-length transform a new estimate of the deviation function is made, thereon setting forth yet another evaluation based on the newly evaluated deviation function. This process continues until the change of deviation is smaller than a prescribed value. Examples have been given in [6.106] for numerically simulated data fields and for experimental measurements of mass transfer at boundary layers.

6.6 Conclusions and Future Trends

Holographic interferometry as a metrological tool offers a lot of advantages, most of which can be exploited sufficiently only with the aid of computers. The rapid development of digital image-processing hardware, concerning increase in storage capacity and processing speed at decreasing prices, enables solutions to problems which were unthinkable some years ago. There is no doubt that this trend will continue in the future. Thus we have no reason to restrict the thinking of new solutions and new problems [6.108].

In developing new software for computer-aided evaluation of holographic interferograms, an important goal should be toward enhancing its flexibility and portability: the software developed today will possibly live through several generations of hardware. The trend will be towards open systems for easy combination of hardware and software modules, but this requires standards which still do not exist. Standardization is a problem to be worked on earnestly in the near future.

The employment of modern hardware, like multiprocessor networks or software concepts like knowledge-based systems and neural networks, has just begun. Future technological developments in the resolution of CCD targets will close the gap between holographic interferometry and TV-holography (ESPI). New holographic materials like electronically addressable spatial light modulators will enable combination of optical and digital processing alternatives to address and attract new applications in the fields of holographic metrology.

This chapter has in detail described not only the interference phase determination for quantitative measurement, which is today the most important application of computers in holographic interferometry, but has additionally addressed other problem fields in qualitative and quantitative evaluation to stimulate further development and new solutions.

Acknowledgements. This chapter could not have been written without the invaluable aid of the coworkers of the laser metrology department of the Bremen Institute for Applied Beam Technology (BIAS). Most of the experiments and all of the figures demonstrated herein have been performed at BIAS. Specific acknowledgement must be made to Professor W. Jüptner, director of BIAS, for his continuous support and stimulation of the work on holographic interferometry and digital image processing.

Portions of the work reported herein were sponsored by research funds of the Commission of the European Communities, the Federal Ministry for Research and Technology and the German Research Council-DFG, especially under contracts Kr 953/1-1, Kr 953/1-2, Kr 953/2-1, and Kr 953/5-1, which is gratefully acknowledged.

References

6.1 N. Abramson: *The Making and Evaluation of Holograms* (Academic, London 1981)
6.2 P. Hariharan: *Optical Holography: Principles, Techniques and Applications* (Cambridge Univ. Press, Cambridge 1984)
6.3 R. Jones, C. Wykes: *Holographic and Speckle Interferometry* (Cambridge Univ. Press, Cambridge, second edition 1989)
6.4 Yu. I. Ostrovsky, M.M. Butusov, G.V. Ostrovskaya: *Interferometry by Holography* (Springer, Berlin, Heidelberg 1980)
6.5 W. Schumann, J.-P. Zürcher, D. Cuche: *Holography and Deformation Analysis* (Springer, Berlin, Heidelberg 1985)

6.6   C.M. Vest: *Holographic Interferometry* (Wiley, New York 1978)
6.7   W. Jüptner, Th. Kreis: Holographic NDT and visual inspection in production line applications, in *Holographic Nondestructive Testing: Status and Comparison with Conventional Methods*, ed. by C. Vest SPIE Proc. **604**, 30–36 (1986)
6.8   G.E. Maddux: Video/computer techniques for static and dynamic experimental mechanics, in *Industrial Laser Interferometry*, ed. by R.J. Pryputniewicz. SPIE Proc. **746**, 52–57 (1987)
6.9   Th. Kreis, W. Jüptner: Determination of defects by combination of holographic interferometry and finite-element-method (in German). *VDI-Reports* **631**, 139–151 (VDI, Düsseldorf 1987)
6.10   W. Jüptner, Th. Kreis (eds): *An External Interface for Processing 3-D Holographic and X-ray Images*. Res. Rep. ESPRIT (Springer, Berlin, Heidelberg 1989)
6.11   J. Balas, J. Sladek, M. Drzik: Exp. Mech. **83**, 196 (1983)
6.12   D.H. Ballard, C.M. Brown: *Computer Vision* (Prentice Hall, Englewood Cliffs 1982)
6.13   K.R. Castleman: *Digital Image Processing* (Prentice Hall, Englewood Cliffs 1979)
6.14   W.K. Pratt: *Digital Image Processing* (Wiley, New York 1978)
6.15   A. Rosenfeld, A.C. Kak: *Digital Picture Processing* (Academic, London 1976)
6.16   J.D. Trolinger: Opt. Eng. **24**, 840 (1985)
6.17   R. Höfling: Opt. Las. Eng. **11**, 49 (1989)
6.18   H.J. Tiziani: Opt. and Quant. Electr. **21**, 253 (1989)
6.19   M.A. Ahmadshahi: Appl. Opt. **30**, 2382 (1991)
6.20   Th. Kreis, H. Kreitlow: Quantitative evaluation of holographic interference patterns under image processing aspects, in *2nd Europ. Cong. on Optics Applied to Metrology*, ed. by P. Meyrueis and M. Grosmann. SPIE Proc. **210**, 196–202 (1979)
6.21   P. Hariharan, B.F. Oreb, C.H. Freund: Appl. Opt. **26**, 3899 (1987)
6.22   S. Nakadate, H. Saito, T. Nakajima: Opt. Acta **33**, 1295 (1986)
6.23   R.J. Pryputniewicz: Opt. Eng. **24**, 843 (1985)
6.24   Th. Kreis: J. Opt. Soc. Amer. A**3**, 847 (1986)
6.25   D.R. Matthys, J.A. Gilbert, T.D. Dudderar, K.W. Koenig: Opt. Las. Eng. **8**, 123 (1988)
6.26   D.R. Matthys, T.D. Dudderar, J.A. Gilbert: Exp. Mech. **28**, 86 (1988)
6.27   P.D. Plotkowski, M.Y.Y. Hung, J.D. Hovanesian, G. Gerhardt: Opt. Eng. **24**, 754 (1985)
6.28   N. Eichhorn, W. Osten: J. Mod. Opt. **35**, 1717 (1988)
6.29   D.A. Tichenor, V.P. Madsen: Opt. Eng. **18**, 469 (1979)
6.30   Th. Kreis: *Evaluation of Holographic Interference Patterns by Spatial Frequency Analysis* (in German). Progr. Rep. VDI, Ser. 8, Vol. 108 (VDI, Düsseldorf 1986)
6.31   D.W. Robinson: Appl. Opt. **22**, 2169 (1983)
6.32   Th. Kreis, W. Osten, R. Biedermann, U. Mieth: *Automatic Detection of Material Flaws in Interference Patterns Using Knowledge-Based Systems and Neural Networks* (in German). Rep. DFG Projekt Kr 953/5-1 (1991)
6.33   S. Nakadate, N. Magome, T. Honda, J. Tsujiuchi: Opt. Eng. **20**, 246 (1981)
6.34   K.A. Stetson: Appl. Opt. **14**, 272 (1975)
6.35   W. Jüptner, K. Ringer, H. Welling: Optik **38**, 437 (1973)
6.36   L. Ek, K. Biedermann: Appl. Opt. **16**, 2535 (1977)
6.37   V. Fossati Bellani, A. Sona: Appl. Opt. **13**, 1337 (1974)
6.38   R.J. Pryputniewicz: J. Opt. Soc. Am. **67**, 1351–1353 (1977)
6.39   K.A. Stetson: Appl. Opt. **14**, 2256 (1975)
6.40   A. Choudry: Automated fringe reduction techniques, in *Interferometric Metrology, Critical Reviews*, ed. by N.A. Massie. SPIE Proc. **816**, 49–55 (1987)
6.41   K. Creath: Phase-measurement interferometry techniques for nondestructive testing, in *2nd Int'l. Conf. on Photomech. and Speckle Metrology*, ed. by F.P. Chiang. SPIE Proc. **1554**, 701–707 (1991)
6.42   R.J. Pryputniewicz: Review of methods for automatic analysis of fringes in hologram interferometry, in *Interferometric Metrology, Critical Reviews*, ed. by N.A. Massie. SPIE Proc. **816**, 140–148 (1987)

6.43 G.T. Reid: Opt. Las. Eng. **7**, 37 (1986)
6.44 Th. Kreis, W. Osten: tm-Techn. Mess. **58**, 235 (1991)
6.45 A.E. Ennos, D.W. Robinson, D.C. Williams: Opt. Acta **32**, 135 (1985)
6.46 W.R.J. Funnell: Appl. Opt. **20**, 3245 (1981)
6.47 F. Becker, G.E.A. Meier, H. Wegner: Automatic evaluation of interferograms, in *Applications of Digital Image Processing*, ed. by A.G. Tescher. SPIE Proc. **359**, 386–393 (1982)
6.48 F. Becker, Y.H. Yu: Opt. Eng. **24**, 429 (1985)
6.49 T. Yatagai, S. Nakadate, M. Idesawa, H. Saito: Opt. Eng. **21**, 432 (1982)
6.50 J. Budzinski: Appl. Opt. **31**, 3109 (1992)
6.51 W. Osten, J. Saedler, H. Rottenkolber: tm-Techn. Mess. **54**, 285 (1988)
6.52 G.A. Mastin, D.C. Ghiglia: Appl. Opt. **24**, 1727 (1985)
6.53 R. Dändliker, B. Ineichen, F.M. Mottier: Opt. Comm. **9**, 412 (1973)
6.54 R. Thalmann, R. Dändliker: Appl. Opt. **26**, 1964 (1987)
6.55 R.J. Pryputniewicz: Opt. Eng. **24**, 849 (1985)
6.56 R. Höfling, E. Lindner: J. Mod. Opt. **35**, 843 (1988)
6.57 K. Creath: Comparison of phase-measurement algorithms, in *Surface Characteriziation and Testing*, ed. by K. Creath. SPIE Proc. **680**, 19–28 (1986)
6.58 P. Hariharan, B.F. Oreb, N. Brown: Appl. Opt. **22**, 876 (1983)
6.59 R. Dändliker, R. Thalmann, J.-F. Willemin: Opt. Commun. **42**, 301 (1982)
6.60 J.E. Greivenkamp: Opt. Eng. **23**, 350 (1984)
6.61 P. Hariharan: Opt. Eng. **24**, 632 (1985)
6.62 B. Breuckmann, W. Thieme: Appl. Opt. **24**, 2145 (1985)
6.63 W. Jüptner: Automatic evaluation of holographic interferograms by the line-scan method (in German), in *Holografische Interferometrie in Technik und Medizin Frühjahrsschule 78*, ed. by H. Kreitlow and W. Jüptner (DPG/DGaO 1978)
6.64 W. Jüptner, Th. Kreis, H. Kreitlow: Automatic evaluation of holographic interferograms by reference beam phase shifting, in *Industrial Applications of Laser Technology*, ed. by W.F. Fagan. SPIE Proc. **398**, 22–29 (1983)
6.65 Th. Kreis: Quantitative evaluation of interference patterns, in *Industrial Optoelectronic Measurement Systems Using Coherent Light*, ed. by W.F. Fagan. SPIE Proc. **863**, 68–77 (1987)
6.66 Th. Kreis: Automatic evaluation of interference patterns, in *Holography Techniques and Applications*, ed. by W. Jüptner. SPIE Proc. **1026**, 80–89 (1988)
6.67 M. Chang, Ch.-P. Hu, P. Lam, J.C. Wyant: Appl. Opt. **24**, 3780 (1985)
6.68 D.J. Bone, H.-A. Bachor, R.J. Sandeman: Appl. Opt. **25**, 1653 (1986)
6.69 D.R. Burton, M.J. Lalor: The precision measurement of engineering form by computer analysis of optically generated contours, in *Industrial Inspection*, ed. by D.W. Braggins. SPIE Proc. **1010**, 17–24 (1989)
6.70 K.A. Nugent: Appl. Opt. **24**, 3101 (1985)
6.71 M. Takeda, H. Ina, S. Kobayashi: J. Opt. Soc. Amer. **72**, 156 (1982)
6.72 M. Takeda: Industr. Metrol. **1**, 79 (1990)
6.73 M. Takeda, Z. Tung: J. Opt. **16**, 127 (1985)
6.74 S. Toyooka, H. Nishida, J. Takezaki: Opt. Eng. **28**, 55 (1989)
6.75 W.W. Macy Jr.: Appl. Opt. **22**, 3898 (1983)
6.76 M. Takeda, M. Kitoh, Spatio-temporal frequency-multiplex heterodyne interferometry, in *Laser Interferometry IV: Computer-Aided Interferometry*, ed. by R.J. Pryputniewicz SPIE Proc. **1553**, 6 76 (1991)
6.77 Th. Kreis, W. Jüptner: Fourier-transform evaluation of interference patterns: the role of filtering in the spatial-frequency domain, in *Laser Interferometry: Quantitative Analysis of Interferograms*, ed. by R.J. Pryputniewicz SPIE Proc. **1162**, 116–125 (1989)
6.78 Th. Kreis: Fourier-transform evaluation of holographic interference patterns, in *Int. Conf. on Photomech. and Speckle Metrology*, ed. by F.P.Chiang. SPIE Proc. **814**, 365–371 (1986)

6.79 Th. Kreis, W. Jüptner: Fourier-transform evaluation of interference patterns: demodulation and sign-ambiguity, in *Laser Interferometry IV: Computer-Aided Interferometry*, ed. by R.J. Pryputniewicz SPIE Proc. **1553**, 263–273 (1991)
6.80 G. Lai, T. Yatagai: J. Opt. Soc. Amer. A**8**, 822 (1991)
6.81 K.-H. Womack: Opt. Eng. **23**, 391 (1984)
6.82 G.W. Johnson, D.C. Leiner, D.T. Moore: Opt. Eng. **18**, 46 (1979)
6.83 K.A. Stetson, W.R. Brohinsky: Opt. Eng. **26**, 1234 (1987)
6.84 K.A. Stetson, W.R. Brohinsky, J. Wahid, T. Bushman: J. Nondestr. Test. **8**, 69 (1989)
6.85 D.J. Bone: Appl. Opt. **30**, 3627 (1991)
6.86 Th.R. Judge, Ch. Quan, P.J. Bryanston-Cross: Opt. Eng. **32**, 533 (1992)
6.87 D.G. Ghiglia, G.A. Mastin, L.A. Romero: J. Opt. Soc. Am. A**4**, 267 (1987)
6.88 Th. Kreis, J. Geldmacher: Evaluation of interference patterns: a comparison of methods, in *2nd Int'l Conf. on Photomech. and Speckle Metrology*, ed. by F.P. Chiang SPIE Proc. **1554**, 718–724 (1991)
6.89 J.B. Schemm, C.M. Vest: Appl. Opt. **22**, 2850 (1983)
6.90 D.R. Burton, M.J. Lalor: Managing some of the problems of Fourier fringe analysis, in *Fringe Pattern Analysis*, ed. by G.T. Reid. SPIE Proc. **1163**, 149–160 (1989)
6.91 N. Ohyama, S. Kinoshita, A. Cornejo-Rodriguez, T. Honda, J. Tsujiuchi: J. Opt. Soc. Amer. A**5**, 2019 (1988)
6.92 K.A. Stetson: Exp. Mech. **7**, 273 (1981)
6.93 R.J. Pryputniewicz: Appl. Opt. **17**, 3613 (1978)
6.94 R.J. Pryputniewicz, K.A. Stetson: Appl. Opt. **15**, 725 (1976)
6.95 Th. Bischof, W. Jüptner: Investigation of the stress distribution in intact bonds by holographic interferometry and finite element method, in *Laser Interferometry IV: Computer-Aided Interferometry*, ed. by R.J. Pryputniewicz. SPIE Proc. **1553**, 326–331 (1991)
6.96 W. Jüptner, Th. Kreis, J. Geldmacher, Th. Bischof: Qualität und Zuverlässigkeit **36**, 417 (1991)
6.97 L.W. Meyer, W. Jüptner, H.-D. Steffens: Fracture toughness investigations using holographic interferometry, in *Laser 75 Opto-electronics*, ed. by W. Waidelich (IPC Science and Technology Press, Guildford 1975) pp. 203–205
6.98 W. Jüptner, K. Grünewald, R. Zirn, H. Kreitlow: Measurement of the stress intensity factor in large specimens by means of holographic interferometry, in *Holographic Data Nondestructive Testing*, ed. by D. Vukicevic. SPIE Proc. **370**, 62–65 (1982)
6.99 P. Will, W. Totzauer, B. Michel: Phys. Stat. Sol. (a) **95**, K113 (1986)
6.100 A. Choudry: Appl. Opt. **20**, 1240 (1981)
6.101 P.V. Farrell, G.S. Springer, C.M. Vest: Appl. Opt. **21**, 1624 (1982)
6.102 T.A.W.M. Lanen, C. Nebbeling, J.L. van Ingen: Opt. Commun. **76**, 268 (1990)
6.103 I. Prikryl, C.M. Vest: Appl. Opt. **21**, 2554 (1982)
6.104 D.W. Sweeney, C.M. Vest: Appl. Opt. **12**, 2649 (1973)
6.105 I.H. Lira, Ch.M. Vest: Appl. Opt. **26**, 3919 (1987)
6.106 S.S. Cha, C.M. Vest: Appl. Opt. **20**, 2787 (1981)
6.107 S.S. Cha, H. Sun: Appl. Opt. **29**, 251 (1990)
6.108 H. Rottenkolber, W. Jüptner: Holographic interferometry in the next decade, in *Laser Interferometry: Quantitative Analysis of Interferograms*, ed. by R.J. Pryputniewicz SPIE Proc. **1162**, 2–15 (1989)

7. Techniques to Measure Displacements, Derivatives and Surface Shapes. Extension to Comparative Holography

P.K. Rastogi

Laboratory of Stress Analysis, Swiss Federal Institute of Technology, CH-1015 Lausanne, Switzerland

The pioneering work of *Stetson* and *Powell* [7.1], in 1965, demonstrating the use of holographic interferometry in vibration analysis, set the tone to the flood of activities devoted to the development and application of the holographic-based methods. A significant amount of research work has been carried out in holographic interferometry [7.2–10]. The work lays down a compact mathematical network to elucidate the relationship between the fringes and the shape contours or changes in shapes of objects. As a result of this investigation a wide range of procedures has been developed which makes it not only possible to measure surface displacements and deformations of engineering structures to an accuracy of a fraction of a micrometer but also to detect material flaws and inhomogeneities having escaped the manufacturing process. The technique is now regularly applied to solve problems in a wide ranging field. The rapid developments in holographic interferometry are significant evidence of the method's popularity.

A holographic interferometer in its basic form provides a means of bringing into interference two diffuse waves which are exactly identical in their broad structure and detail. For the condition of identity of the two waves to be satisfied it is imperative that these waves originate from the same point P, on the object surface, in its two different states before and after the object is deformed. The wave fronts diffracted from points other than the set of identical points do not contribute to fringe formation at the point P. The concept of corresponding points is of prime importance in holographic interferometry. The interference fringes are localized in that region of the space where the homologous rays intersect. The interference pattern is formed due to the changes in the optical path length which is dependent on object deformation and the configuration geometry during the recording and reconstruction of the hologram. Based upon the configuration geometries, a class of holographic techniques has emerged, which is capable of addressing a wide range of specific measurement problems. This family of techniques has been employed to obtain displacement, strain, slope-change and curvature contours of a deformed structure. These techniques have also been used to measure shape variations of a 3-D specimen. More recently, methods have been developed which allow one to visualize the contours of the difference in displacements of two similarly stressed objects. The basic technique, better known as comparative or difference displacement holographic interferometry, has an important potential in nondestructive testing.

The goal of this chapter is to review the line of holographic techniques applied to deformation analysis, difference displacement and shape measurements. The emphasis will be on the description of techniques based on whole field and real-time approach. This approach is in large part responsible for the success of holographic interferometry.

7.1 Measurement of Out-of-Plane Displacements

The measurement of out-of-plane displacements has admittedly been one of the most wide-spread techniques in the tool box of holographic interferometry. Simple to implement, the technique owes its popularity to the ease with which fringes corresponding to the out-of-plane displacements are displayed on the examined object surface.

7.1.1 Basic Configurations

Consider a point P(r) on the surface to be illuminated and observed along the propagation vectors (K_e, K_o). The space vectors M_o and N_e define the positions of the point of observation and the source of illumination with respect to the coordinate system (x, y, z) shown in Fig. 7.1. When the object is deformed, the point P(r) moves through a displacement given by the vector $L(r)$

$$L(r) = \hat{i}L_x + \hat{j}L_y + \hat{k}L_z. \tag{7.1.}$$

The optical phase difference between the rays emanating from P and P' is given by

$$\varphi(P) = (K_e - K_o) \cdot L(r). \tag{7.2}$$

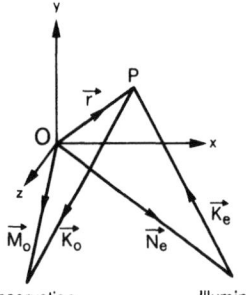

Fig. 7.1. Illumination and observation geometry

7.1 Measurement of Out-of-Plane Displacements

The propagation vectors K_e and K_o are, respectively, given by

$$K_e = k\hat{K}_e = k\frac{r - N_e}{|r - N_e|} \quad \text{and} \quad K_o = k\hat{K}_o = k\frac{M_o - r}{|M_o - r|}, \tag{7.3}$$

where $\hat{K}_e$ and $\hat{K}_o$ are unit vectors along their respective illumination and observation directions; k is the magnitude of the propagation vectors

$$k = |K_e| = |K_o| = \frac{2\pi}{\lambda}.$$

For the sake of simplicity, let us consider a plane illumination and observation parallel to the x–z plane. Equation (7.2) reduces to

$$\varphi(P) = \frac{2\pi}{\lambda}[(K_{ex} - K_{ox})L_x - (K_{ez} + K_{oz})L_z], \tag{7.4}$$

where (K_{ex}, K_{ey}, K_{ez}) and (K_{ox}, K_{oy}, K_{oz}) are Cartesian components of the illumination and observation vectors, respectively. For the contribution of the in-plane component to cancel out in (7.4), it suffices that $K_{ex} = K_{ox}$. The corresponding fringe equation becomes

$$L_z = \frac{n\lambda}{2\cos\theta}, \tag{7.5}$$

where n is the fringe number and θ is the angle which the illumination beam makes with the surface normal. Two systems of measurements emerge from (7.5): one corresponding to the illumination and observation directions normal to the object surface, and the other corresponding to oblique illumination and observation directions making equal and opposite angles with the surface normal (Fig. 7.2). The sensitivity of the method decreases with the increase in the obliquity of the pair of illumination and observation beams. The sensitivity becomes $\lambda/2$ for the configuration $\theta = \theta_0 = 0$ and ∞ for the configuration $\theta = \theta_0 = \pi/2$. The schematic of an optical set-up to observe fringes corresponding to out-of-plane displacements is shown in Fig. 7.3. The holographic plate is placed inside an immersion tank mounted on a universal translation and rotation stage. The reconstructed interference image is observed in real-time on the TV monitor. An example of fringe pattern depicting the out-of-plane displacements in a rectangular aluminium plate – clamped along the edges and drawn out at the center by means of a bolt – subjected to two point loading is depicted in Fig. 7.4(a). Figure 7.4(b) refers to an illustration taken from the study of the embryonic behaviour of embryos during incubation. The photograph shows a specific limb movement near the air space.

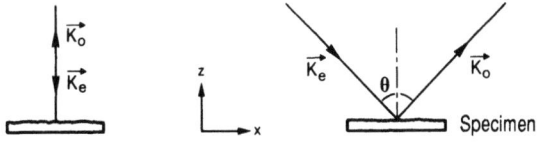

Fig. 7.2. Schematic of the illumination and observation configurations generally used for measuring out-of-plane displacements

216   7. Techniques to Measure Displacements, Derivatives and Surface Shapes

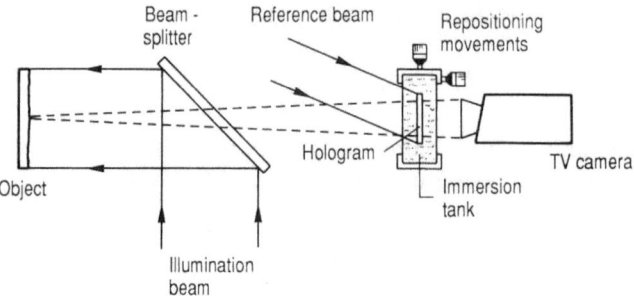

Fig. 7.3. Layout of an optical set-up to measure out-of-plane displacements

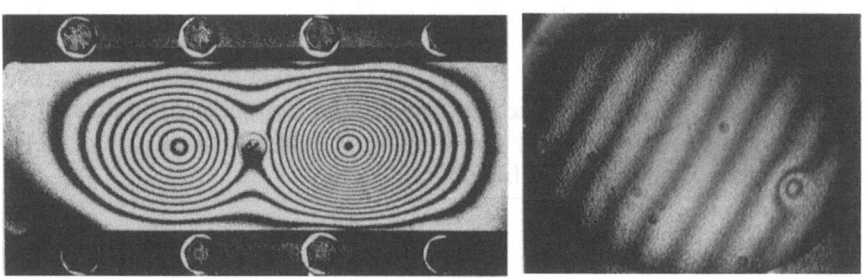

Fig. 7.4. (a) An example of out-of-plane displacement fringes produced on a rectangular aluminum plate – clamped along the edges and drawn out at the center by means of a bolt – subjected to two point loading. (b) Illustration of a specific limb movement near the air space of a hen embryo during incubation

A popular system for studying large objects, undergoing predominantly out-of-plane displacements, is to illuminate and observe a point $(x, y, 0)$ on the surface from the same point $(0, 0, \rho)$ lying on the optical axis of the measurement set-up. A characteristic of such a system is that the pair of illumination and observation rays, propagating towards and away from the surface, always remain aligned irrespective of the position of the running point on the object surface. When applied to a deformation problem where in-plane displacements $L_{x,y}$ can be neglected, the out-of-plane displacement component in terms of Cartesian components is given by

$$L_z(x, y) = \frac{n\lambda}{2[1 + (x^2 + y^2)/\rho]^{-1/2}}. \tag{7.6}$$

The interpretation of the fringes as contours of equal out-of-plane displacements is thus valid only in the neighbourhood of the optical axis. As the fringe sensitivity varies over the object surface, a correction at each object point is called for when analyzing the fringes. Finally, this interferometer cannot be used for the evaluation of out-of-plane displacements in the presence of in-plane

7.1 Measurement of Out-of-Plane Displacements

deformations. The contribution of in-plane displacements to fringe formation can, however, be considerably reduced by positioning the illumination source and observation point at a large distance from the object.

Yamaguchi and *Saito* [7.11] have demonstrated that the technique could be used to determine Poisson's ratio, a material property. Subjecting a prismatic bar with a rectangular cross-section, to pure bending in one of its principal planes, transforms the surface of the bar to a saddle-like shape. Choosing the origin of the coordinate system to be at the centroid of the cross-section and the x–z plane as the principal plane of bending, the contour lines mapping the out-of-plane deformation of a plate subjected to four-point bending appear in the form of a family of hyperbolas [7.12] given by

$$x^2 - vy^2 = \text{constant}. \tag{7.7}$$

The equation of the asymptotes is

$$x^2 - vy^2 = 0, \tag{7.8}$$

which yields the Poisson's ratio of the material

$$v = \tan^2 \alpha_1, \tag{7.9}$$

where α_1 is the half of the smaller angle between the asymptotes.

7.1.1.1 Remarks on the Removal of Phase Ambiguities and Measurement of Small Deformations

A problem encountered in holographic interferometry when applied to displacement analysis is related to the unambiguous determination of fringe orders and directions [7.13–15]. A way to overcome the problem of phase ambiguities consists of superposing a set of linear parallel fringes or carrier to the deformation pattern. The changes in the direction and spacing of the original pattern with respect to the added phase help to identify the direction of displacement undergone by the object. If the carrier dominates, the modulated pattern displays monotonically increasing fringe orders, which allow the latter to be determined at any point without ambiguity. The phase change due to deformation is obtained by subtracting the known linear carrier from the modulated fringe pattern composed of the linear and deformation-induced phase. Recently, circular [7.16] and cylindrical [7.17] fringe carrier techniques have been proposed with a view to eliminating the ambiguities in holographic fringe interpretation. The determination of the sign of displacement is greatly facilitated by employing real-time or sandwich holographic interferometry for measurements. An example of fringe pattern obtained by adding a linear phase to the deformation fringes is depicted in Fig. 7.5. The type of loading is similar to that performed to obtain Fig. 7.4a, with the exception that the bolt is slightly released after recording the hologram.

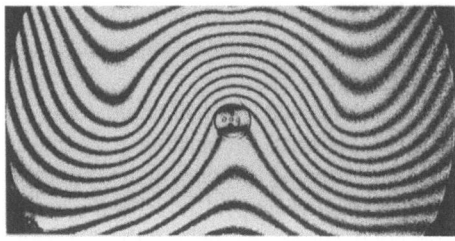

Fig. 7.5. An example of fringe pattern obtained by adding a linear phase to the deformation fringes

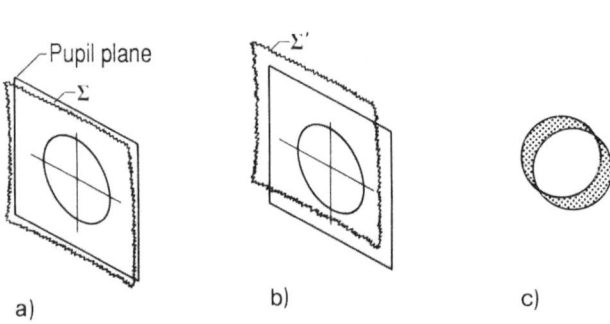

Fig. 7.6a–c. Description of the pupil plane decorrelation: (a) and (b) illustrate the relative shift of the two reconstructed wavefronts in the pupil plane; (c) size of the effective pupil (front view) is given by the overlapping region of the pupil contours. The shaded regions are uncorrelated

The addition of a linear phase carrier to the phase change induced by very small object deformations has been used to measure these small deformations [7.18–20], which can be of the order of 1/50th of a wavelength. The phase difference due to object deformation is retrieved by subtracting the linear phase from the modulated phase on a point by point basis. *Toyooka* [7.18] has proposed a whole field approach based on double exposure, addition of linear carrier phase and post-processing of the resulting interferogram to obtain phase-difference amplification in holographic interferometry. The maximum sensitivity reported by him is $\lambda/16$ per fringe in comparison to the $\lambda/2$ achievable by conventional means.

7.1.2 Speckle Decorrelation and Fringe Visibility

In general, an arbitrary surface deformation gives rise to a relative displacement between the reconstructed wave fronts in the pupil plane. The effect of this displacement is that during reconstruction the pupil samples slightly different portions of the wave fronts Σ and Σ'. This is illustrated in Fig. 7.6. Since the diffracted amplitudes are random in nature, only the identical parts interfere, the rest presenting themselves in the form of a noise lead to a fall in fringe visibility. This one-to-one correspondence, to a first order, between the interfering speckles constitutes an essential feature of fringe formation in holographic interferometry. The speckle fields participating in the fringe formation are given by the unshaded region of the pupil contour cut on each of the two wave fronts. The shaded portions are uncorrelated and act as noise diminishing the fringe visibility. The mean intensity of the reconstructed fringes is the sum of the mean intensity of the modulated part and the mean intensity of the unmodulated part of the speckle field. It can be written in the form

$$\langle I(\text{P}) \rangle = \langle I_0 \rangle (1 + V_1 \cos \varphi) + \langle I_0' \rangle. \tag{7.10}$$

The visibility of the reconstructed fringes is given by

$$V = \frac{\langle I_{\max} \rangle - \langle I_{\min} \rangle}{\langle I_{\max} \rangle + \langle I_{\min} \rangle} = V_1 V_2, \tag{7.11a}$$

with

$$V_2 = \frac{\langle I_0 \rangle}{\langle I_0 \rangle + \langle I_0' \rangle}. \tag{7.11b}$$

The factor V_2, being the ratio of the unshaded surface area of the pupil to the total surface area, could be recognized as the normalized autocorrelation function of the pupil. Equation (7.11a) reveals that the fringe visibility not only varies with decorrelations due to object displacements and strains, but depends also on the shift between the speckle fields in the pupil plane.

The decorrelations arising from object displacements and strains account for fringe delocalization. In the presence of the lateral displacement component L_t the fringes tend to move away from the object surface. The fringes are localized (that is to say, their contrast is maximum) in that region of the space where the distance between the homologous rays [7.21–24] is minimum (Fig. 7.7). The coherent superposition of two speckle patterns which are related to one another by a difference of phase and a lateral shift L_t, of modulus L_t, gives rise to a fringe

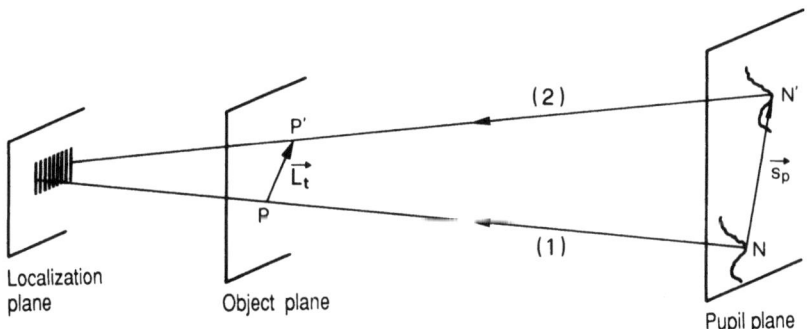

Fig. 7.7. The fringes are localized where the distance between the homologous rays (1) and (2) is the shortest

system modulating the speckle field. The fringe visibility is given by the autocorrelation function of the speckle for the argument L_t [7.4–6, 22–24]. The latter is given by the Fourier transform of the pupil function limiting the spectral extent of the object. A consideration of the influence of both shifts, L_t in the object plane and s_p in the pupil plane, leads to the following expression for fringe visibility V for a circular pupil of radius a [7.25, 26]

$$V = \frac{1}{\pi} \int_{\theta'=0}^{2\pi} a'^2(\theta') \left(\operatorname{sinc} \chi - \frac{1}{2} \operatorname{sinc}^2 \frac{\chi}{2} \right) d\theta', \tag{7.12}$$

where

$$a'(\theta') = \sqrt{(1 - A^2 \sin^2 \theta')} - A\sqrt{(1 - \sin^2 \theta')},$$

$$\chi = 3.83 \, Ba'(\theta') \cos(\theta' - \eta'), \quad A = \frac{|s_p|}{2a}, \quad B = \frac{|L_t|}{\sigma},$$

σ being the mean lateral speckle size in the hologram image plane.

7.1.3 Compensation of Rigid-Body Movements and Fringe Control

The use of holographic interferometry is often limited by its sensitivity to rigid-body movements. These movements cause the fringe pattern to alter completely or lose their contrast to the extent of even vanishing from the object surface. As a result fringe control techniques [7.27–34] have been developed which, in combination with a real-time procedure, provide a highly satisfactory performance in deformation applications. Sandwich holography [7.35–41] has been the other preferred mode of operation to overcome these disadvantages.

As indicated in Sect. 7.1.2, the object and pupil plane decorrelations accompanying object deformation are important factors affecting the fringe visibility. The speckle shifts are schematically decomposed in splitting the vectorial displacement field in terms of rigid-body movements (T_r, R_r) and deformations (T_d, R_d). T and R are the translation and rotation vectors:

$$s_p = F_1 \cdot (R_r + R_d) + F_2 \cdot (T_r + T_d), \tag{7.13a}$$

$$L_t = T_r + T_d, \tag{7.13b}$$

$F_{1,2}$ being 3×3 matrices which are dependent solely on the recording geometry. For a known recording geometry and displacement components, one can calculate these shifts by using the explicit formulations contained in [7.42–45].

An inspection of (7.13) indicates that the fringe visibility could be optimized by compensating the rigid-body movements (T_r, R_r), which are generally predominant on loading the model. The compensation of T_r is of prime importance, as this movement tends to delocalize the fringes, resulting in a rapid decrease in the fringe visibility on the object surface. Localizing the fringes on the object surface, or its immediate neighbourhood, is sufficient to eliminate T_r. The rigid-body rotations R_r principally affect the spatial frequency of the fringe patterns. The reduction or removal of this spatial frequency is essential to a distinct observation of defects in the specimen.

The advantage of real-time operation lies essentially in its execution stage. For the compensation of translation T_r, one can easily sense the direction and magnitude of the movement to be given to the three translation micrometers, while observing the TV monitor for fringe localization and visibility. The unwanted fringes are eliminated by an adequate manipulation of the reconstructed wave front. In double exposure, the trial and error procedure being prohibited, three operations are absolutely called for before the second exposure: measurement of the six components of rigid-body movements, calculation of the compensatory movements of different system elements (hologram, reference, illumination) and implementing these compensations in dark. In real-time the effort reduces uniquely to the last operation. Another technique for eliminating the influence of rigid-body movements from the fringe patterns is based on reflection holography [7.46, 47]. The requirements on the stability of the set-up are relaxed considerably when using this technique. Finally, *Kreitlow* et al. [7.48] have proposed that with a reference mirror fastened to the object and undergoing the same displacements as the object, the demands for stability of the set-up are much less than in usual holographic interferometry.

7.2 Measurement of In-Plane Displacements

The measurement of in-plane components of displacement using holographic interferometry has been investigated actively by a number of researchers. The first works in this line, especially by *Aleksandrov* and *Bonch-Bruevich* [7.49], *Ennos* [7.50], and later by *Fossati-Bellani* and *Sona* [7.51] vividly put forward the difficulties involved in the measurement. Their techniques, based on a point-by-point approach, require observing the object from different directions. The in-plane information follows from the analysis of the resulting interferograms. The first attempts at whole field visualization are due to *Boone* [7.52] and *Butters* [7.53]. By illuminating the object along two directions symmetric to the surface normal and observing along a common direction, they could make apparent moiré fringes representing the projection of the displacement vector on the object surface in the direction containing the two beams. Although the approach works in theory, low-density fringes of variable spacing and localization make it difficult to observe the moiré. *Schlüter* and *Nowatzyk* [7.54] reported that changing the reference-beam phase and using a video-electronic processing helps to overcome the problem. Another solution consists of generating an auxiliary system of fringes localized on the object surface [7.55, 56]. *Sciammarella* and *Gilbert* [7.57] proposed modifying the position of the hologram according to a given law. Since then the method has seen a rapid growth of interest and development [7.25, 26, 58–70]. Of the other methods of interest,

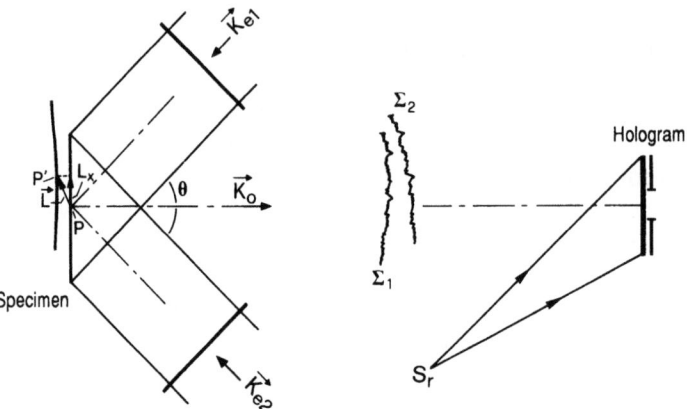

Fig. 7.8. Schematic of the holographic moiré set-up to measure in-plane displacement component L_x

Katzir and *Glaser* [7.71], and *Katzir* et al. [7.72] proposed recovering the in-plane displacement information by reconstructing the reference wave of a double-beam-illuminated object. *Pirodda* [73–75] proposed the use of conjugate-wave holographic interferometry to obtain fringes corresponding to in-plane displacements. More recently, *Rastogi* and *Denarié* [7.76] extended the use of the phase shifting technique to facilitate the quantitative analysis of fringe patterns obtained with holographic moiré.

7.2.1 Holographic Moiré

7.2.1.1 Basic Relations

The configuration of the interferometric set-up for in-plane displacement measurement is shown in Fig. 7.8. The specimen is illuminated by two beams defined by the wave vectors K_{e1} and K_{e2}. The beams are supposed to lie in the x–z plane. Wave field Σ_1 received by the hologram results from the superposition of the two wave fronts arising, respectively, from the diffraction of the two beams illuminating the object; Σ_2 is the wave field scattered toward the hologram after the object is deformed. The observation is carried along wave vector K_o. The first phase of the method consists of exposing the holographic plate H to the wave field Σ_1. The holographic plate is processed and returned to its original position. The object is deformed. If A_1 and A_1', A_2 and A_2', represent the corresponding amplitudes at the point P of the image before and after object deformation, the intensity observed at a point P in the image plane is given by

$$\langle I(\mathrm{P})\rangle = \langle |e^{i\pi}(A_1 + A_2) + A_1' + A_2'|^2\rangle, \tag{7.14}$$

where

$$A'_1 = A_1 e^{i\varphi_1}, \tag{7.14a}$$

$$A'_2 = A_2 e^{i\varphi_2}, \tag{7.14b}$$

φ_1 and φ_2 being the corresponding phase differences introduced by the object deformation, express

$$\varphi_1(P) = (\mathbf{K}_{e1} - \mathbf{K}_o) \cdot \mathbf{L}, \tag{7.15a}$$

$$\varphi_2(P) = (\mathbf{K}_{e2} - \mathbf{K}_o) \cdot \mathbf{L}. \tag{7.15b}$$

Since Σ_1 and Σ_2 are coherent and uncorrelated, only the amplitudes corresponding to scattered field pairs (Σ_1, Σ'_1) and (Σ_2, Σ'_2) are correlated.

$$\langle A_1 A_2^* \rangle = \langle A_1 A_2'^* \rangle = \langle A_1' A_2^* \rangle = \langle A_1' A_2'^* \rangle$$
$$= \langle A_2 A_1^* \rangle = \langle A_2 A_1'^* \rangle = \langle A_2' A_1^* \rangle = \langle A_2' A_1'^* \rangle = 0, \tag{7.16a}$$

$$\langle |A_1|^2 \rangle = \langle |A_1'|^2 \rangle = \langle |A_2|^2 \rangle = \langle |A_2'|^2 \rangle = \langle I_1 \rangle, \tag{7.16b}$$

where $\langle \cdots \rangle$ represents the mean over the ensemble, and $*$ indicates the complex conjugate. The intensity in the reconstructed image becomes

$$\langle I(P) \rangle = 4 \langle I_1 \rangle [1 - \cos\varphi \cos\psi], \tag{7.17}$$

where we have set $\varphi = (\varphi_1 - \varphi_2)/2$ and $\psi = (\varphi_1 + \varphi_2)/2$. The reconstructed image exhibits a granular random distribution of mean intensity $4\langle I_1 \rangle$, modulated by a slowly varying system of fringes. In holographic interferometry one is generally concerned with fringe patterns whose spacing is much larger than the speckle grain. Thus the measured intensity will be in good accordance with (7.17) if it is averaged over a region large compared to the mean speckle size and small compared to fringe spacing. In terms of the notations used in Fig. 7.8,

$$\varphi(P) = \frac{\mathbf{K}_{e1} - \mathbf{K}_{e2}}{2} \cdot \mathbf{L} = \frac{2\pi}{\lambda} L_x \sin\theta, \tag{7.18a}$$

$$\psi(P) = \left(\frac{\mathbf{K}_{e1} + \mathbf{K}_{e2}}{2} - \mathbf{K}_o \right) \cdot \mathbf{L} = \frac{2\pi}{\lambda} L_z (1 + \cos\theta). \tag{7.18b}$$

The moiré fringe loci represent the contour lines of the projection of the displacement vector $\mathbf{L}$ on the object plane in the direction containing the two beams. The spacing of the moiré fringes corresponds to an incremental displacement of $\lambda/2\sin\theta$.

There are some difficulties which must be overcome. For a good visualization of the moiré fringes, it is necessary that the two interacting patterns have their maximum contrast on the same plane, are of high density and are oriented in nearly the same direction. From a practical viewpoint, these conditions can be met by introducing artificially an additional phase difference much larger than that introduced by the object deformation. This additional phase difference gives rise to the formation of parallel equidistant fringes of high density localized on the object surface. In the presence of these so-called auxiliary fringes (7.17) becomes

$$\langle I(P)\rangle = 4\langle I_1\rangle[1 - \cos\varphi\cos(\psi + \psi_a)], \tag{7.19}$$

where ψ_a is the auxiliary phase difference introduced between the two exposures. The term $\cos(\psi + \psi_a)$ represents the modulation corresponding to the primary fringes. The primary fringes are the result of the combination of the phase differences due to out-of-plane displacements and deformations of the object surface and the auxiliary displacement introduced between the two exposures.

Having assembled the conditions for obtaining good moiré formation, the question arises of how to produce large phase differences capable of generating fringes localized on the object surface. We will now concentrate on this aspect, not only because of its importance in obtaining good-quality moiré fringes but also in procedures such as enhancement of a method's sensitivity, fringe control or in the determination of the sign of displacement. Last but not least, the generation of auxiliary fringes is of great importance in all techniques based on the concept of holographic moiré.

7.2.1.2 Generation of Auxiliary Fringes

We shall consider only those methods of producing auxiliary fringes which require a relative displacement of the holographic plate and reference beam between the two exposures. The methods based on imparting displacement to the model are not considered here due to their lack of interest from a practical standpoint.

7.2.1.2.1 Rotation of the Holographic Plate Around an Axis Parallel to its Plane. This method [7.57] proposes generating auxiliary fringes by rotating the holographic plate around an axis parallel to its plane (Fig. 7.9a). In this case the

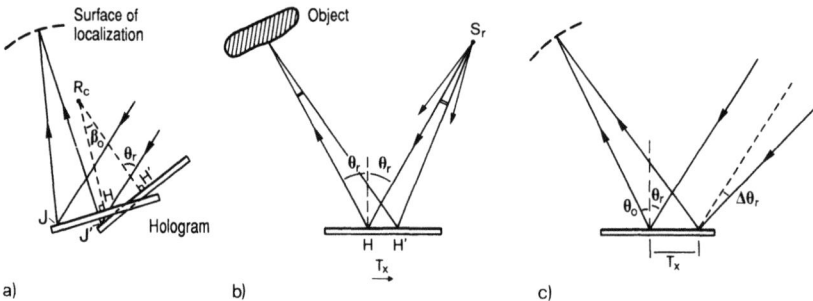

Fig. 7.9a–c. Generation of auxiliary fringes by **(a)** rotation of the holographic plate around an axis passing by R_c and parallel to its plane; **(b)** translation of the holographic plate in its own plane; and **(c)** rotation of the reference beam and translation of the holographic plate

localization plane depends on the coordinates of the rotation center C and on the angle of illumination of the reference beam

$$\rho_a = \frac{z_c}{1 - \cos\theta_r}. \tag{7.20}$$

The fringe spacing of the auxiliary fringes is given by

$$P_A = \frac{\lambda}{\beta_0(1 - \cos\theta_r)}. \tag{7.21}$$

The apparent difficulty for the user is that for a stage holder with a fixed movement of rotation the method loses much flexibility of operation. While using a stage holder possessing four degrees of freedom (ω_{px}, ω_{py}, T_x, T_y), this handicap can be overcome and the rotation axis varied at will to meet any experimental requirement.

7.2.1.2.2 Translation of the Hologram in its Plane. This method [7.24, 25, 66] consists of translating the holographic plate in a configuration in which the object and reference sources, placed at equal distances, are oriented symmetrically with respect to the normal to the plate (Fig. 7.9b). *Monneret* [7.24] has shown that under these conditions the fringes are localized on the object surface independent of the magnitude of translation. The fringe spacing of the auxiliary fringes is given by

$$P_A = \frac{\lambda\rho_r}{T_x \cos\theta_r}, \tag{7.22}$$

where ρ_r, the distance from the hologram to the reference source, is equal to ρ_a, the distance from the hologram to the object.

7.2.1.2.3 Rotation of the Reference Beam and Translation of the Holographic Plate. This method [7.77] generates auxiliary fringes by rotating the reference beam and providing an adequate lateral translation to the holographic plate (Fig. 7.9c). The lateral translation must ensure the intersection of the homologous rays on the object surface. If the reference beam is tilted around an axis perpendicular to the plane of the paper, the interfringe is given by

$$P_A = \frac{\lambda}{\Delta\theta_r \cos\theta_r}, \tag{7.23}$$

where θ_r is the angle of incidence of the reference beam and $\Delta\theta_r$ is the induced rotation.

7.2.1.2.4 Rotation of the Reference Beam. This method [7.25, 62, 68] for the generation of auxiliary fringes is applicable to the systems based on image-plane holography. In image-plane holography a lens is used to image the object on a holographic plate. The method consists of tilting the collimated reference beam. In this case the fringes are localized on the plate. The interfringe is given by (7.23).

7.2.1.3 Localization

The geometrical definition of the surface of localization as the loci of points, where the distance between the homologous rays is minimum, lends itself well to the study of the fringe localization for any given illumination and recording geometry. However, a more analytical approach would consist of applying the concept of the stationary conditions of the path differences [7.78–82]. It is this latter formalism which is used here to compute fringe localization. If ε denotes the path difference at any point A of the field diffracted by the hologram

$$\varepsilon = S_r J'A - S_r JA,$$

the fringes will localize in the neighbourhood of A if ε is independent of the position of any point J on the hologram (Fig. 7.10). In other words, the surface of localization is given by

$$\frac{\partial \varepsilon}{\partial \xi} = 0; \quad \frac{\partial \varepsilon}{\partial \eta} = 0 \tag{7.24}$$

subject to their being satisfied simultaneously.

The coordinate system $[H, \xi \eta \zeta]$ is chosen so that the $[H, \xi \eta]$ plane coincides with the hologram plane, H being the center of the illuminated region of the

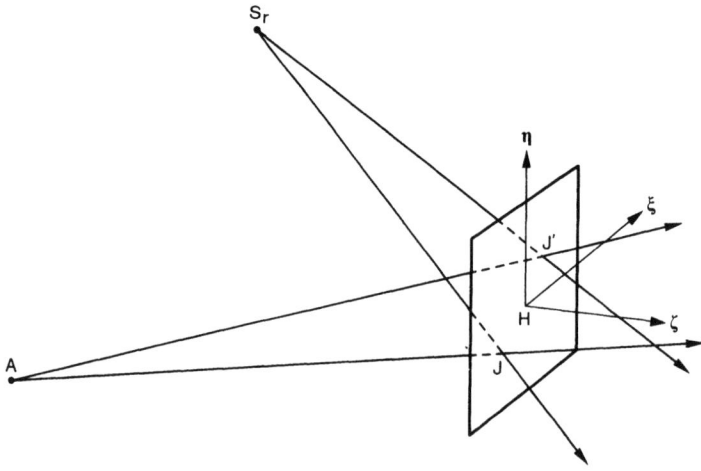

Fig. 7.10. Geometry of hologram reconstruction

hologram. The reference source S_r is placed in the $[H\xi\zeta]$ plane. The direction cosines of the directions AH and S_rH are respectively given by (l_a, m_a, n_a) and (l_r, m_r, n_r); $AH = \rho_a$, $S_rH = \rho_r$. The displacement of the holographic plate between the two exposures can be decomposed into a rotation $\omega_p(\omega_{px}, \omega_{py}, \omega_{pz})$ around an axis passing through H and into a translation $T(T_x, T_y, T_z)$. The displacement ΔJ of a variable point $J(\xi, \eta, 0]$ is given by

$$\Delta J \begin{vmatrix} \Delta \xi = T_x - \omega_{pz}\eta, \\ \Delta \eta = T_y + \omega_{pz}\xi, \\ \Delta \zeta = T_z + (\omega_{px}\eta - \omega_{py}\xi). \end{vmatrix} \quad (7.25)$$

Limiting the development to second-order terms, the path difference is given by

$$\varepsilon = \Delta\xi \left[l_r - l_a + \xi\left(\frac{1}{\rho_r} - \frac{1}{\rho_a}\right) + \frac{l_a \mu_a}{\rho_a^2} - \frac{l_r \mu_r}{\rho_r^2} \right]$$

$$+ \Delta\eta \left[m_r - m_a + \eta\left(\frac{1}{\rho_r} - \frac{1}{\rho_a}\right) + \frac{m_a \mu_a}{\rho_a^2} - \frac{m_r \mu_r}{\rho_r^2} \right]$$

$$+ \Delta\zeta \left[n_r - n_a + \zeta\left(\frac{1}{\rho_r} - \frac{1}{\rho_a}\right) + \frac{n_a \mu_a}{\rho_a^2} - \frac{n_r \mu_r}{\rho_r^2} \right], \quad (7.26)$$

with

$$\mu_i = (l_i \xi + m_i \eta + n_i \zeta)\rho_i \quad (i = a, r).$$

Let us limit our consideration to first-order terms only. The higher-order terms contribute to the evaluation of fringe visibility but their influence is much less compared to the first-order terms. The first-order terms determine the optimal localization. The zero-order terms give an indication of the position and angular spacing of the fringes. If one considers the case of auxiliary fringes perpendicular to the plane of incidence,

$$T_y = T_z = 0, \quad \omega_{px} = \omega_{pz} = 0, \quad \Delta\xi = T_x, \quad \Delta\eta = 0, \quad \Delta\zeta = -\xi\omega_{py}, \quad m_r = 0$$

and the left-hand side of the localization condition (7.24) becomes

$$\frac{\partial \varepsilon}{\partial \xi} = T_x \left(\frac{1 - l_r^2}{\rho_r} - \frac{1 - l_r^2}{\rho_a} \right) - \omega_{py}(n_r - n_a) + \omega_{py} R_\xi(\xi, \eta, \zeta), \quad (7.27a)$$

$$\frac{\partial \varepsilon}{\partial \eta} = \frac{l_a m_a}{\rho_a} T_x + \omega_{py} R_\eta(\xi, \eta, \zeta), \quad (7.27b)$$

where $R_\xi(\xi, \eta, \zeta)$ and $R_\eta(\xi, \eta, \zeta)$ correspond to the derivation of second-order terms of ε. In (7.27) only the terms independent of (ξ, η) are relevant for determining the surface of localization. The existence of the other two terms only expresses the fact that the visibility on the localization plane is not maximum. The condition for localization becomes

$$T_x\left(\frac{1-l_r^2}{\rho_r} - \frac{1-l_r^2}{\rho_a}\right) = \omega_{py}(n_r - n_a), \tag{7.28a}$$

$$\frac{l_a m_a}{\rho_a} T_x = 0. \tag{7.28b}$$

It is evident that (7.28b) is satisfied only for the points lying on the $[H, \xi\zeta]$ and $[H, \eta\zeta]$ planes. Hence the sections of the surface of localization in these two planes are given by (7.28a) with m_a or $l_a = 0$.

Considering the case of auxiliary fringes parallel to the plane of incidence, the sections of the surface of localization are obtained similarly by inserting

$$T_x = T_z = 0, \quad \omega_{py} = \omega_{pz} = 0, \quad \Delta\xi = 0, \quad \Delta\eta = T_y, \quad \Delta\zeta = \eta\omega_{px}, \quad m_r = 0$$

in (7.24). Table 7.1 presents the sections of the surfaces of localization obtained in the configurations described in Figs. 7.9a and b. As an example, Fig. 7.11 depicts the sections of the surface of localization in the $[H, \xi\zeta]$ and $[H, \eta\zeta]$ planes corresponding to the configuration shown in Fig. 7.9a. The fringes are supposed to be orthogonal to the plane of incidence. For the points in the field not belonging to the $[H, \xi\zeta]$ or $[H, \eta\zeta]$ planes, the homologous rays do not intersect and in this condition the localization plane (defined as the plane where the fringe contrast is maximum but not equal to unity) is situated in the neighborhood of the shortest distance between them. Hence to increase contrast, the aperture has to be reduced.

Table 7.1. A comparative study of the sections of the surface of localization in the $[H, \xi\eta]$ and $[H, \zeta\eta]$ planes corresponding to the configurations shown in Figs. 7.9a, b

Fringe directions	Planes	Configuration in Fig. 7.9a	Fig. 7.9b
⊥ to the plane of incidence	$[H, \xi\zeta]$	$\rho_a = \dfrac{T_x}{\omega_{py}} \dfrac{n_a^2}{n_a - n_r}$	$\rho_a = \rho_r \dfrac{n_a^2}{n_r^2}$
	$[H, \eta\zeta]$	$\rho_a = \dfrac{T_x}{\omega_{py}} \dfrac{1}{n_a - n_r}$	$\rho_a = \dfrac{\rho_r}{n_r^2}$
∥ to the plane of incidence	$[H, \xi\zeta]$	$\rho_a = \dfrac{T_y}{\omega_{px}} \dfrac{1}{n_r - n_a}$	$\rho_a = \rho_r$
	$[H, \eta\zeta]$	$\rho_a = \dfrac{T_y}{\omega_{px}} \dfrac{n_a^2}{n_r - n_a}$	$\rho_a = n_a^2 \rho_r$

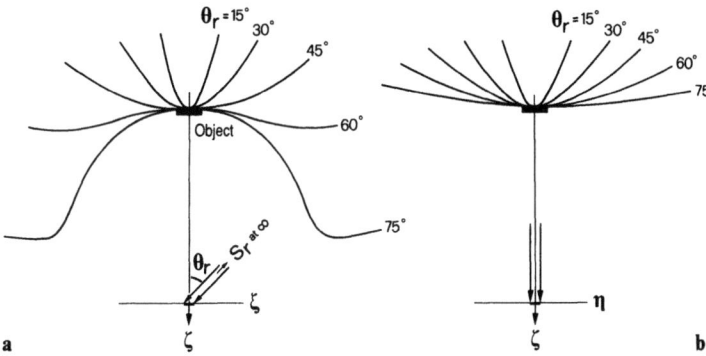

Fig. 7.11a, b. Surface of localization of the auxiliary fringes for the method 'Rotation of the holographic plate around an axis parallel to its plane': **(a)** trace in the $[H, \xi\zeta]$ plane; **(b)** trace in the $[H, \eta\zeta]$ plane

7.2.1.4 Some Remarks Concerning Imaging Parameters

Optimum experimental conditions at the outset aid considerably in the successful implementation of the method. The two parameters in imaging viz. aperture and magnification have an important influence on the contrast of the primary fringe and hence by implication on the quality of the moiré fringes. The three relevant parameters which characterize an aperture are its shape, size and orientation. The use of an aperture of appropriate shape and orientation can help to satisfy several requirements for good fringe formation. For example, the choice of a rectangular aperture can help to dampen object-plane decorrelation by orienting its larger side perpendicular to the maximum lateral displacements and, at the same time, increase resolution by generating the carrier fringes parallel to the shorter side of the imaging aperture. Different rigid-body rotations may desorient the carrier fringes but these could be reoriented along a preferred direction by providing appropriate compensations in a real-time set-up. The choice of aperture size is delicate; while an aperture of a small size slows down the decorrelation in the object plane it favorises the decorrelation in the pupil plane. However, a compromise is generally possible by adequately designing an optical system which endeavours to keep the ratios $|L_{t\,max}|/\sigma$ and $|s_p|/q'$ reasonably low so as to be able to maintain the fringe visibility in an acceptable range; q' is the length of the longer side of the rectangular aperture. *Sciammarella* and *Chawla* [7.62] have suggested that working with small magnifications is more advantageous, as it increases the tolerance on the focusing of the localization planes.

Figure 7.12a shows an example of a fringe pattern obtained on a thin aluminum sheet with a hole in it and loaded in uniaxial tension. Figure 7.12b exhibits another example obtained on a rubber membrane. The contrast of the moiré fringes has been enhanced by optically filtering the carrier fringes.

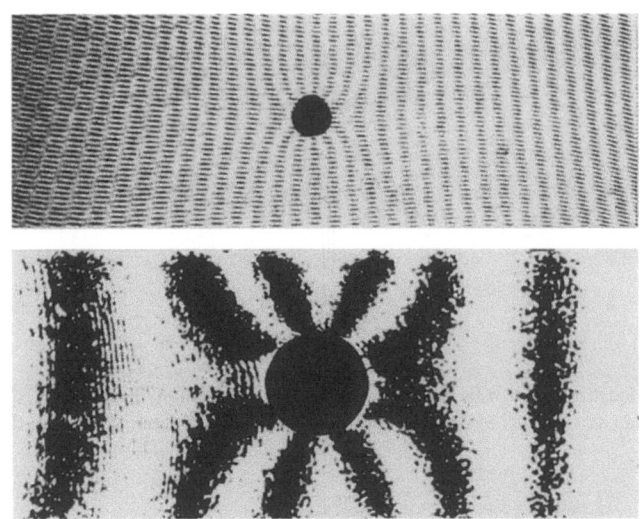

Fig. 7.12. (a) An example of fringe pattern corresponding to horizontal displacements in an aluminum sheet with a hole under tension. (b) Idem for a rubber membrane. The contrast of the moiré fringes has been improved by optically filtering the carrier fringes

Fig. 7.13. An example of in-plane displacement fringes along x direction obtained in the study of the mechanical response of an external fixation device

Figure 7.13 presents an example of a fringe pattern obtained in the study of the mechanical behaviour of an external fixation device widely used in the healing of bone fractures. These fringe patterns correspond to the L_x displacement component.

7.2.2 Reconstructed Reference Wave Holographic Interferometry

Katzir and *Glaser* [7.71] demonstrated an alternative approach of measuring in-plane displacements using an image plane holographic set-up, as shown in Fig. 7.14. The object is illuminated by two beams, oriented symmetrically with respect to the surface normal. The diffracted waves are recorded on a holographic plate using a collimated reference beam. The holographic plate is developed and repositioned in its initial position. The readout of the hologram is performed by lighting the object with its two illumination beams and observing the hologram plane in the direction of the reference wave. The reference beam is blocked during the readout step. By calculating the amplitude transmittance of

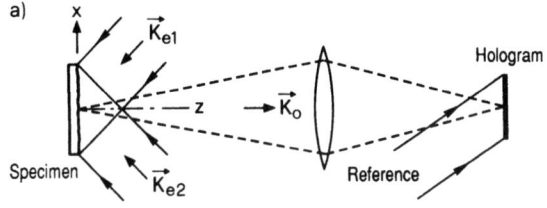

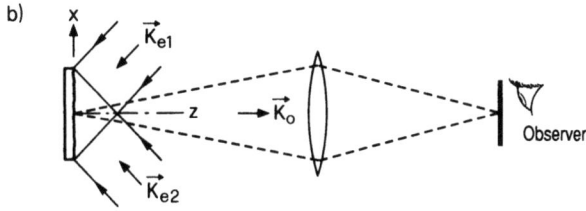

Fig. 7.14a,b. Schematic of the (a) recording and (b) reconstruction arrangements of the reconstructed reference wave holographic interferometry

the hologram and multiplying it by the sum of the amplitudes of the object beams, the term of interest corresponding to the amplitude of the reconstructed reference wave is

$$A_c = (|A_1|^2 + |A_2|^2)A_r, \tag{7.29}$$

where A_r denotes the amplitude of the reference wave in the hologram plane. The term (7.29) represents a wave identical to the reference wave.

The object is now deformed. The corresponding amplitudes A'_1 and A'_2 after deformation are given by (7.14a, b). The reference wave presently emerging from the hologram can be expressed as

$$A'_c = (A_1^* A'_1 + A_2^* A'_2)A_r. \tag{7.30}$$

Equation (7.30) shows that the fringes arising from the terms within the bracket modulate the reference wave. Ignoring for simplicity the reference and object intensity terms, the intensity distribution of the reconstructed reference wave can be shown to be proportional to

$$|A'_c|^2 \propto \cos^2\left(\frac{K_{e1} - K_{e2}}{2}\right) \cdot L. \tag{7.31}$$

This expression reveals that the reconstructed reference wave is modulated by fringes of equal in-plane displacement. Each fringe corresponds to an incremental in-plane displacement of $\lambda/2 \sin \theta$.

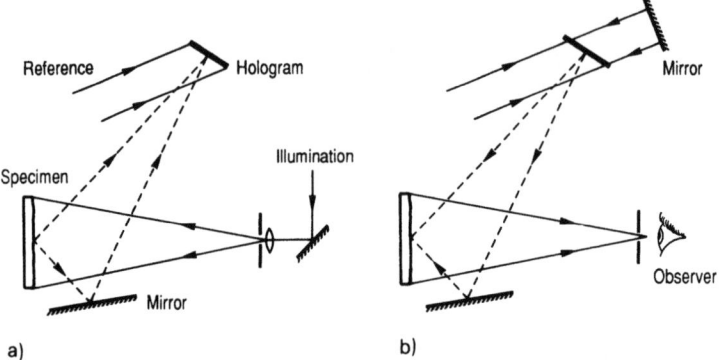

Fig. 7.15a, b. Schematic of the (a) recording and (b) reconstruction arrangements of the conjugate wave holographic interferometry

7.2.3 Conjugate-Wave Holographic Interferometry

Pirodda [7.73] has demonstrated a novel approach capable of measuring in-plane displacements of a diffuse object surface. The approach consists of reconstructing the illumination wave by projecting on the object its real image. The real image of the object is obtained by illuminating the hologram by a plane wave travelling anti-parallel to the original reference beam.

The optical set up (Fig. 7.15) consists of illuminating the object with a plane or slightly diverging wave in a direction normal to the object surface. The object is viewed holographically along two directions at equal angles to the surface normal. A plane reference beam strikes the holographic plate. A hologram of the object in its initial state is recorded and after processing replaced in its original position. The illumination beam is turned off and the hologram is lit by a beam travelling in a direction opposite to the original reference beam. The reconstructed waves propagate along the same path from which the two scattered waves emerged toward the hologram. These two waves form two real images of the object upon itself. The light the object scatters reconstructs the conjugate of the illuminating wave. The illumination source is removed and the object is now deformed. An observer looking through an aperture placed just in front of the illumination source sees a system of interference fringes modulating the object surface. These fringes contour path length variations related to an in-plane component of the displacement vector. The measurement is independent of the out-of-plane component of displacement.

Let us suppose that the object is illuminated by a laser beam travelling in the z direction normal to its surface. The observation is carried along two directions symmetrical to the surface normal and making an equal angle θ_0 with the z axis. The beams are supposed to lie in the x–z plane. The amplitudes of the two

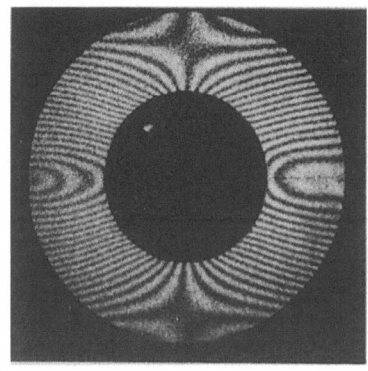 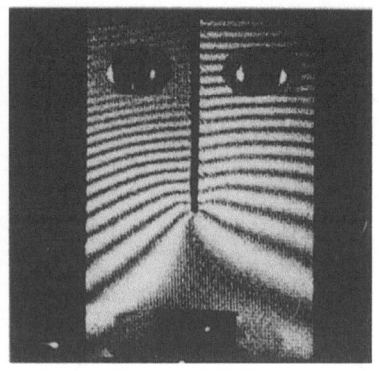

Fig. 7.16a, b. Examples of in-plane fringe contours L_x produced on (**a**) a ring subjected to a diametral compression and (**b**) a notched carbon fiber composite plate loaded in its plane by forces tending to open the notch. (Courtesy of L. Pirodda)

scattered waves contributing to fringe formation are given by

$$A'_1 = \exp\left\{i\frac{2\pi}{\lambda}[L_x \sin\theta_0 + L_z(1+\cos\theta_0)]\right\}, \tag{7.32a}$$

$$A'_2 = \exp\left\{i\frac{2\pi}{\lambda}[-L_x \sin\theta_0 + L_z(1+\cos\theta_0)]\right\}. \tag{7.32b}$$

The mean intensity in the image is given by the relation

$$\langle I(P)\rangle = \langle(A'_1 + A'_2)(A'^*_1 + A'^*_2)\rangle, \tag{7.33a}$$

which yields for the irradiance of the fringe pattern

$$\langle I(P)\rangle = 2\left[1 + \cos\left(\frac{4\pi}{\lambda}L_x\sin\theta_0\right)\right]. \tag{7.33b}$$

The in-plane displacement L_x between two consecutive fringes is given by $\lambda/2\sin\theta_0$. Figure 7.16 shows examples of fringe patterns corresponding to horizontal displacements obtained on (a) a ring subjected to diametral compression and (b) a notched carbon fiber composite plate loaded in its plane by forces tending to open the notch.

7.2.4 Phase-Shifted Holographic Moiré

Rastogi and *Denarié* [7.76] have reported the use of phase shifting [7.83–89] and computer image-processing facilities to obtain a highly improved visualization of the phase distribution corresponding to the in-plane displacements of an object subjected to stress. A review of phase-shifting holographic

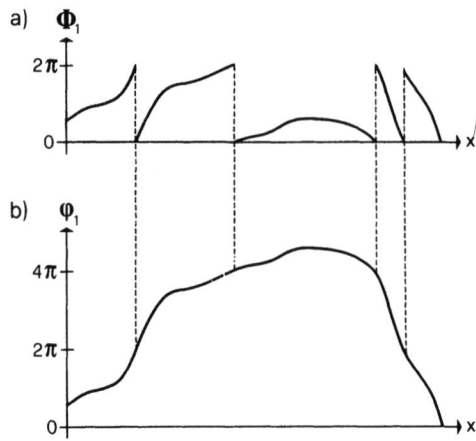

Fig. 7.17. (a) Modulo 2π phase Φ_1 calculated by phase shifting interferometry; (b) reconstruction of the actual phase φ_1

interferometry is presented in Chap. 5. Referring to (7.15), the moiré fringes are given by

$$\varphi_1(P) - \varphi_2(P) = 2n_m\pi,$$

where n_m is the moiré fringe number. A very useful way of implementing (7.15) is provided by the phase-shifting technique: given three or more digitized phase-shifted versions of an interferogram, the calculation of the fringe phase function φ_1 or φ_2 can be achieved within a few seconds with the help of a microcomputer. The operation of calculating a fringe phase function, say φ_1, is performed in two steps. The first step results in a modulo 2π function Φ_1, which is displayed as a saw-tooth, fringe-like intensity pattern, as shown in Fig. 7.17a. The saw-tooth pattern is free from contrast and background intensity variations. In the next step, Fig. 7.17b, φ_1 is reconstructed by means of a simple algorithm, provided that the sampling requirements are fulfilled during the digitization process.

The first step of the procedure, carried out separately for the interference patterns defined by (7.15a, b), leads to modulo 2π phase functions Φ_i:

$$\Phi_i = \varphi_i - 2\pi \operatorname{Int}\left(\frac{\varphi_i}{2\pi}\right), \tag{7.34}$$

where Int is the integer-part function, and i represents 1 and 2. These functions conceal enough information for one to be able to extract the holographic moiré order n_m. Since the moiré phase function φ is provided by the difference between the two phase sets φ_1 and φ_2, Φ the saw-tooth version of φ can be written as [7.90]

$$\Phi = \varphi_2 - \varphi_1 - 2\pi \operatorname{Int}\left(\frac{\varphi_2 - \varphi_1}{2\pi}\right). \tag{7.35}$$

7.2 Measurement of In-Plane Displacements

Replacing φ_1 and φ_2 with the help of (7.34) yields

$$\Phi = \Phi_2 - \Phi_1 + 2\pi G, \tag{7.36}$$

where

$$G = \text{Int}\left(\frac{\varphi_2}{2\pi}\right) - \text{Int}\left(\frac{\varphi_1}{2\pi}\right) - \text{Int}\left(\frac{\varphi_2 - \varphi_1}{2\pi}\right). \tag{7.37}$$

It follows

$$G = \begin{cases} 1 & \text{if } \Phi_2 - \Phi_1 < 0, \\ 0 & \text{otherwise.} \end{cases}$$

Figure 7.18 presents a graphical description of the above process. The phase functions $\varphi_{1,2}$ have been chosen to be linear, for the sake of simplicity. Detecting pixels where $G = 1$ and subsequently adding 2π to $\varphi_2 - \varphi_1$ can then be achieved with an elementary routine. The result is the modulo 2π phase function Φ.

The schematic of the optical set-up is similar to that in Fig. 7.8 with the exception that the reference beam is reflected from a mirror driven by a piezoelectric transducer (PZT). Three images are recorded and digitized using the phase-stepping technique. The phase φ_i is calculated pixel by pixel with the help of the following algorithm:

$$\varphi_i = \tan^{-1}\frac{\sqrt{3}(I'_1 - I'_3)}{2I'_2 - I'_1 - I'_3},$$

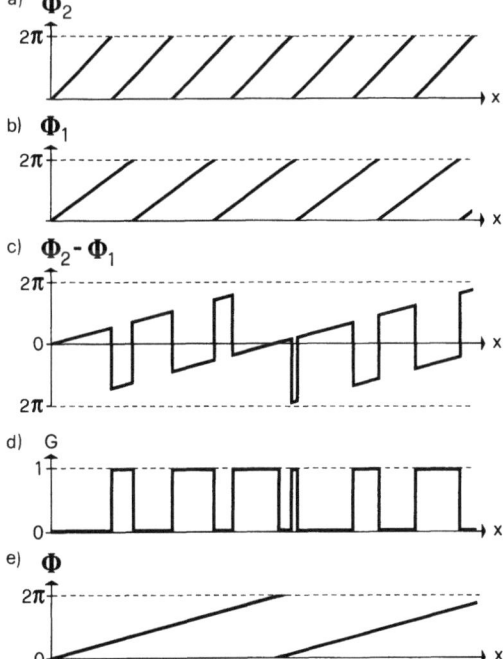

Fig. 7.18a–e. Schematic of the method used to obtain the phase function Φ: (a) modulo 2π phase distribution Φ_2; (b) modulo 2π phase distribution Φ_1; (c) difference between (a) and (b); (d) variation of G; and (e) reconstruction of Φ obtained from (c) and (d)

Fig. 7.19. An example of moiré phase distribution corresponding to in-plane displacement L_x

where I'_1, I'_2, and I'_3 are the intensities corresponding to the phase shifts of $-2\pi/3, 0$, and $2\pi/3$, respectively. The result is the modulo 2π phase map of the original interferogram.

Figure 7.19 shows an example of the moiré phase distribution corresponding to in-plane displacements. This pattern type, being no different from the one obtainable from the usual phase shifting procedure, naturally enjoys all the possibilities inherent in the phase shifting interferometry for analysis and quantitative measurements. The loading consists of splitting a notched fiber-reinforced concrete specimen by means of a wedge pressed between rollers that are placed on the top of the tested specimen.

The display of results in a pseudocolor representation is both attractive and significant from a user's point of view. Plate 7.1(a) reveals a reconstruction of the true phase corresponding to the in-plane displacement L_x. As an example of data analysis, Plate 7.1(b) shows a three-dimensional plot of the in-plane phase distribution obtained from the reconstruction shown in Plate 7.1(a).

7.3 Measurement of the Derivatives of Displacements

In spite of being equipped with the ability to address opaque objects in real time and on a whole field basis, until recently holographic interferometry showed little promise to address those engineering applications which require an experimental study of flexural deformations of structures such as plates and beams. This shortcoming arises from the fact that, whereas holographic interferometry records displacements, the study of flexural deformations needs a computation of their derivatives. Similarly, another significant physical quantity which has yet to find a promising solution using holographic interferometry is in-plane

7.3 Measurement of the Derivatives of Displacements

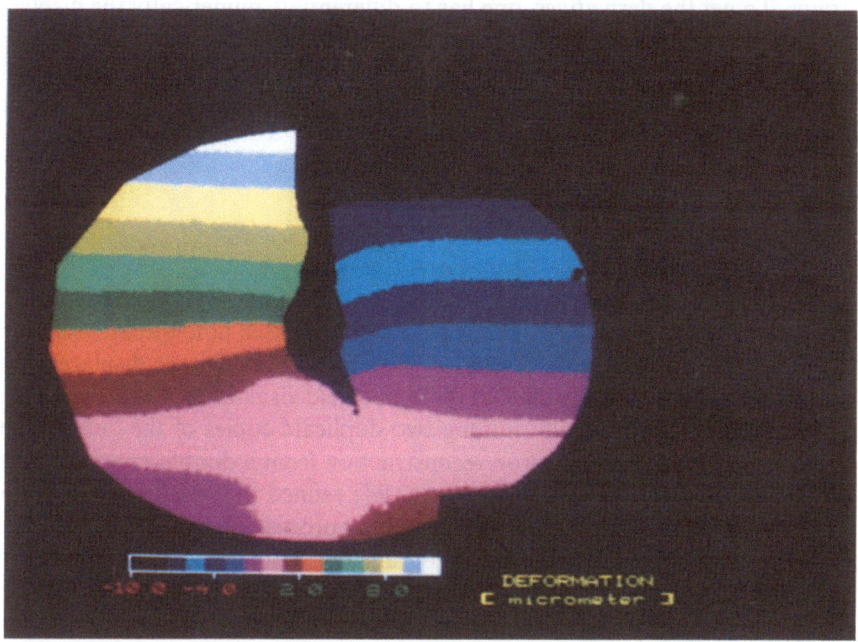

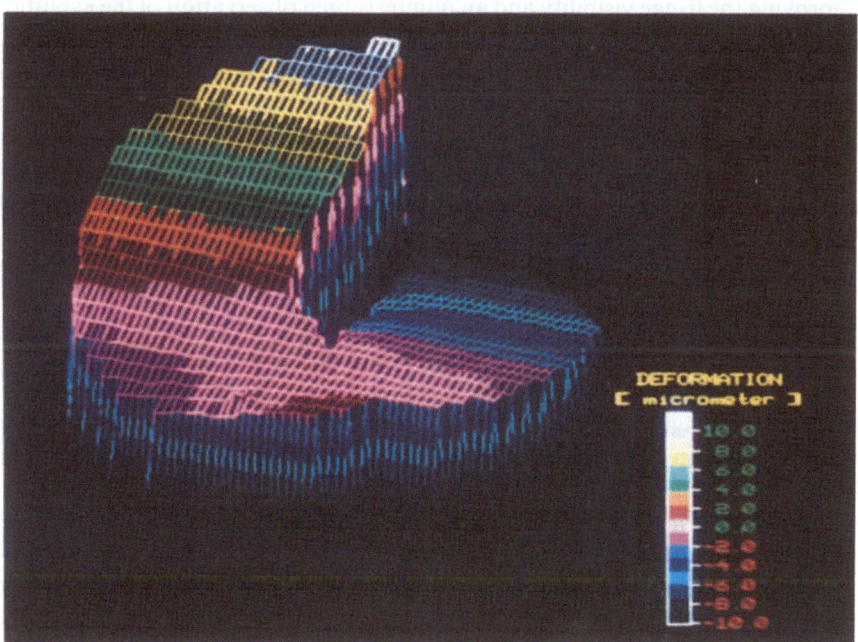

Plate 7.1. Examples of results in pseudocolor representation. **a** Reconstruction of the true in-plane phase distribution; **b** its corresponding three-dimensional plot

strains. To get the derivatives, one has to differentiate numerically the displacement pattern at each point on the object surface. The measurement of curvature (second derivatives of the deflection) presents a major interest, since it is directly related to the bending moment [7.91] and, in consequence, to the stresses induced in the loaded structural element.

7.3.1 Measurement of Slope Change and Curvature

Stetson [7.92] was the first to demonstrate a moiré-based holographic technique to obtain information on the second derivatives of out-of-plane displacements. *Boone* and *Verbiest* [7.93] obtained the whole field visualization of slope change by superposing and mutually shifting two duplicate copies of the out-of-plane displacement fringes obtained on reconstruction from a double-exposure holographic interferometry set-up. *Cadoret* [7.94] refined the technique further by proposing an optical superposition and shifting process. His proposal consists of reconstructing a double-exposure hologram by illuminating it with two mutually incoherent reference waves, making an angle between them. However, in both these techniques the displacement information for a given loading is first frozen and the derivation is achieved as a second step only. The possibility of improving the fringe visibility and an uninterrupted observation of the evolution of the slope distribution with increasing loads is thus not possible. Holographic moiré [7.95, 96] alleviates these problems by enabling one to obtain in real time the patterns of slope-change contours by shearing wave fronts diffracted from the object surface. The formation of moiré fringes is obtained in a one-step process. The combination of holographic moiré and phase shifting techniques [7.97] has been reported to determine contours of constant values of the second derivatives of displacements.

7.3.1.1 Holographic Moiré for Slope-Change Measurement

A schematic diagram illustrating the principle used for slope change measurement is shown in Fig. 7.20. A flat object is illuminated by means of a coherent beam. An optical device interposed in the observation arm of the interferometer is adjusted to bring two adjacent points P_1 and P_2, symmetrically situated to P, to contribute to a single point in the image plane. The points P_1 and P_2 are described by the position vectors $r_1 - \delta r_1/2$ and $r_1 + \delta r_1/2$ referred to the origin of the coordinate system (Fig. 7.20). The points P_1 and P_2 are each illuminated and observed along propagation vectors (K_{e11}, K_{o11}) and (K_{e12}, K_{o12}), respectively. The broken lines show the mean illumination and observation directions (K'_{e1}, K'_{o1}) of the element P_1 and P_2. If the object now deforms such that the

7.3 Measurement of the Derivatives of Displacements

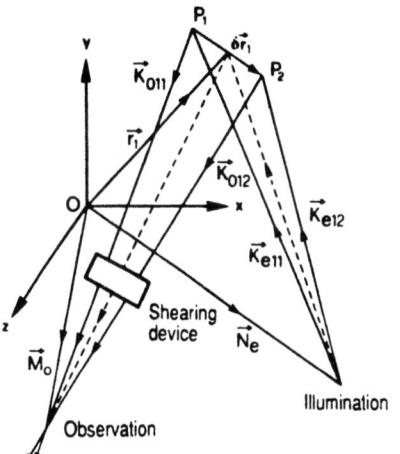

Fig. 7.20. Illumination and observation geometry in holographic moiré for slope measurement

point $P(r_1)$ undergoes a displacement $L(r_1)$ and its neighbouring points P_1 and P_2 move through displacements given by the vectors $L\left(r_1 - \dfrac{\delta r_1}{2}\right)$ and $L\left(r_1 + \dfrac{\delta r_1}{2}\right)$

$$L\left(r_1 - \frac{\delta r_1}{2}\right) = L(r_1) - \frac{\delta L}{2}, \tag{7.38a}$$

$$L\left(r_1 + \frac{\delta r_1}{2}\right) = L(r_1) + \frac{\delta L}{2}, \tag{7.38b}$$

where δL is the vector representing the relative increment in displacements between the points P_1 and P_2

$$\delta L = \hat{i}\delta L_x + \hat{j}\delta L_y + \hat{k}\delta L_z. \tag{7.39}$$

The wave fields now observed through the hologram are composed of the wave fields scattered by the pair of points P_1 and P_2 in their initial and final states. The phase differences introduced at P_1 and P_2 due to object deformation are given by

$$\varphi_1 = (K_{e12} - K_{o12}) \cdot L\left(r_1 + \frac{\delta r_1}{2}\right), \tag{7.40a}$$

$$\varphi_2 = (K_{e11} - K_{o11}) \cdot L\left(r_1 - \frac{\delta r_1}{2}\right). \tag{7.40b}$$

The reconstructed image displays a slowly varying fringe modulation related to the difference in phase $\varphi_1 - \varphi_2$. The beat arises from the interaction of the interference patterns formed due to wave front pairs scattered by points P_1 and P_2. The corresponding moiré is governed by the phase equation:

$$(K'_{e1} - K'_{o1}) \cdot (\delta r_1 \cdot \nabla) L(r_1) + (\Delta K_{e1} - \Delta K_{o1}) \cdot L(r_1) = 2 n_m \pi, \tag{7.41}$$

where we have made use of the Taylor series expansion of the displacement-vector difference to first-order terms; ΔK_{e1} and ΔK_{o1} represent the vector increments between the illumination and observation-vector pairs (K_{e12}, K_{e11}) and (K_{o12}, K_{o11}), respectively. Expression (7.41) relates the derivatives of displacement to the moiré fringes for any shearing, illumination and observation geometry. The method is sensitive to the displacement components as well. The mean illumination and observation vectors are given by

$$K'_{e1} = k \frac{r_1 - N_e}{|r_1 - N_e|}; \quad K'_{o1} = k \frac{M_o - r_1}{|M_o - r_1|}. \tag{7.42}$$

The vector increments ΔK_{e1} and ΔK_{o1} can be represented as

$$\Delta K_{e1} = \frac{k}{\rho_{e1}} \delta r_1 - \frac{\delta r_1 \cdot K_{e11}}{\rho_{e1}} K_{e11}, \tag{7.43a}$$

$$\Delta K_{o1} = \frac{\delta r_1 \cdot K_{o11}}{\rho_{o1}} K_{o11} - \frac{k}{\rho_{o1}} \delta r_1. \tag{7.43b}$$

The corresponding resolved components follow as

$$\Delta K_{e1x} = \frac{\Delta x_1 - \Delta \rho_{e1} K_{e11x}}{\rho_{e1}} \quad \Delta K_{o1x} = \frac{\Delta \rho_{o1} K_{o11x} - \Delta x_1}{\rho_{e1}},$$

$$\Delta K_{e1y} = \frac{\Delta y_1 - \Delta \rho_{e1} K_{e11y}}{\rho_{e1}} \quad \Delta K_{e1y} = \frac{\Delta \rho_{o1} K_{o11y} - \Delta y_1}{\rho_{e1}}, \tag{7.44}$$

$$\Delta K_{e1z} = \frac{-\Delta \rho_{e1} K_{e11z}}{\rho_{e1}} \quad \Delta K_{e1z} = \frac{\Delta \rho_{o1} K_{o11z}}{\rho_{e1}},$$

where $\Delta \rho_{e1}$ and $\Delta \rho_{o1}$ are the respective increments in the illumination and observation directions induced by shearing δr_1. Knowledge of (7.42, 43) allows for calculating explicitly (7.41) and thereby the fringe dependence for any specific recording and observation geometry. If we define K_{s1} and K_{s2} as the sensitivity vectors lying along the direction bisecting the illumination and viewing directions at the points P_1 and P_2, i.e.,

$$K_{s1} = K_{e11} - K_{o11}$$

$$K_{s2} = K_{e12} - K_{o12}$$

(7.41) takes the form

$$K_s \cdot (\delta r_1 \cdot \nabla) L(r) + \Delta K_s \cdot L(r) = 2 n_m \pi, \tag{7.45}$$

7.3 Measurement of the Derivatives of Displacements

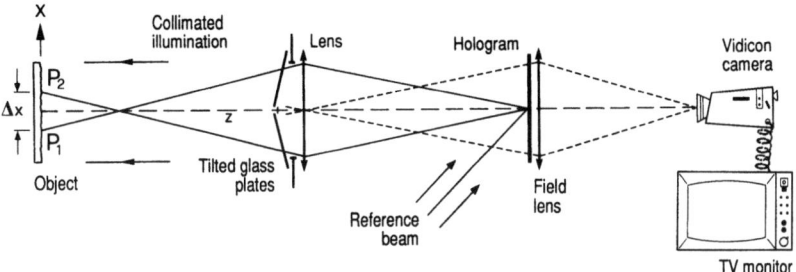

Fig. 7.21. Schematic of the real-time holographic moiré set-up for the observation of the slope change. The optical elements leading the collimated light on the object surface are not shown in the figure

where $\Delta K_s = K_{s2} - K_{s1}$; or, in terms of Cartesian components

$$\Delta K_s = k[\hat{i}(\Delta K_{e1x} - \Delta K_{o1x}) + \hat{j}(\Delta K_{e1y} - \Delta K_{o1y})$$
$$+ \hat{k}(\Delta K_{e1z} - \Delta K_{o1z})]. \tag{7.46}$$

Equation (7.45) leads to a useful moiré-fringe interpretation. The observed moiré-fringe pattern yields information relating to the projection of the change in the displacement vector $L(r)$ – corresponding to a change in position δr_1 – on the sensitivity vector plus the projection of the displacement vector $L(r)$ on the vector representing the difference between the sensitivity vector pair (K_{s1}, K_{s2}) corresponding to a change in position δr_1. Fortunately, in most practical set-ups the contribution of this latter term to fringe formation is quite small and could be eliminated by illuminating the object by a parallel beam and viewing it by means of a telecentric system. Ignoring this term, (7.45) becomes

$$K_{s1} \cdot (\delta r_1 \cdot \nabla) L(r) = 2n_m \pi. \tag{7.47}$$

Since in holographic interferometry it is more convenient to point the sensitivity vector in a direction normal to the object surface, the measurement of the derivatives of the out-of-plane movements is more immediate as compared to the measurement of the derivatives of in-plane displacements.

The schematic of an optical set-up is depicted in Fig. 7.21 where a plane object is imaged on the holographic plate by means of a lens. The object is illuminated along a direction normal to the object surface. Two similar, thick, parallel, glass plates are placed inclined in front of the lens in such a way that each plate covers half of the lens aperture. This gives rise to two laterally-sheared speckle modulated images along the x direction. The magnitude of shearing Δx_1 can be varied by changing the angle between the plates. The reconstructed image is observed on a TV monitor. If the observation is carried by means of a telecentric system, (7.47) reduces to

$$2\Delta x_1 \frac{\partial L_z}{\partial x} = n_m \lambda. \tag{7.48}$$

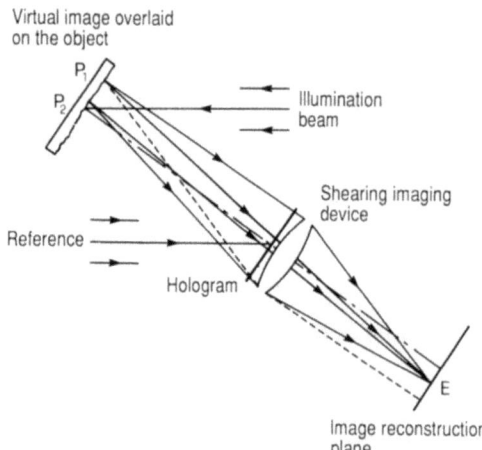

Fig. 7.22. Layout of the optical system incorporated with the facility to vary in live the measurement sensitivity and slope direction

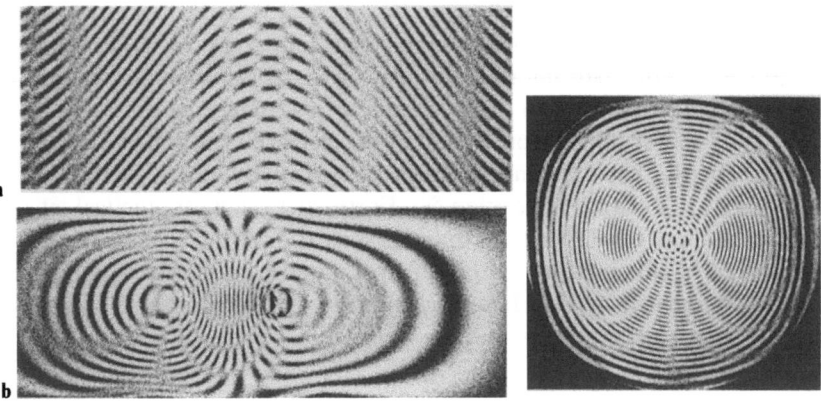

Fig. 7.23a–c. Examples of fringe contours corresponding to $\partial L_z/\partial x$ produced on (**a**) an aluminum beam fixed at both ends and deflected at its center, (**b**) the specimen shown in Fig. 4a, loaded by releasing the bolt and deflecting the beam at a point on the left of the bolt and (**c**) a centrally loaded circular aluminum plate clamped along its boundary

The inclined glass-plate shearing device suffers from the drawback of shear variation over the object surface and introduces aberrations which make it only worthwhile for studying small objects. The advantages of the method can be extended to include the possibility to change the direction of derivation and fringe contour sensitivity during the same experiment. The main hindrance in providing freedom to vary the sensitivity and direction of the slopes comes from the difficulty felt in changing the magnitude and direction of δr_1 during an experiment. The difficulty is linked to the fact that the shearing mechanism is incorporated in the recording process. This makes any a posteriori manipulation of the vector δr_1 impossible. However, by delinking the wave-front shearing

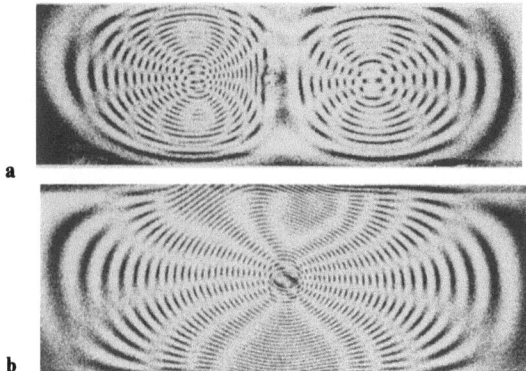

Fig. 7.24a, b. Examples of fringe contours corresponding to $\partial L_z/\partial y$ produced on (a) the specimen shown in Fig. 4a submitted to two point loading one on each side of the bolt and (b) a rectangular aluminum plate clamped along the edges and drawn out at the center by means of a bolt. The loading is introduced by slightly releasing the bolt

from the recording stage but still incorporating it in the image reconstruction, it becomes possible to render to the holographic slope measuring set-up the option to change at will the sensitivity and the direction of slopes. A schematic of the recording and reconstruction of the hologram is shown in Fig. 7.22. A split lens is placed behind the hologram. The light contribution to the reconstructed field at a point E in the image plane comes from the neighbouring points P_1 and P_2 in the object plane. The capability of the split lens to change in permanence the direction and magnitude of δr_1 not only allows to vary, during each loading step, the slope direction but also the method's sensitivity in function of the object's response to the applied load.

Examples of fringe contours corresponding to slope distribution $\partial L_z/\partial x$ are shown in Fig. 7.23. Figure 7.23a illustrates the case of an aluminum beam fixed at both ends and deflected at its center. Figure 7.23b corresponds to the specimen shown in Fig. 7.4a with the difference that the loading now consists of releasing the bolt and deflecting the plate at a point on the left of the bolt. Figure 7.23c depicts the case of a centrally-loaded square aluminum plate clamped along its boundary. Examples of fringe contours corresponding to slope distribution $\partial L_z/\partial y$ are exhibited in Fig. 7.24. The model and loading type in Fig. 7.24a are similar to that shown in Fig. 7.4a. Figure 7.24b depicts the case of a rectangular aluminum plate clamped along the edges and drawn out at the center by means of a bolt. The bolt is slightly released after recording the hologram.

7.3.1.2 Measurement of Curvature

Schematic displaying the underlying principle used for the measurement of second-order derivatives is displayed in Fig. 7.25. The method employed for wave-front shearing is used in its more sophisticated form to obtain the second derivatives of displacements. An image-shearing device placed behind the holographic plate enables the observation of a point on the object along two distinct neighbouring directions. A lateral wave-front shearing ensues and, as a result,

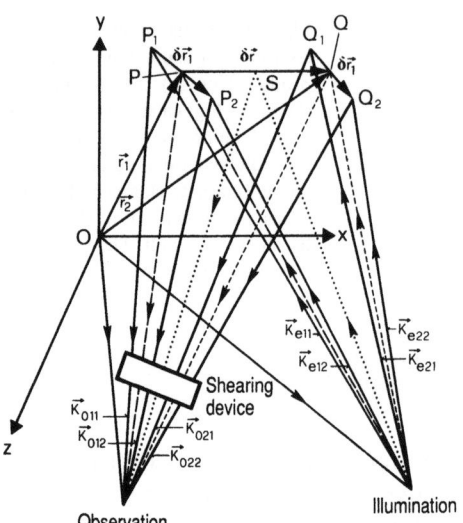

Fig. 7.25. Schematic diagram of the underlying principle used for fringe formation corresponding to slope and curvature. The object is assumed to be in its initial stage

brings two adjacent points P_1 and P_2 on the object plane to coincide in the image plane. The amplitude distribution in the image plane results from the superposition of the wave fronts diffracted from the points P_1 and P_2. Let us now go a step further and envision the introduction of another shearing, whose role is to induce a lateral translation of the ensemble of the pairs of mutually shifted images. As a result of this displacement, two adjacently situated points Q_1 and Q_2, mutually shifted by δr_1, are brought to contribute to the same image point as the pair (P_1, P_2) in the image plane. This contribution is, however, sought on a sequential and not on a simultaneous basis. The points P and Q are laterally shifted by δr. It is evident that the points P and Q on the object could be restored from the final image by tracing back along their pair of viewing directions to their respective intersections. The points Q_1 and Q_2 are described by position vectors $r_2 - \delta r_1/2$ and $r_2 + \delta r_1/2$. The points Q_1 and Q_2 are each illuminated and observed along propagation vectors (K_{e21}, K_{o21}) and (K_{e22}, K_{o22}), respectively. The broken lines (short dashes) show the mean illumination and observation directions (K'_{e2}, K'_{o2}) of the element $Q_1 Q_2$.

The pair of points P and Q are situated symmetrically to S. The dotted lines show the mean illumination and observation directions (K'_e, K'_o) of the element PQ. The relative phase change due to the displacement between points Q_1 and Q_2 is given by

$$\varphi' = (K'_{e2} - K'_{o2}) \cdot (\delta r_1 \cdot \nabla) L(r_2) + (\Delta K_{e2} - \Delta K_{o2}) \cdot L(r_2), \qquad (7.49)$$

where ΔK_{e2} and ΔK_{o2} represent the vector increments between the illumination and observation vector pairs (K_{e12}, K_{e11}) and (K_{o12}, K_{o11}), respectively. Responses to deformation of the two elements centered at P and Q are respectively

7.3 Measurement of the Derivatives of Displacements

given by (7.41 and 49). The calculation of the relative phase change between the two elements leads to the phase equation

$$(K'_e - K'_o) \cdot (\delta r_1 \cdot \nabla)(\delta r \cdot \nabla) L(r_1) + (\Delta K'_e - \Delta K'_o) \cdot (\delta r_1 \cdot \nabla) L(r_1) +$$
$$\frac{\Delta K'_e - \Delta K'_o}{2} \cdot (\delta r_1 \cdot \nabla)(\delta r \cdot \nabla) L(r_1) + (\Delta K_{e1} - \Delta K_{o1}) \cdot (\delta r \cdot \nabla) L(r_1) = 2 N_m \pi,$$
(7.50)

where N_m is the second-order moiré fringe number and, $\Delta K'_e$ and $\Delta K'_o$ are the vector increments between the illumination and observation vector pairs (K'_{e2}, K'_{e1}) and (K'_{o2}, K'_{o1}), respectively. Equation (7.50) relates the moiré fringes to the second derivative of displacements for any shearing, illumination, and observation geometry. This relationship displays the influence of divergent-beam illumination and field angles, and explains the part of each first-order and second-order displacement derivative component to fringe formation. An examination of these splitted parts permits one to assess the effectiveness and scope of the technique in a particular measurement problem.

The vector increments $\Delta K'_e$ and $\Delta K'_o$ can be expressed by the relations

$$\Delta K'_e = \frac{k}{\rho_e} \delta r - \frac{\delta r \cdot K'_{e1}}{\rho_e} K'_{e1}, \tag{7.51a}$$

$$\Delta K'_o = \frac{\delta r \cdot K'_{o1}}{\rho_o} K'_{o1} - \frac{k}{\rho_o} \delta r. \tag{7.51b}$$

The corresponding resolved components follow as

$$\Delta K'_{ex} = \frac{\Delta x - \Delta \rho_e K'_{e1x}}{\rho_e} \qquad \Delta K'_{ox} = \frac{\Delta \rho_o K'_{o1x} - \Delta x}{\rho_o},$$

$$\Delta K'_{ey} = \frac{\Delta y - \Delta \rho_e K'_{e1y}}{\rho_e} \qquad \Delta K'_{oy} = \frac{\Delta \rho_o K'_{o1y} - \Delta y}{\rho_o}, \tag{7.52}$$

$$\Delta K'_{ez} = \frac{-\Delta \rho_e K'_{e1z}}{\rho_e} \qquad \Delta K'_{oz} = \frac{\Delta \rho_o K'_{o1z}}{\rho_o},$$

where $\Delta \rho_e$ and $\Delta \rho_o$ are the respective increments in the illumination and observation directions induced by shearing δr. Equations (7.44, 52) show that the influence of ΔK terms in (7.50) could be eliminated for objects illuminated and observed by means of a collimated beam and a telecentric system respectively. In such a case (7.50) reduces to

$$(K'_e - K'_o) \cdot (\delta r_1 \cdot \nabla)(\delta r \cdot \nabla) L(r_1) = 2 N_m \pi. \tag{7.53}$$

The observed fringe pattern gives information relative to the projection of the change in the vector $\delta r_1 \cdot \nabla L$ (corresponding to a change δr in position) on the sensitivity vector defined by $K'_e - K'_o$. ∇L is a dyadic. If the object is illuminated and observed along a direction normal to its surface, (7.53) becomes independent of the in-plane derivative terms. The fringes can now be interpreted as the rate of change of slope, $\delta r_1 \cdot \nabla L_z$, in the direction δr.

Equation (7.53) enables the observation of the contours of constant values of the second-order derivatives with respect to the direction of shearing. If one choses $\delta r_1 = i\Delta x_1$ and $\delta r = i\Delta x$, (7.53) reduces to

$$\frac{\partial^2 L_z}{\partial x^2} = \frac{N_m \lambda}{2\Delta x_1 \Delta x}. \tag{7.54a}$$

Choosing $\delta r_1 = \hat{j}\Delta y_1$ and $\delta r = \hat{j}\Delta y$ provides fringes corresponding to

$$\frac{\partial^2 L_z}{\partial y^2} = \frac{N_m \lambda}{2\Delta y_1 \Delta y}. \tag{7.54b}$$

The patterns (7.54a, b) show the family of fringes corresponding to whole field curvature distributions along x and y directions, respectively. Twist, defined as the rate of change in slope in a direction orthogonal to that of slope, is obtained by choosing $\delta r_1 = \hat{j}\Delta y_1$ and $\delta r = \hat{i}\Delta x$,

$$\frac{\partial}{\partial x}\left(\frac{\partial L_z}{\partial y}\right) = \frac{N_m \lambda}{2\Delta y_1 \Delta x}. \tag{7.54c}$$

Equation (7.54c) displays the family of fringes corresponding to the contours of constant twist.

From the discussion it is obvious that the curvature contours are produced in the form of second-order moiré fringes obtained by the interference of a pair of fringe patterns corresponding to slope distributions. The fringes corresponding to the slope distribution themselves appear as additive moiré fringes modulating a pair of carrier fringes. The presence of several moiré patterns modulating the interference image makes the viewer unable to discern moiré contours corresponding to curvature. Hence a convenient way to obtain curvature fringe contours is to combine phase shifting technique with the concept described above.

The layout of the experimental arrangement is shown in Fig. 7.26. A shearing-image reconstruction device is placed behind the hologram. An observer reconstructing the hologram through the split lens receives wave fields from the sheared holographic virtual images of the object superposed on the wave fields scattered by the corresponding points on the object surface. A field lens is placed in the reconstructed image plane. The image of this plane is formed by means of another image-splitting device. A second field lens then enables one to obtain the final image on a television screen. By setting the second shearing device to give zero shift, $\delta r = 0$, the method enables the evolution of the slope fringes with increased object deformation to be observed. The phase-shifting technique is applied separately on the two interference patterns being reconstructed through

7.3 Measurement of the Derivatives of Displacements   247

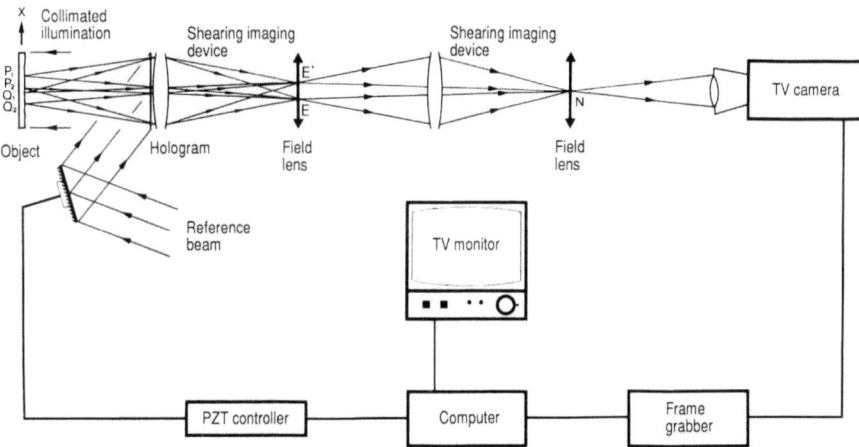

Fig. 7.26. Layout of the experimental arrangement to obtain fringes corresponding to the spatial derivatives of the out-of-plane displacement fields

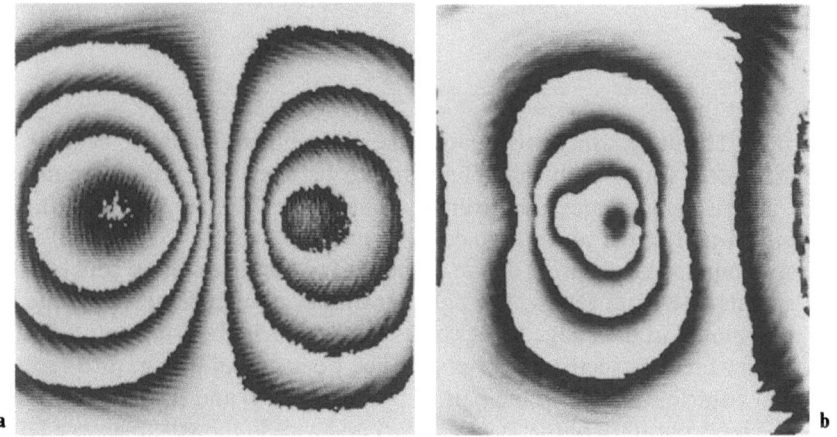

Fig. 7.27a, b. The phase maps representing contours of **(a)** constant slope change and **(b)** constant curvature along the x direction. The model is a square aluminum plate in which a flaw is simulated by drilling a small untraversing hole on its rear surface

each half of the split lens. The process gives rise to modulo 2π phase maps. The modulo 2π moiré phase is then computed from the pair of the phase maps stored in the microcomputer. One such example is shown in Fig. 7.27a. The model is an aluminum square plate 1 mm thick in which a flaw is simulated by drilling a small blind hole on its rear surface. The plate is clamped along the edges and submitted to centrally concentrated loads. The phase map corresponds to the contours of partial slopes. The presence of the flaw is only weakly detected in the

248   7. Techniques to Measure Displacements, Derivatives and Surface Shapes

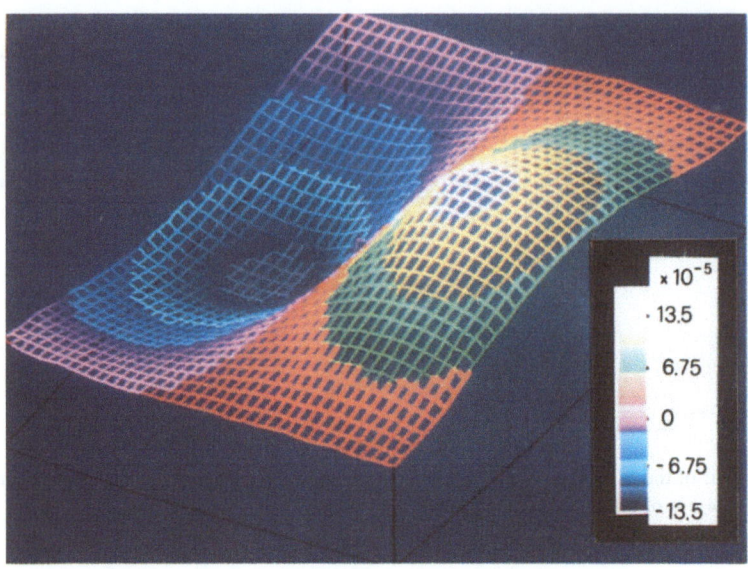

Plate 7.2. Three-dimensional plot of the phase distribution corresponding to $\partial L_z/\partial x$

phase map. The uncertainty surrounding the phase function relative to slopes is one fiftieth of a fringe or better. The distance between two adjacent contour planes in the slope pattern is given by 3.86×10^{-5}. Plate 2 gives a three-dimensional plot of the phase map corresponding to slopes.

The second image-shearing device is adjusted to obtain a shift δr between the two elements $P_1 P_2$ and $Q_1 Q_2$. The light contribution to the final reconstructed field at a point N in the image plane comes from the neighbouring points E and E′ in the first reconstructed image plane. The points E and E′ in turn receive their light contributions from the pair of points (P_1, P_2) and (Q_1, Q_2), respectively. The incorporation of the second shearing device in the set-up gives a large amount of flexibility to the system. The direction and magnitude of shearing in view of obtaining the second derivative can be varied at will. Using modulo 2π moiré fringe computed for the two elements, the modulo 2π second-order moiré fringe is computed in an analogous manner from the pair of slope phase maps stored in the microcomputer. An example is shown in Fig. 7.27b. The precision of the system is one twenty-fifth of a fringe or better for the measurement of phase function relative to curvature. The phase map represents contours of constant curvature with a contour interval of 5.78×10^{-6} mm^{-1}. The phase display, in contrast with that in Fig. 7.27a clearly and unambiguously outlines the presence and location of the flaw. The access to the isophase maps corresponding to curvature contours has considerably extended the flexibility of holographic interferometry in non-destructive inspection.

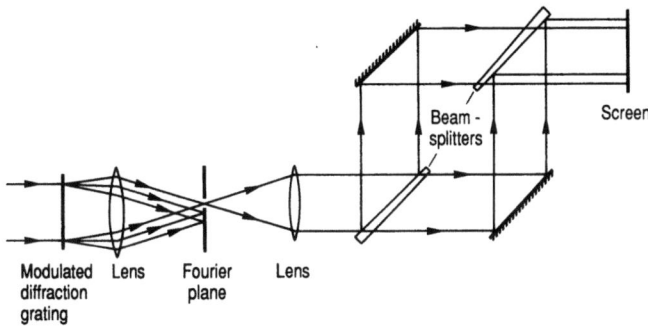

Fig. 7.28. Schematic of the optical arrangement for spatial filtering and for shearing inteferometry

7.3.1.3 Measurement of Slope and Curvature using Modulated Diffraction Gratings

The configuration of the optical set-up is similar to Fig. 7.3. The object is illuminated and observed in a direction normal to its surface. A double-exposure hologram of the object in its unstressed and stressed states is recorded. Before exposing the plate the second time, a linear system of carrier fringes is added to the deformation fringes by providing a slight tilt $\Delta\theta_e$ to the object beam. The reconstructed image is recorded on a photographic plate. The photographic transparency acts as a modulated diffraction grating [7.98]. Assuming unit magnification, the transmission function of the developed plate is given by

$$t_1(x, y) = \sum_{m=-\infty}^{\infty} b_m \exp\left[imk(2L_z + x \sin \Delta\theta_e)\right], \quad (7.55)$$

where b_m is the amplitude of the mth-order diffraction spectrum. The photographic transparency consists of a dense system of carrier fringes modulated by a term proportional to $L_z(x, y)$. This transparency is placed at the input plane of the set-up shown in Fig. 7.28 and illuminated with a plane wave. The image is filtered and its pth-diffracted component is led through a shearing interferometer in-built in the system. The shearing interferometer divides the incoming wave front into two laterally sheared wave fronts which are superposed on a screen. Using Taylor series expansion and neglecting the terms higher than the second order, the intensity observed on the screen is proportional to

$$I = c \cos^2\left(pk \frac{\partial L_z(x, y)}{\partial x} \Delta x_1\right), \quad (7.56)$$

where c is a constant, and Δx_1 represents the lateral shear between the two shifted images. Equation (7.56) represents the irradiance of the fringe pattern corresponding to the slope change. Fringe sensitivity is proportional to the

250   7. Techniques to Measure Displacements, Derivatives and Surface Shapes

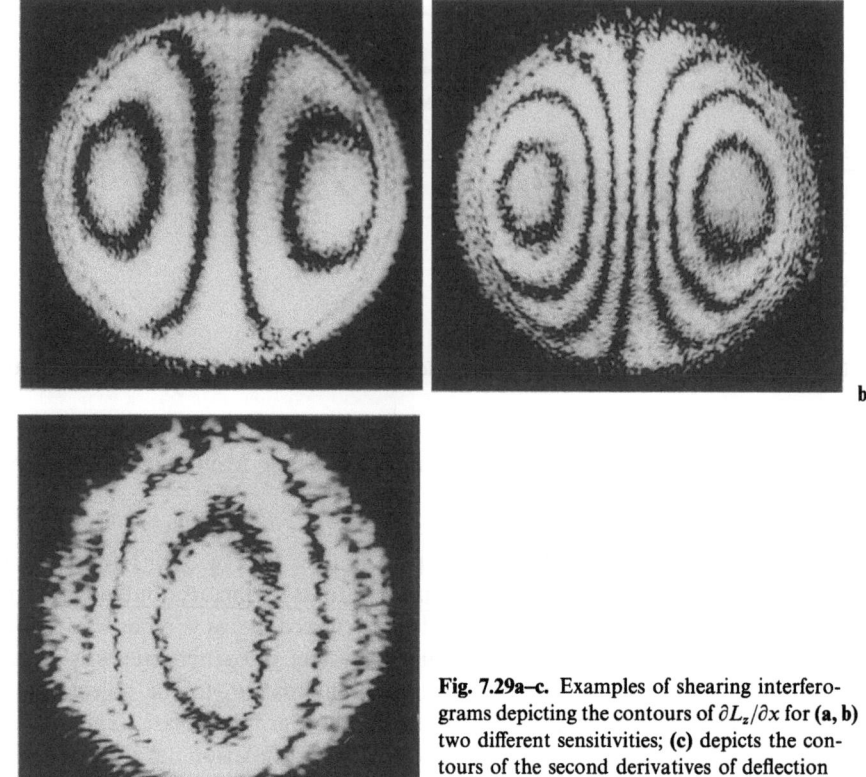

Fig. 7.29a–c. Examples of shearing interferograms depicting the contours of $\partial L_z/\partial x$ for (**a, b**) two different sensitivities; (**c**) depicts the contours of the second derivatives of deflection $\partial^2 L_z/\partial x^2$. (Courtesy of S. Toyooka)

diffraction order p. Examples of fringe patterns are shown in Fig. 7.29. Figures 7.29a, b depict the contours of the first derivatives of the out-of-plane displacements undergone by a uniformly loaded circular plate clamped along its edges. The fringe sensitivity is two times higher in Fig. 7.29b as compared to that shown in Fig. 7.29a.

Toyooka [7.98] proceeded along similar lines to obtain contours corresponding to the second derivatives of deflection. One of the wave fronts passing through the shearing interferometer is tilted by a small angle $\Delta \alpha_1$ by slightly rotating the beam splitter. This results in the addition of a linear system of carrier fringes to the slope-related fringes. The reconstructed image is recorded on a photographic plate. The transmission of the plate after development is given by

$$t_2(x,y) = \sum_{l=-\infty}^{\infty} d_l \exp\left\{ilk\left[2p\frac{\partial L_z}{\partial x}(\Delta x_1 + \varepsilon') + x\sin\Delta\alpha_1\right]\right\}, \quad (7.57)$$

where d_l is the amplitude of the lth-order diffraction spectrum, and ε' is the lateral shear introduced by the tilt of the beam-splitter. This transparency constitutes the second modulated diffraction grating. The grating is now placed at the input plane of the set-up shown in Fig. 7.28. The image is filtered and its qth diffracted component is led through the shearing interferometer producing a lateral shift of Δx between the two shifted images. The intensity of the qth-order diffracted wave observed on the exit screen is given by

$$I = c' \cos^2 pqk \frac{\partial^2 L_z}{\partial x^2}(\Delta x_1 + \varepsilon')\Delta x, \tag{7.58}$$

where c' is a constant. Equation (7.58) represents the irradiance of the fringe pattern corresponding to the second derivative of deflection. Figure 7.29c shows the contours of the second derivatives of deflection $\partial^2 L_z/\partial x^2$. Similarly, if one of the two wave fronts in the interferometer is sheared by Δy in the y direction, the intensity pattern on the screen changes to

$$I = c' \cos^2 pqk \frac{\partial^2 L_z}{\partial x \partial y}(\Delta x_1 + \varepsilon')\Delta y. \tag{7.59}$$

Equation (7.59) represents the contour lines of the twist. Fringe sensitivity in (7.58, 59) is proportional to the product pq.

7.3.1.4 Relationship Between the Second Derivatives, Moments and Stresses

On the assumption of small deformations and elastic behaviour, the relationships between the moments M_x, M_y, M_{xy} per unit length and the plate deflection L_z for a plate under transverse bending are given by [7.91]

$$M_x = -D\left(\frac{\partial^2 L_z}{\partial x^2} + v\frac{\partial^2 L_z}{\partial y^2}\right),$$

$$M_y = -D\left(\frac{\partial^2 L_z}{\partial y^2} + v\frac{\partial^2 L_z}{\partial x^2}\right),$$

$$M_{xy} = 2D(1-v^2)\frac{\partial^2 L_z}{\partial x \partial y},$$

where M_x and M_y are known as flexural moments and M_{xy} as torsional moments. The relationships between the moments M_i and stresses σ_i are given by

$$\sigma_i = \frac{6M_i}{h^2},$$

where i represents x, y or xy; h depicts the thickness and D the flexural rigidity of the plate whose Poisson's ratio is given by v. σ_x and σ_y, are known as bending stresses and σ_{xy} as twisting stresses on the plate. Considering the case of the bending of beams and supposing linearly elastic conditions, the relationship for M_x, for example is given by

$$M_x = D\frac{\partial^2 L_z}{\partial x^2}.$$

7.3.2 Measurement of In-Plane Strains

Studies by *Dubas* and *Schumann* [7.82, 99], and *Schumann* and *Dubas* [7.5] have amply demonstrated the dependence of the position of fringe localization on the derivatives of displacements. Localization conditions can be stated explicitly in terms of rotation and strain. This remarkable property has been used to deduce strains from the measurement of the fringe localization positions. *Stetson* [7.100–102], and *Pryputniewicz* and *Stetson* [7.103], and *Pryputniewicz* [7.104, 105] presented methods based on fringe-vector analysis for the determination of strains of three-dimensional objects. *Charmet* and *Montel* [7.106], and *Ebbeni* and *Charmet* [7.107] proposed the use of contrast measurements to determine the in-plane strain state of the object surface. *Sciammarella* and *Gilbert* [7.108] combined the multiple-hologram technique and the least-squares method [7.109, 110] to measure the strain components. *Sciammarella* and *Chawla* [7.60–62], and *Gilbert* [7.63] presented methods based on the set-up described in Sect. 7.2.1. Two densified shifted patterns are obtained corresponding to each of the illumination beams. These patterns are spatially filtered and the fringe order relative to strain is thereupon computed by subtracting the fringe orders of the individual patterns at each point on the object surface. Based on the superposition and shifting of identical fringe patterns, *Rastogi* et al. [7.26] suggested another approach, which consists of first producing a multiplicative holographic moiré pattern corresponding to in-plane displacements. In-plane strain fringe contours are displayed on the video monitor. *Toyooka* [7.111] devised a method using a holographic modulated diffraction grating. A method based on similar lines was presented recently by *Fang* and *Dai* [7.112].

7.3.2.1 Measurement of In-Plane Strains using Modulated Diffraction Gratings

The optical set-up for the method [7.111] is same, as shown in Fig. 7.8. A double-exposure hologram of the object in its unstressed and stressed states is recorded. In between the two exposures two linear sets of carrier fringes are added, one each to the two sets of deformation fringes existing in the system. These fringes are added by changing the directions of incidence of the two illuminating beams, one by an angle $\Delta\theta_{ey}$ around the y axis and the other by an angle $\Delta\theta_{ex}$ around the x axis. The irradiance observed in the reconstructed image is made up of two groups of interference fringes approximately perpendicular to each other. The phase differences corresponding to the two patterns can be written as

$$\varphi_1 = (\boldsymbol{K}_{e1} - \boldsymbol{K}_o) \cdot \boldsymbol{L} + \Delta \boldsymbol{K}_1 \cdot \boldsymbol{r}, \tag{7.60a}$$

$$\varphi_2 = (\boldsymbol{K}_{e2} - \boldsymbol{K}_o) \cdot \boldsymbol{L} + \Delta \boldsymbol{K}_2 \cdot \boldsymbol{r}, \tag{7.60b}$$

where $\Delta \boldsymbol{K}_1 = \hat{\boldsymbol{i}} \Delta K_{ex}; \Delta \boldsymbol{K}_2 = \hat{\boldsymbol{j}} \Delta K_{ey}$.

7.3 Measurement of the Derivatives of Displacements

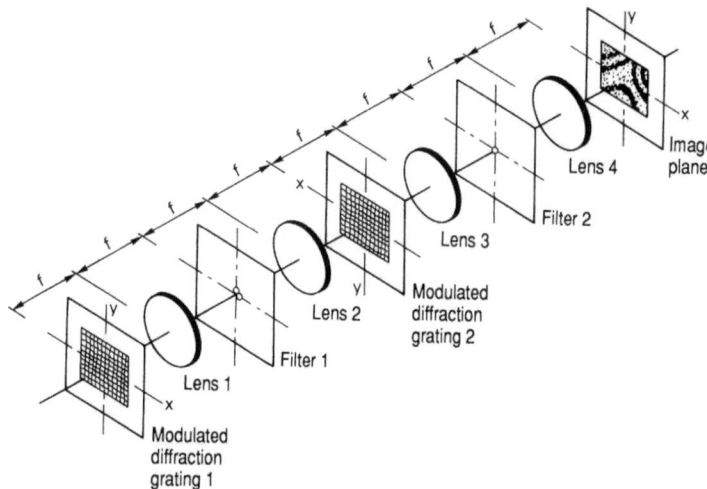

Fig. 7.30. Schematic of the coherent double diffraction spatial filtering system for obtaining fringes corresponding to the derivatives of the in-plane displacement fields

The reconstructed image is recorded on a photographic plate. The photographic transparency is a modulated diffraction grating. An identical copy of this image is made on another plate. One of the plates is placed at the input plane of the coherent double-diffraction system shown in Fig. 7.30. The other plate is placed in the back focal plane of lens 2 so as to be in exact coincidence with the projected image of the first plate. The far-field diffraction of the grid structure recorded on the plate is displayed in the first frequency plane. Two diffraction orders, lying approximately on orthogonal axes, are selected by placing an adequate two slot filter in the frequency plane. The complex amplitudes of the two filtered waves in the back focal plane of lens 2 are proportional to $\exp(i\varphi_1)$ and $\exp(i\varphi_2)$. The plate contained in this plane is now illuminated by these two waves. Let this plate be shifted by an amount Δx along the x axis. If a filter with a small aperture is now introduced at the origin of the second frequency plane, a two beam interference pattern is observed on a screen placed at the back focal plane of lens 4. The intensity distribution in the interferogram is given by

$$I \propto |\exp\{i[\varphi_1(x + \Delta x, y) - \varphi_1(x, y)]\}$$
$$+ \exp\{i[\varphi_2(x + \Delta x, y) - \varphi_2(x, y)]\}|^2. \tag{7.61}$$

The phase term of interest giving rise to fringe formation could thus be noted as

$$\varphi = [\varphi_1(x + \Delta x, y) - \varphi_1(x, y)] - [\varphi_2(x + \Delta x, y) - \varphi_2(x, y)],$$

which with the help of Taylor series expansion reduces to

$$\varphi = \frac{2\pi}{\lambda}\left(2\Delta x \sin\theta \frac{\partial L_x}{\partial x} + \Delta\theta_{ey}\Delta x \cos\theta\right). \tag{7.62a}$$

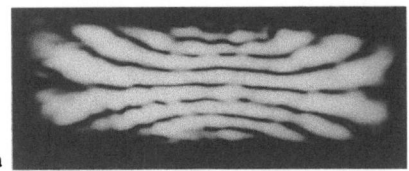 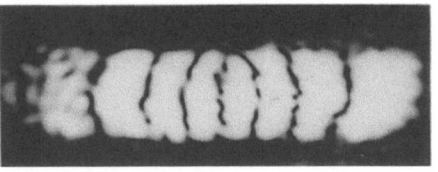

Fig. 7.31a, b. Examples of fringe patterns corresponding to (a) $\partial L_x/\partial x$ and (b) $\partial L_x/\partial y$ obtained in the case of a rectangular steel beam subjected to three point bending. (Courtesy of S. Toyooka)

The interference pattern gives contours of phase variations related to the in-plane strain. Due to the presence of a constant additional term $\Delta\theta_{ey}\Delta x\cos\theta$, the absolute amount of strain is not proportional to the absolute fringe numbers. The strain increment between two consecutive fringes is equal to

$$\Delta\left(\frac{\partial L_x}{\partial x}\right) = \frac{\lambda}{2\Delta x \sin\theta}. \tag{7.62b}$$

Similarly, if the second plate were shifted by an amount Δy along the y axis, the expression for the phase difference changes to

$$\Delta\varphi = \frac{2\pi}{\lambda}\left(2\Delta y \sin\theta \frac{\partial L_x}{\partial y} - \Delta\theta_{ex}\Delta y\right) \tag{7.63a}$$

and the strain increment per fringe interval is given by

$$\Delta\left(\frac{\partial L_x}{\partial y}\right) = \frac{\lambda}{2\Delta y \sin\theta}. \tag{7.63b}$$

Examples of fringe patterns corresponding to (a) $\partial L_x/\partial x$ and (b) $\partial L_x/\partial y$ are shown in Fig. 7.31. The specimen is a rectangular steel beam supported on two rods and subjected to a concentrated load placed on the middle of the beam.

7.4 Contouring of Three-Dimensional Objects

Holographic interferometry encompasses a group of techniques that provide information on the relief variations of a three dimensional object surface [7.113, 114]. A practical way to display this information consists of obtaining contour maps showing the intersection of an object with a set of equidistant planes perpendicular to the line of sight. The wavelength-change [7.115–120] and immersion [7.121–123] methods apart, holographic techniques for contouring applications are, in general, the fruit of modifications in the illumination and observation directions [7.115, 124–127]. While using his well-known sandwich holography *Abramson* [7.124] demonstrated that it was possible to obtain

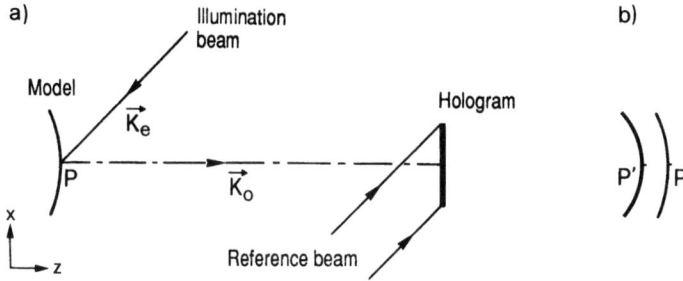

Fig. 7.32. Schematic of two wavelength holographic contouring

contour maps by first tilting a plane-parallel glass plate in the illumination arm of the interferometer during recording and then rotating the sandwich hologram during reconstruction. *Yonemura* [7.126] presented a method in which demodulation was obtained by translation of the holographic plate in a double-exposure experiment. *Rastogi* and *Pflug* [7.128, 129] used a phase organization technique to obtain a significant improvement in the methods performances. *Rastogi* and *Pflug* [7.130–132] reported methods based on holographic moiré and suggested a phase-management solution to improve their performances. These methods combine the advantage of variable sensitivities with those of a variable orientation of the standard planes.

7.4.1 Wavelength-Difference Contouring

Hildebrand and *Haines* [7.115] proposed a contouring method based on the use of two wavelengths. The schematic of a real-time contouring set-up is shown in Fig. 7.32a. A point on the model surface is illuminated and observed along the propagation vector pair (K_e, K_o). Let λ_1 be the wavelength utilized in the recording of the hologram. The object point and hologram are now illuminated by a light of slightly different wavelength and the reconstructed image is viewed through the hologram. As a result of the wavelength change the reconstructed image is shifted relative to the real object in both lateral and longitudinal directions. The new propagation vectors of the illumination and observation directions are given by (K'_e, K'_o). The lateral displacements can be eliminated by means of plane recording and reconstructing reference beams, and shifting back the reconstructed image by tilting the reference beam by an appropriate amount such that the condition

$$\sin \theta'_r = \frac{\lambda_2}{\lambda_1} \sin \theta_r \tag{7.64}$$

is satisfied. θ_r and θ'_r represent the initial and final angular positions of the reference beam. Another way of eliminating the lateral shift requires bringing the object very close to the photographic plate. This is achieved by recording an image plane hologram of the object.

The longitudinal shift gives rise to a change in magnification between the reconstructed and object images. The two images, no longer perfectly coincident, interfere with each other (Fig. 7.32b) to produce fringes corresponding to the contours of constant altitude. If A_1 and A'_1, A_r and A'_r, represent the corresponding amplitudes of the object and reference waves at a point on the holographic plate before and after the wavelength change, the intensity of the interference image can be written as

$$\langle I \rangle = \langle |p't \exp(i\pi) A'_r A_1 A_r^* + \upsilon A'_1|^2 \rangle \tag{7.65}$$

where

$$A'_1 = A_1 \exp(i\varphi_1), \tag{7.66a}$$

$$A'_r = A_r \exp(i\varphi'_1), \tag{7.66b}$$

and p' is the slope of the amplitude transmittance versus exposure curve of the photographic plate, and t is the exposure. Factor υ accounts for the beam weakening when passing through the hologram, and φ_1 and φ'_1 are the corresponding phase differences introduced by the change of wavelength in the object and reconstructed waves. We can express the phase difference on the object waves as

$$\varphi_1 = (\Delta \mathbf{K}_e - \Delta \mathbf{K}_o) \cdot \mathbf{r}, \tag{7.67}$$

where the position vector $\mathbf{r}$ defines the position of a point on the object, and $\Delta \mathbf{K}_e$ and $\Delta \mathbf{K}_o$ represent the change of propagation vectors induced by the wavelength change. Writing in a more explicit form

$$\varphi_1 = 2\pi \frac{\lambda_2 - \lambda_1}{\lambda_1 \lambda_2} [-x \sin \theta - (1 + \cos \theta) z], \tag{7.68}$$

where θ is the angle which the illumination beam makes with respect to the optical axis. The observation is carried along the direction of the optical axis by using a telecentric viewing system. Similarly, the phase introduced on the reconstructed wave is

$$\varphi'_1 = 2\pi \frac{\lambda_2 - \lambda_1}{\lambda_1 \lambda_2} x \sin \theta_r, \tag{7.69}$$

where the reference beam is supposed to lie in the x–z plane. Supposing that the equality $\upsilon = p't\langle I_r \rangle$ is satisfied, the irradiance distribution in the interference image becomes

$$\langle I \rangle = 2p'^2 t^2 \langle I_r \rangle^2 \langle I_1 \rangle [1 - \cos(\varphi_1 - \varphi'_1)]. \tag{7.70}$$

The phase corresponding to contour information is given by

$$\varphi_1 - \varphi'_1 = 2\pi \frac{\lambda_2 - \lambda_1}{\lambda_1 \lambda_2} [-x(\sin \theta + \sin \theta_r) - (1 + \cos \theta) z]. \tag{7.71}$$

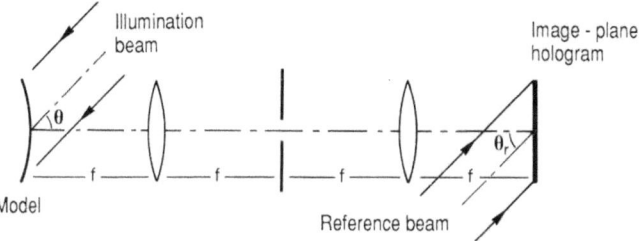

Fig. 7.33. An optical arrangement with telescopic viewing to obtain contour maps by two wavelength holographic contouring

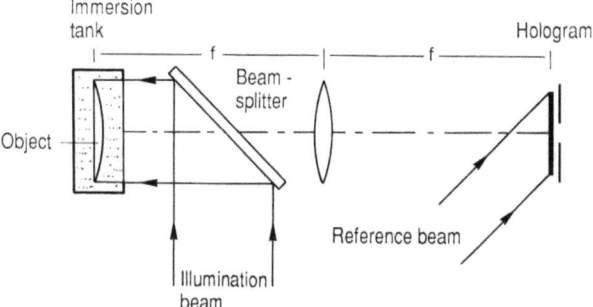

Fig. 7.34. Schematic of the immersion method for determining shapes of three-dimensional objects

The linear phase term can be eliminated by arranging the object and reference beams such that $\theta = -\theta_r$ (Fig. 7.33). The contour sensitivity per fringe issuing from a change in the phase difference of 2π then follows as

$$\Delta z = \frac{\lambda_1 \lambda_2}{(\lambda_1 - \lambda_2)(1 + \cos\theta)}. \tag{7.72}$$

The fringe planes intersect the object in a direction parallel to the hologram plane.

7.4.2 Immersion Method

The schematic of the immersion method is illustrated in Fig. 7.34. The model is placed in a glass tank filled with a liquid (or gas) having refractive index n'_1. The window facing the incident light is made of plane glass. A hologram of the object is recorded and the liquid contained in the tank replaced by another one having a refractive index n'_2. An observer looking through the hologram will see the object surface modulated by a set of interference fringes. These fringes arise from the change of optical path in the light rays traversing the two liquids.

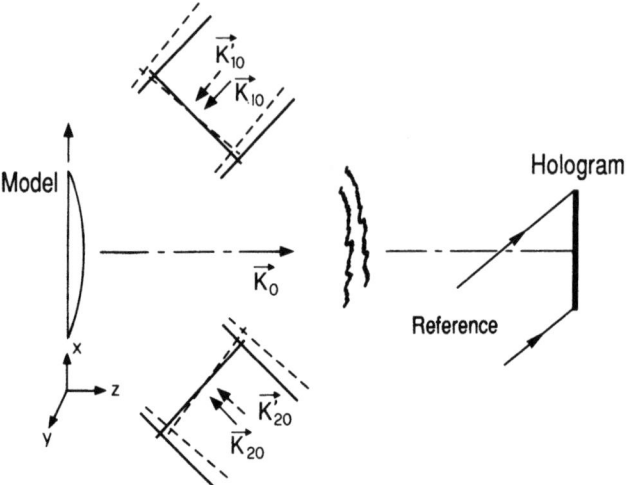

Fig. 7.35. Schematic diagram of the holographic moiré multiple sources contouring set-up

The object is assumed to be illuminated by a plane wave in a direction perpendicular to the glass plane. Observation is carried out using a telecentric imaging system. The rays propagating towards the hologram will thus not be perceptibly deviated by a change in the refractive index of the medium. In the absence of any lateral shifts, the fringes remain localized on the object surface. The optical phase difference giving rise to fringe formation is

$$\varphi = \frac{4\pi}{\lambda}(n'_1 - n'_2)z, \qquad (7.73)$$

where z is the distance between the object surface and the front glass window. The contour interval arising from a change in the phase difference of 2π is given by

$$\Delta z = \frac{\lambda}{2(n'_1 - n'_2)}. \qquad (7.74)$$

7.4.3 Holographic Moiré Multiple-Sources Contouring

7.4.3.1 Basic Technique

A simple representation of the optical system is shown in Fig. 7.35. The surface to be contoured is illuminated by means of two coherent collimated beams situated symmetrically with respect to the optical axis. The two beams are supposed to be in the x–z plane. The method is based on the principle of four-wave-front mixing in real time. Of these wave fronts, two are recorded on the holographic plate. These are played back during the reconstruction of the hologram. The two wave fronts are due to one illumination each. The other two wave fronts, also due to one illumination each, correspond to the live wave fronts issuing from the object surface and observed through the holographic plate. Controlled phase differences are introduced on the two live wave fronts and the resulting response is observed in real time on a TV screen.

Consider a point P(r) on the surface to be illuminated and observed along propagation vector pairs (K_{10}, K_0) and (K_{20}, K_0), respectively. Controlled phase modifications are introduced on the two arms of the interferometer. From the underlying properties of correlation in holographic interferometry, it follows that only wave fronts owing their origin to the same illumination beam fulfil the condition to participate in the interference process. The interferometer thus generates two patterns, one corresponding to each of the two illumination beams. If both the illumination beams are tilted by the same amount and in the same direction around an axis perpendicular to the plane of the paper, the equation of the resulting moiré fringe pattern becomes

$$\varphi = (\Delta K_{10} - \Delta K_{20}) \cdot r = 2n_m \pi, \tag{7.75}$$

where ΔK_{10} and ΔK_{20} define the changes in the illumination wave vectors resulting from the rotation of the two beams.

The contribution of the small tilts provided to the illumination beams is to encode the interference patterns by the relief of the object. The simultaneous observation of the two encoded patterns results in an optical subtraction, which enables one to decode the relief of the object in the form of a low frequency modulation whose significance is described by (7.75). Each moiré fringe in the interferogram can be described as the locus of all points on the surface that lie at a constant height from a fixed plane. The increment of height Δz between two consecutive contour planes is given by

$$\Delta z = \frac{\lambda}{2 \sin \theta \sin \Delta \theta_{eo}}. \tag{7.76}$$

The set of intersecting planes are situated in a direction perpendicular to the line of sight. Table 7.2 lists sensitivities attainable by the method for two sets of fringe frequencies in the reconstructed image and whose components f_x along x direction are, respectively, 1 line/mm and 1.5 line/mm. These frequencies lie in the upper range of frequencies with which one is usually accustomed to work in holography. The scope of the method thus remains limited due to its relatively low sensitivity. The range within which its sensitivity for a given illumination angle can be varied is quite limited.

It might seem paradoxical that the upper limit on contour sensitivity is approached while any degradation on the visibility of the carrier fringes is hardly discernible. The observation of moiré necessitates, however, that the carrier fringes are resolved by the observation system. Hence this apparent paradox is more due to the difficulty in discerning moiré impressed on a dense granular carrier fringe system rather than due to any significant fall in the fringe

Table 7.2. Maximum values of sensitivity Δz [mm] attainable by the method

θ	f_x	
	1 line/mm	1.5 lines/mm
30°	0.87	0.58
40°	0.60	0.40
50°	0.42	0.28

visibility. These remarks point out that potential for enhanced sensitivity measurements is present in the interferometer. The use of the method to suit finer varying surfaces is described below.

7.4.3.2 Phase Management Solution

The term corresponding to the frequency component along x direction can be written as

$$\psi_o(P) = \left(\frac{\Delta K_{10} + \Delta K_{20}}{2} - K_o\right) \cdot r. \tag{7.77}$$

This equation is independent of variations in shape. A convenient way to enhance the technique's potential is to drain off phase from the ψ_o term in (7.77), and fill-in the vacancy so created by the shape sensitive term φ. The phase pull-out corresponding to ψ_o is subject to two indispensable conditions. It should be carried out in a manner firstly to leave the phase of contour fringes unaffected, and secondly leave the plane of localization of the remaining carrier fringe system in the vicinity of the object surface. The retreat of phase from ψ_o can be performed in a number of ways, two of which are (i) reference beam rotation and holographic plate translation, and (ii) holographic plate rotation and its translation. In an image-plane configuration, the phase retreat could be carried out by rotation of the reference beam. Consider that procedure (i) is chosen to perform the phase pull out from the ψ_o term. Assuming the reference beam to be collimated and lie in the x–z plane, the condition for the remaining carrier fringes to continue to localize in the immediate vicinity of the object surface requires that

$$\frac{T_{x1}}{\sin \Delta \theta_{r1}} = \frac{\rho_a \cos \theta_r}{\cos^2 \theta_o}. \tag{7.78}$$

Under the assumption that (7.78) is satisfied, the spatial frequency of the carrier fringes leaving the parent fringe system is given by

$$s_{r1} = \lambda^{-1} \cos \theta_r \sin \Delta \theta_{r1}. \tag{7.79}$$

The value of ψ'_o left in ψ_o, after the first phase of withdrawal is over, can be expressed as

$$\psi'_o = (1 - \chi_1)\psi_o, \tag{7.80}$$

where

$$\chi_1 = \frac{\cos\theta_r \sin\Delta\theta_{r1}}{\sin\theta \sin\Delta\theta_{e0}}.$$

The first stage of phase withdrawal accomplished, two sets of fringes are reintroduced in a manner similar to obtain (7.75). The incoming phases have two components: one component contributes towards increasing the sensitivity of the method to contour measurements, and the other serves to reinforce the presence of remaining carrier frequency on the object surface. If ΔK_{11} and ΔK_{21} are the new representations of the induced changes in the illumination wave vectors:

$$\Delta K_{11} = K_{11} - K_{10}, \qquad K_{21} = K_{21} - K_{20},$$

the equation of the ensuing beat pattern becomes

$$(\Delta K_{11} - \Delta K_{21}) = 2n_m\pi, \tag{7.81}$$

which means a new contour sensitivity per moiré fringe, given by

$$\Delta z = \frac{\lambda}{2\sin\theta \sin\Delta\theta_{e1}}, \tag{7.82}$$

where

$$\Delta\theta_{e1} = \Delta\theta_{e0} + \Delta\theta'_{e1},$$

$\Delta\theta'_{e1}$ being the additional increment of rotation provided to the illumination beams. The phase term related to carrier fringes now becomes

$$\psi_{p1} = (1 - \chi_1)\psi_o + \psi_1, \tag{7.83}$$

where

$$\psi_1 = \left(\frac{\delta K_{11} + \delta K_{21}}{2} - K_0\right) \cdot r,$$

and relative increments δK_{11} and δK_{21} are given by

$$\delta K_{11} = K_{11} - K'_{10}, \qquad \delta K_{21} = K_{21} - K'_{20}.$$

If a proper implementation of the process is achieved, the contour sensitivity attained in (7.82) is augmented by a factor of two or more of that depicted in (7.76). If this sequence is repeated n times, the moiré observed at the end of the nth sequence would be

$$(\Delta K_{1n} - \Delta K_{2n}) \cdot r = 2n_m\pi. \tag{7.84}$$

The distance separating two adjacent contour planes is now given by

$$\Delta z = \frac{\lambda}{2 \sin \theta \sin \Delta\theta_{en}}, \qquad (7.85)$$

where

$$\Delta\theta_{en} = \Delta\theta_{e0} + \Delta\theta'_{e1} + \Delta\theta'_{e2} + \cdots + \Delta\theta'_{en}.$$

The phase term relative to the remaining carrier fringes can be expressed as

$$\psi_{pn} = (1 - \chi_n)\{(1 - \chi_1)(1 - \chi_2) \ldots (1 - \chi_{n-1})\psi_o + (1 - \chi_2)(1 - \chi_3) \ldots$$
$$(1 - \chi_{n-1})\psi_1 + (1 - \chi_3)(1 - \chi_4) \ldots (1 - \chi_{n-1})\psi_2 + \cdots + \psi_{n-1}\} + \psi_n. \qquad (7.86)$$

Assuming $\chi_n = 1$, (7.86) takes on a much simpler form

$$\psi_{pn} = \psi_n = \left(\frac{\delta K_{1n} + \delta K_{2n}}{2} - K_o\right) \cdot r, \qquad (7.87)$$

where

$$\delta K_{1n} = \Delta K_{1n} - \Delta K_{1(n-1)},$$
$$\delta K_{2n} = \Delta K_{2n} - \Delta K_{2(n-1)}.$$

In the final count, a total withdrawal of carrier frequency amounting to

$$s_{rn} = \lambda^{-1} \cos \theta_r \sin \Delta\theta_{rn} \qquad (7.88)$$

is effected over n steps and under the assumption that the equality

$$\frac{T_{xn}}{\sin \Delta\theta_{rn}} = \frac{\rho_a \cos \theta_r}{\cos^2 \theta_0} \qquad (7.89)$$

is satisfied, where

$$\Delta\theta_{rn} = \Delta\theta_{r1} + \Delta\theta'_{r2} + \cdots + \Delta\theta'_{rn},$$
$$T_{xn} = T_{x1} + T'_{x2} + \cdots + T'_{xn}.$$

A note of caution must be sounded before we move on to evaluate the method's performance. Bear in mind that the number of times these sequences could be repeated depends upon the degradation in the visibility of the carrier fringes which can be tolerated. The fall in fringe visibility is a direct consequence of these underlying sequences enhancing the sensitivity of the method. The plot in Fig. 7.36 shows the variation of contour sensitivity as a function of the illumination angle for several different values of the tilt $\Delta\theta_{en}$ of the illumination beams. The visibility of the carrier fringes falls to zero if the changes in the angles of illumination exceed

$$\Delta\theta_{en} \geq \sin^{-1} \frac{\lambda}{\sigma \cos \theta}. \qquad (7.90)$$

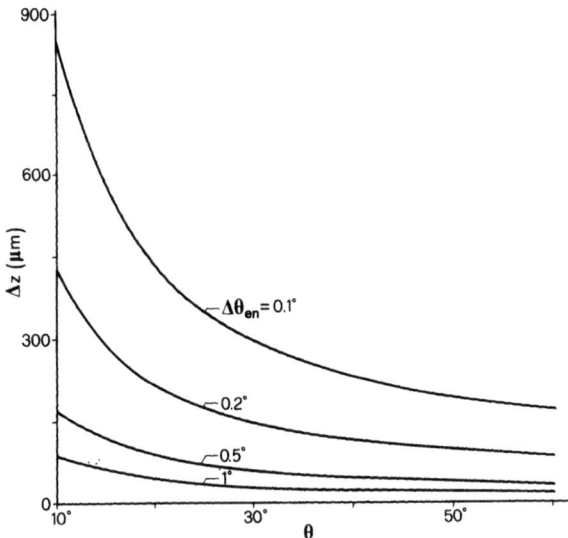

Fig. 7.36. Plot of the variation of contour sensitivity as a function of illumination angle for few selected values of $\Delta\theta_e$

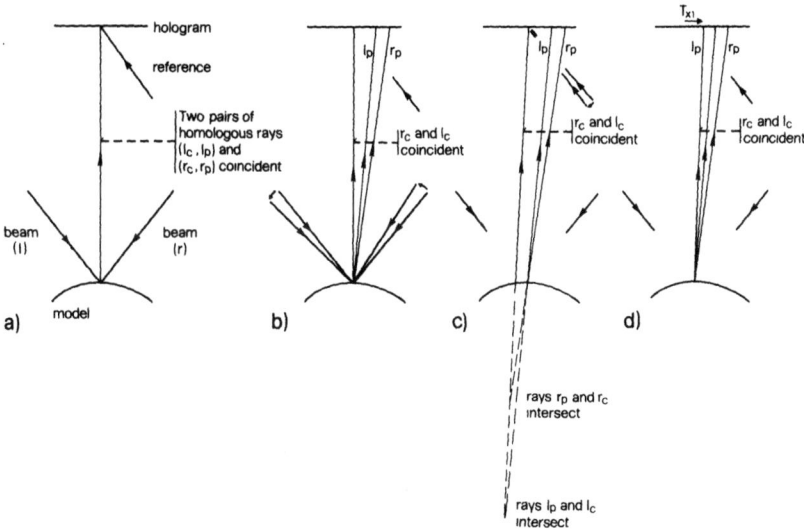

Fig. 7.37a–d. Schematic diagram showing the mechanism of fringe formation

7.4.3.3 Mechanism of Fringe Formation

The contour fringes are formed in a four-wave-front mixing interferometer. This type of interferometric mixing is presented in Fig. 7.37a. The pairs of homologous rays (l_c, l_p) and (r_c, r_p) are shown to be in a state of coincidence. The two beams are then subjected to small tilts. As a result of these tilts the rays l_p and r_p undergo differential angular displacements. The differential displacement results in the formation of a moiré pattern, which is directly related to the three dimensional features of the object. The fringes are localized on the object surface. These aspects are displayed in Fig. 7.37b.

The process of phase withdrawal is shown in Fig. 7.37c–d. A small rotation of the reference beam results in an angular movement of the rays (l_c, r_c). This movement also causes a delocalization of the fringe pattern from the model surface. An adequate lateral translation of the photographic plate helps to bring back the runaway fringe pattern on the object surface. As a result of these manipulations the fringes reappear on the object but only after having undergone a reduction in their frequencies. Although Figs. 7.37b and d correspond to the same beat defined by (7.75), their corresponding carrier fringes are respectively described by (7.77, 80). A repetition of these procedures provides the ability to vary continuously and in a wide range the contour sensitivity of the method.

7.4.3.4 Performance Evaluation

The performance of the phase management approach can best be evaluated in terms of the gain in the capacity of the method to generate finer contour planes intersecting the object. The reason underlying the enhancement of the method's capacity lies in the capability of the phase management approach to amplify the contribution of the shape-induced term in a fringe pattern phase shifted by the topographic variations. Such an approach is a substantial departure from conventional techniques using moiré. In conventional techniques, a moiré pattern related to contour information is, in general, formed by superimposing a grid pattern on its own replica modified by the shape of the object. Sensitivities, in general, are enhanced by choosing finer pitches for the grid or by changing other geometrical parameters related to the configuration. The notion of amplifying the strength of phase relative to shape contribution in a *given* distorted fringe pattern – phase shifted by the topography of the object – is an interesting approach in interferometry. The amplification is obtained at the expense of ψ_o and related terms.

A measure of the efficiency of the proposed approach is obtained by defining here a quantity M_a, the modulation aspect factor, as the ratio of the spatial frequencies in the z and x directions. This quantity enables the evaluation of the relative strength of the shape-induced phase contribution in the overall phase contained in each arm of the interferometer. The value of M_a before the application of the phase management approach is given by

$$M_{ai} = \tan \theta, \tag{7.91}$$

where a subscript i has been added to identify this value as the initial value of M_a. Equation (7.91) shows that M_{ai} is independent of the amount of phase that can be introduced individually in each arm, and has a unique predetermined

Table 7.3. Values of M_a obtained before and after the application of the phase-management approach for $f_x = 11$ mm and $\sigma = 20$ μm. Parenthesis enclose corresponding contour sensitivities Δz(μm)

θ	Method	Method integrating the phase-management approach V					
		0.9	0.8	0.7	0.6	0.5	0.4
30°	0.58	2.89	5.77	8.66	11.55	14.43	17.32
	(870)	(173)	(87)	(58)	(43)	(35)	(29)
50°	1.19	5.96	11.92	17.90	23.84	29.80	35.75
	(420)	(84)	(42)	(28)	(21)	(17)	(14)

value for each θ. Application of the phase management approach leads to the following formula for M_a

$$M_a = M_{ai} \frac{\sin \Delta\theta_e}{\sin \Delta\theta_e - \sin \Delta\theta_r}, \tag{7.92}$$

where, for ease of writing, θ has been considered to be equal to θ_r. It is obvious that for comparative analysis of M_a's to have a quantitative justification, the evaluation should be carried out for a constant value of fringe frequency f_x. On the other hand, if we were to consider values of M_a, M_{a1}, and M_{a2}, for two different fringe frequencies, f_1 and f_2, then the ratio between these two values is obtained as

$$\frac{M_{a1}}{M_{a2}} = \frac{f_1}{f_2},$$

where we have used the fact that M_{ai} is constant for a given θ. Although the moiré distribution does not modify with a change in carrier frequency, a change in the quality of their observation is noted due to a contribution of a different number of carrier fringes in the formation of a moiré fringe.

Table 7.3 lists a series of values for M_a obtained before and after the application of the phase management approach to the contouring technique. The mean speckle size and the minimum acceptable value of visibility are taken to be 20 μm and 0.4, respectively, and f_x is supposed to be equal to 1 line/mm on the reconstructed interferograms. The table reveals that the shape-modulation strengthening capacity of the method is very high. For a frequency f_x of 1 line/mm and visibility $V = 0.4$, the capacity generated is almost 30 times larger than the technique based on conventional means. A gain in fringe sensitivity is 30 times better than what would have been achieved by the method operating without the integrated approach.

An example of results obtained by applying the phase-management approach is exhibited in Fig. 7.38. The management approach is repeated many times over to obtain a gradual increase in the sensitivity of the method. Figures

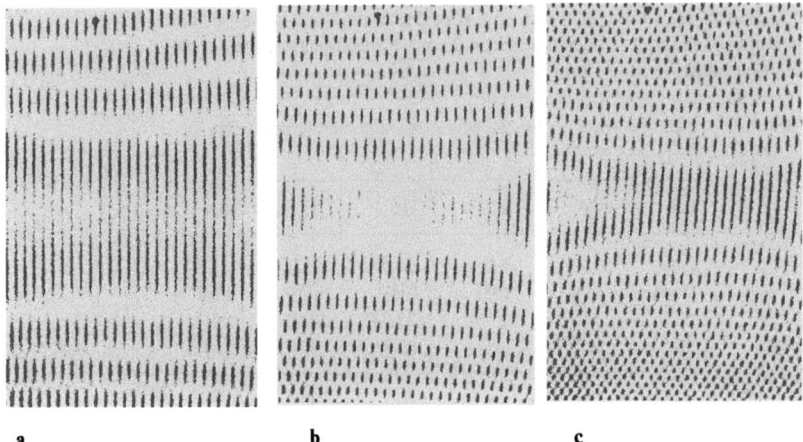

Fig. 7.38a–c. A result to show the continuous generation of sensitivities obtained by the application of the phase management approach. The sensitivities generated in (a), (b) and (c) are given by 475, 230 and 126 μm, respectively

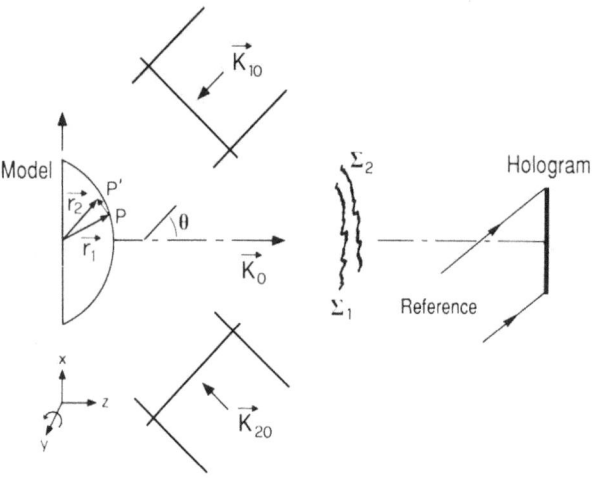

Fig. 7.39. Schematic of the optical configuration of the dual-illumination object tilt holographic contouring method

7.38a–c show some contouring patterns obtained in the sequence. The sensitivities are 345, 177 and 126 μm, respectively. The method also offers the possibility to change in real-time the orientation of the contour planes.

7.4.4 Holographic Moiré Object Tilt Contouring

In this method relief sensitivity is obtained by slightly rotating the object surface in a direction perpendicular to the plane containing the two beams. The

7.4 Contouring of Three-Dimensional Objects

resulting moiré represents a topographical map of the tested surface. The phase difference corresponding to the moiré fringes is given by

$$\varphi(P) = (K_{10} - K_{20}) \cdot (r_2 - r_1), \tag{7.93}$$

where the points P and P', before and after object displacement, are defined by the position vectors r_1 and r_2 referred to the origin of a coordinate system (Fig. 7.39). If the two wave vectors K_{10} and K_{20} make an equal angle θ with the z axis, the contour sensitivity per moiré fringe issuing from an increment of the difference in phase $\varphi_{10}(P)$ equal to 2π is given by

$$\Delta z = \frac{\lambda}{2\sin\theta \sin\Delta\theta_o}. \tag{7.94}$$

The contour planes intersect the object in a direction perpendicular to the line of sight. The implementation of this method would have little practical utility if one were constrained to sensitivity limits echoed in (7.94). A breakdown in the method's potentialities arises from a saturation of the phase carrying capacity of the interferometer. Since the fringe frequency in the x direction, f_x, is normally two to four and a half times much higher, within the range $53° > \theta > 25°$, than in the z direction, the phase demodulation computed by the method naturally suffers from a weak sensitivity. The application of the phase management approach, enables a substantial improvement of the method's performance. At the end of the nth step, the sensitivity of contouring is given by

$$\Delta z = \frac{\lambda}{2\sin\theta \sin\Delta\theta_{on}} \tag{7.95}$$

where

$$\Delta\theta_{on} = \Delta\theta_o + \Delta\theta'_{o1} + \Delta\theta'_{o2} + \cdots + \Delta\theta'_{on},$$

$\Delta\theta'_{oi}$ ($i = 1, 2, \ldots, n$) being the increment of rotation provided to the object in the ith step of the phase-management approach. The upper limit to measurements is set by

$$\sin^{-1}\frac{\sigma}{h'} \leq \Delta\theta_{on} \geq \sin^{-1}\frac{\lambda}{\sigma(1 + \cos\theta)}, \tag{7.96}$$

where h' denotes the object depth. Whereas the inequality on the left originates from the relative displacement of the homologous rays at the observation plane, the one on the right accounts for the shift between the interfering wave fronts at the pupil plane. At any given instant the fall in visibility is the result of the combined effect of both these factors. Figure 7.40 shows a 3-D plot of the variation of fringe visibility on a contour plane.

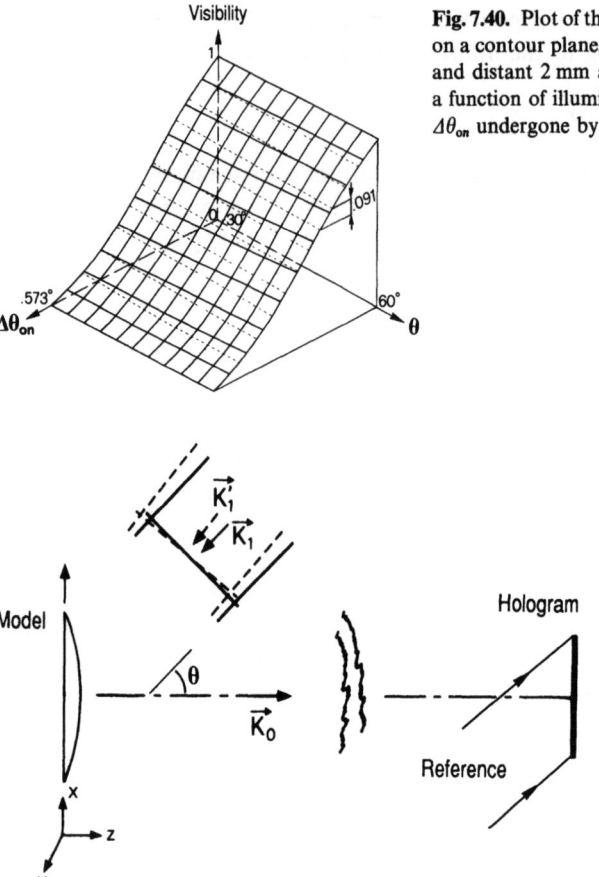

Fig. 7.40. Plot of the variation of fringe visibility on a contour plane, parallel to the x and y axes, and distant 2 mm away from the origin, as a function of illumination angle θ and rotation $\Delta\theta_{on}$ undergone by the model

Fig. 7.41. Schematic representation of the multiple sources holographic contouring arrangement

7.4.5 Multiple-Sources Contouring

7.4.5.1 Basic Technique

The optical arrangement of the method is shown in Fig. 7.41. The object is illuminated by a parallel beam of light making an angle θ with respect to the optical axis. The wave front originating from the object is recorded on the holographic plate. The specimen and its reconstruction are observed simultaneously through the hologram lit by its reference beam. The principle of contour generation consists of projecting Young's fringes obliquely onto the object surface by providing a tilt to the illumination beam. The fringe pattern is modulated by the three-dimensional variations of the object surface. The demodulation of these fringes is obtained by reshuffling the phase terms in the interferometer in a manner to eliminate all terms except those directly dependent on the relief of the object surface. In moiré such a subtraction is normally achieved by the combination of grids, corresponding to a regular grid and its counterpart modified by the relief of the object. Contrary to the formation of moiré fringes the demodulated fringe contours obtained by reshuffling the phase terms are exempt from the carrier fringes. A proper management of the method's phase resources permits one to improve the contour exploration capacity of the method by several orders of magnitude.

Let a point P(r) on the surface be initially illuminated and observed along the propagation vector pair (K_1, K_0). A suitable amount of phase is injected into the interferometer by tilting the illumination beam to its new position, defined by K_1'. The equation of the resulting fringe pattern can then be expressed as

$$\varphi_0 = \Delta K_1 \cdot r = 2n\pi, \tag{7.97}$$

where ΔK_1 defines the change in the illumination wave vector resulting from the rotation of the beam.

The contribution to φ_0 is composed of two types of phase components: one, φ_1' owing it presence to the relief variations of the object and the other, φ_2', a consequence of the linear phase variations along x direction of the object surface. The step to extract the phase corresponding solely to the depth information consists of eliminating the term φ_2' from the total phase difference developed in the interferometer. This is carried out by generating in real time in the interferometer an additional phase factor capable of neutralizing φ_2'. An indispensable requirement on this phase factor is that it should be capable of neutralizing φ_2' without either generating any new phase component related to depth or disturbing the localization of the fringes from on, or near, the object surface. A complete phase neutralization is obtained if the contribution of the corresponding phase φ_n is equal and opposite in sign to that of φ_2'. In this case (7.97) reduces to

$$\varphi_0 = \frac{2\pi}{\lambda} \sin \theta \sin \Delta\theta_e z, \tag{7.98}$$

where $\Delta\theta_e$ is the small angle by which the beam is rotated around the 0y axis. The gradual reduction of phase $\varphi_n - \varphi_2'$ is monitored in real time on a plane object surface placed beside the specimen. The situation of zero fringe on the plane surface confirms the complete elimination of the φ_2' term. The corresponding fringe pattern represents true depth contours whose planes are oriented in a direction perpendicular to the line of sight. The contour sensitivity per fringe is given by

$$\Delta z = \frac{\lambda}{\sin \theta \sin \Delta\theta_e}. \tag{7.99}$$

A way to generate the additional phase term needed to obtain phase neutralization is to tilt slightly the reference beam illuminating the hologram in an image-plane configuration. The phase introduced on the reconstructed wave is given by

$$\varphi_n = \frac{2\pi}{\lambda} x \sin \Delta\theta_r \cos \theta_r \tag{7.100}$$

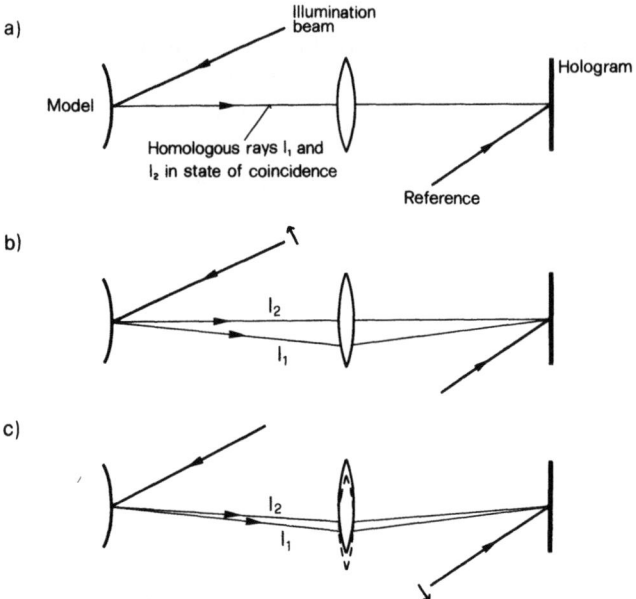

Fig. 7.42a–c. Schematic drawing showing the mechanism of fringe formation

where $\Delta\theta_r$ is the small tilt provided to the reference beam supposed to lie in the x–z plane. The introduction of the phase difference φ_n gives rise to the generation of fringes which are well localized on the holographic plate.

7.4.5.2 Mechanism of Fringe Formation

The way the method works is explained briefly by means of Fig. 7.42a showing a conventional image plane set-up in real-time operation. A hologram of the object is recorded and repositioned in its original position. The pair of homologous rays, l_1 and l_2, originating from a point P on the reconstructed and specimen surfaces respectively are in a state of coincidence. The tilt of the illumination beam results in an angular displacement of the ray l_1 (Fig. 7.42b). The phase difference corresponding to this displacement is given by (7.97). The ray l_2 is now shifted angularly by providing an appropriate tilt to the reference beam. This results in the lateral shift of the reconstructed pupil with respect to its initial position (Fig. 7.42c). An observer looking through the hologram sees two identical but slightly shifted images of the pupil distribution in the pupil plane. The aim of this manipulation is to develop enough phase in the interferometer required to neutralize the contribution of the phase φ'_2.

Table 7.4. Sensitivities Δz [mm] generated by the method for three pairs of values (f_x, θ)

θ	f_x	
	1 line/mm	1.5 line/mm
30°	1.73	1.16
40°	1.19	0.79
50°	0.84	0.56

7.4.5.3 Enhancement of Contouring Sensitivity

In contouring investigations, using conventional fringe projection techniques, the sensitivity is very low. Hence the application of these methods has been limited mainly to the investigation of relatively coarse objects. Table 7.4 gives an idea of the mediocre performance available from the technique. An on-line management of the phase resources inherent in the interferometer, however, permits one to considerably improve upon these performance levels.

The source of the problem resides in the relief encoding part of the method. The capacity of the interferometer to receive phase is relatively low. The introduction of more phase above this capacity could only be to the detriment of the quality of the relief modulated fringe pattern. Thus most of the phase reserves lying within the interferometer remain in a dormant state. Moreover, the ratio of the topographical phase information impressed on the fringe pattern to the total phase introduced in the interferometer is relatively small. The problem can be alleviated by working out means to improve this ratio in the interferometer. One such approach is based on a selective elimination of the phase terms of type φ'_2 from the interferometer. The phase vacancies so created are progressively reallocated to the phase terms of type φ'_1.

Equation (7.98) describes the interference phase with the corresponding contour interval given by (7.99). The elimination of the phase φ'_2 from the interferometer creates a phase vacancy which can now be filled up by an additional phase term of type φ_0. This modification of the phase balance in the interferometer gives rise to a grid pattern but where the ratio of the topographical to the total phase contribution now stands enhanced as compared to that obtained at the end of the demodulation part of the method. The new phase situation φ_{01} in the interferometer is

$$\varphi_{01} = \varphi'_1 + \Delta\varphi_{01},$$

where $\Delta\varphi_{01}$ is the phase increment added to the interferometer. It is given by

$$\Delta\varphi_{01} = \Delta\varphi'_{11} + \Delta\varphi'_{21},$$

where $\Delta\varphi'_{11}$ and $\Delta\varphi'_{21}$ are the increments corresponding to relief and linear phase variations, respectively. The phase increment $\Delta\varphi'_{21}$ is neutralized by introducing an additional phase shift $\Delta\varphi_n$, of the type φ_n, in the interferometer. The corresponding phase position in the interferometer becomes

$$\varphi_{01} = \varphi_1 + \Delta\varphi'_{11},$$

which leads to a set of contour planes separated by

$$\Delta z = \frac{\lambda}{\sin\theta \sin\Delta\theta_{e1}}, \qquad (7.101)$$

where $\Delta\theta_{e1} = \Delta\theta_e + \Delta\theta'_{e1}$; $\Delta\theta'_{e1}$ being the angle of rotation provided to the illumination beam to inject the phase shift $\Delta\varphi_{01}$ in the interferometer. Suppose that the steps of phase removal and subsequent reinforcement are repeated n times over. At the end of the nth step, the phase term corresponding to the fringe pattern can be written as

$$\varphi_{0n} = \varphi_1 + \Delta\varphi'_{11} + \Delta\varphi'_{12} + \cdots + \Delta\varphi'_{1n}, \qquad (7.102)$$

where $\Delta\varphi'_{1i}$ is the relief related phase increment introduced in the interferometer during the ith step. Writing (7.102) in its present form requires that

$$\frac{\sin\Delta\theta_{en}}{\sin\Delta\theta_{rn}} = \frac{\cos\theta_r}{\cos\theta}, \qquad (7.103)$$

where

$$\Delta\theta_{en} = \Delta\theta_e + \Delta\theta'_{e1} + \Delta\theta'_{e2} + \cdots + \Delta\theta'_{en},$$

$$\Delta\theta_{rn} = \Delta\theta_r + \Delta\theta'_{r1} + \Delta\theta'_{r2} + \cdots + \Delta\theta'_{rn}.$$

The development of (7.102) shows that the distance separating two adjacent contour planes is now given by

$$\Delta z = \frac{\lambda}{\sin\theta \sin\Delta\theta_{en}}. \qquad (7.104)$$

Equation (7.104) reports a substantial increase in the measurement sensitivity. Figure 7.43a presents plots of the variation of fringe visibility as a function of contour sensitivity. The curves are drawn for sensitivities higher than 300 µm, as the visibility of these fringes is relatively higher toward the lower end of the sensitivity scale. The upper limit to measurements is reached for

$$\Delta\theta_{en} \geq \sin^{-1}\frac{\lambda}{\sigma\cos\theta}.$$

Although we have discussed the type of phase movements, nothing has so far been said about the magnitude of these movements taking place in the interferometer. The extent of phase movements taking place in the interferometer is illustrated by means of plots shown in Fig. 7.43b. Drawn for two values of illumination angles, these curves indicate the fringe-frequency dependence of the contouring sensitivity developed by the interferometer. To consider a numerical example, for an object supposed to be illuminated at 30° to the z axis, a fringe

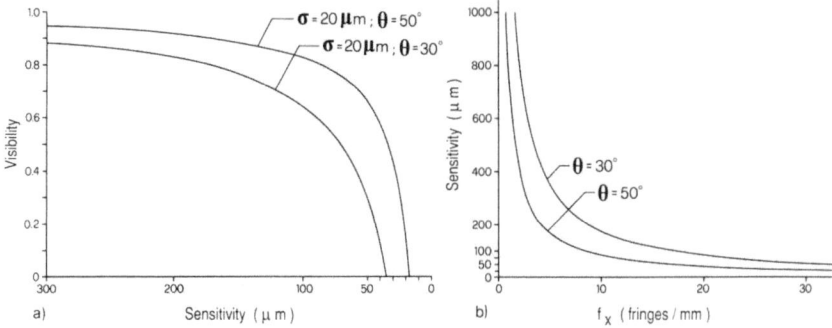

Fig. 7.43. (a) Plot of visibility as a function of contour sensitivity for two selected values of the pair (σ, θ). (b) Plots showing the dependence of the contour sensitivity developed as a function of the generation of fringe frequency in the interferometer for two selected values of the illumination angle

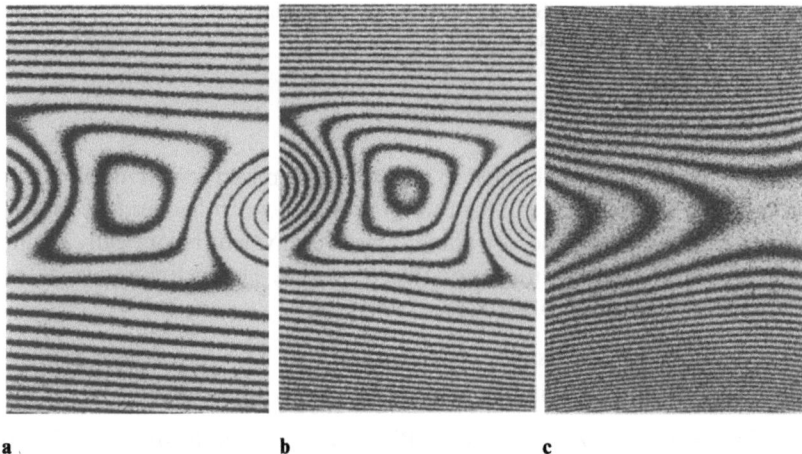

Fig. 7.44a–c. Illustration of the capacity of the interferometer to generate in real-time contour intervals of progressively increasing sensitivities. The sensitivities generated in (**a**) and (**b**) are given by 408 and 236 µm respectively. (**c**) Finer sensitivity of 30 µm per fringe interval is shown for a more smoothly varying surface

frequency of 43.3 lines/mm would necessarily be generated if one were to develop in the interferometer the capacity to deliver a contour sensitivity of 40 µm. That the method has the capacity to handle such large phase movements in a controlled manner speaks in itself of the high performance levels which the method can possibly attain. Figure 7.44 demonstrates the capacity of the interferometer to generate in real-time contour intervals of progressively increasing sensitivities. The sensitivities generated in Figs. 7.44a, b are given by 408 and 236 µm, respectively. Figure 7.44c shows a contouring pattern obtained on a specimen of relatively smoothly varying relief. The distance between two adjacent contouring planes is 30 µm.

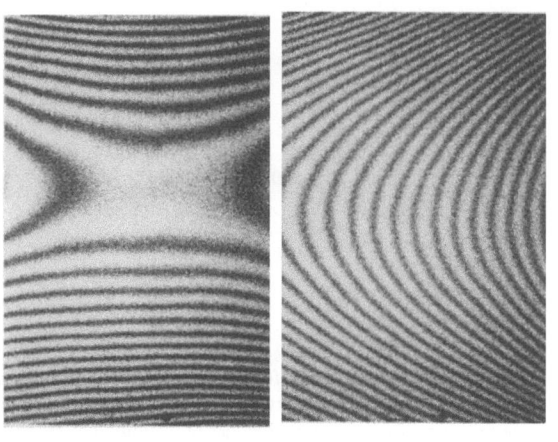

Fig. 7.45a, b. Illustration of the possibility of obtaining oblique slicing of the models. Contour plane is inclined around (**a**) $0x$ and (**b**) $0y$ axes

7.4.5.4 Orientation of the Contour Planes

The method's flexibility is not limited to creating customized sensitivities only. It also offers the means to orient the contour planes in a controlled and a continuous manner with respect to the x–y plane. This is equivalent to optically slicing three dimensional surfaces by means of oblique planes. The inclination, α, of the contour plane with respect to the x axis is given by

$$\alpha = \tan^{-1}\left(\cot\theta \pm \frac{\cos\theta_r \sin\Delta\theta_{ri}}{\sin\theta \sin\Delta\theta_{ei}}\right). \tag{7.105a}$$

This equation shows that $\alpha = 0$ only if (7.103) is satisfied. On the other hand, and on the assumption that the contour planes perpendicular to the line of sight have already been generated, these planes can be oriented around the y axis by an angle β given by

$$\beta = \tan^{-1}\left(\frac{\sin\Delta\theta_{ry}}{\sin\theta \sin\Delta\theta_{ei}}\right), \tag{7.105b}$$

where $\Delta\theta_{ry}$ is the component of the tilt provided to the reference beam along the y direction. Fringe patterns in Fig. 7.45 illustrate the flexibility associated with the interferometer to pivot in real-time the orientation of the contour planes.

7.5 Comparative Holographic Interferometry

Comparative holographic interferometry [7.77, 133–152] is a relatively recent development in non-destructive testing. The potential of the technique is widely recognized [7.9, 10, 153]. The special feature of the technique is that it provides the contours of path variations related to the difference in displacements or shapes of two objects. The possibility of instantaneously comparing two objects considerably broadens the scope of holographic interferometry in non-destructive inspection. One of the difficulties associated with the development of the method is inherent in the task itself, which aims at comparing two macroscopically similar but physically different surfaces. The technique, by way of its ability to compare (i) the mechanical responses of two nominally identical specimens subjected to the same loadings or (ii) the shapes of two nominally identical specimens, offers a versatile and practical means to detect the anomalies in the test piece with respect to the flaw free master piece. The first effort to code the variation of the differences in phase of two objects subjected to stresses, in the form of holographic fringes, was made by *Neumann* [7.134]. Work along similar lines was reported by *Füzessy* and *Gyimesi* [7.136–146]. *Rastogi* [7.77, 147–149] proposed a real-time technique sensitive to the measurement of the difference in displacement components. A similar approach was later presented by *Simova* and *Sainov* [7.150–152] in an attempt to accelerate the method's application in non-destructive testing. More recently, *Rastogi* et al. [7.90] extended the use of phase shifting interferometry to comparative holography.

7.5.1 Equation of the Difference Displacement Vector of Two Surface Elements

When a body deforms, its surface shape changes. Thus the comparison of two nominally identical bodies reduces to the extraction of the difference in the change in surface shapes at each point of the two deformed bodies. In order to acquire the true contours of the difference in surface shape changes, it is indispensable that only the corresponding points on the two deformed bodies be compared.

Let us associate the two bodies to the cartesian coordinate system (x, y, z). The undeformed states of both the bodies are taken to be their initial states. The bodies are supposed to be in their final states after having been subjected to loading. In Fig. 7.46a, a point P_M on a small unstrained element on the surface of the master specimen occupies a position (x, y, z). In its final state the point P_M attains the position $P_{M1}(x', y', z')$. Let (L'_x, L'_y, L'_z) be the components of the displacement vector of the point P_M. In Fig. 7.46b, a point P_T on an element on the surface of the test specimen, corresponding to the master specimen, occupies a position (x, y, z) in its initial state and the position $P_{T1}(x'', y'', z'')$ in its final state. Let (L_x, L_y, L_z) be the components of the displacement vector of the point P_T. Inasmuch as the corresponding points on the two deformed bodies are compared, we obtain

$$P_T P_{T1} - P_M P_{M1} = P_{M1} P_{T1} \tag{7.106a}$$

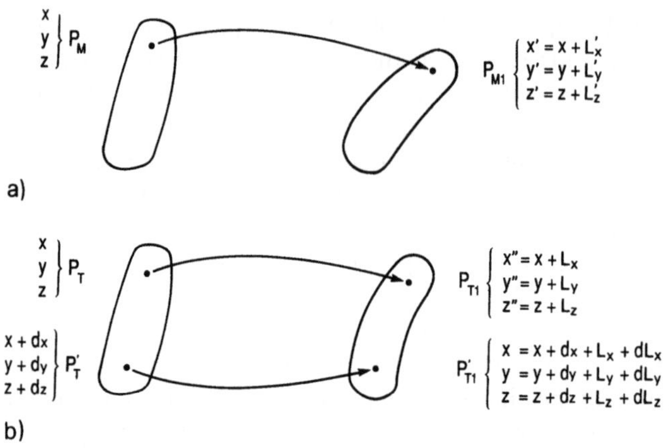

Fig. 7.46a, b. Geometry of small elements on the surface of two bodies in their undeformed and deformed states

or, in terms of displacement components

$$\begin{bmatrix} L_x \\ L_y \\ L_z \end{bmatrix} - \begin{bmatrix} L'_x \\ L'_y \\ L'_z \end{bmatrix} = \begin{bmatrix} \Delta L_x \\ \Delta L_y \\ \Delta L_z \end{bmatrix}, \tag{7.106b}$$

where $(\Delta L_x, \Delta L_y, \Delta L_z)$ denote the components of difference displacement vector between the two corresponding points. The fact that, due to system errors, non-corresponding points could get compared, brings with it an increased complexity. A point P'_T lying in the neighbourhood of P_T occupies in its initial state a position $(x + dx, y + dy, z + dz)$ and in its final state the position P'_{T1}. Let $(L_x + dL_x, L_y + dL_y, L_z + dL_z)$ be the components of the displacement vector of the point P'_T. The displacement vector can be written as

$$P'_T P'_{T1} = P_T P_{T1} + (P_M P'_T \cdot \nabla) P_T P_{T1} \tag{7.107a}$$

or, in matricial form the equation becomes

$$\begin{bmatrix} L_x + dL_x \\ L_y + dL_y \\ L_z + dL_z \end{bmatrix} = \begin{bmatrix} L_x \\ L_y \\ L_z \end{bmatrix} + \begin{bmatrix} \dfrac{\partial L_x}{\partial x} & \dfrac{\partial L_x}{\partial y} & \dfrac{\partial L_x}{\partial z} \\ \dfrac{\partial L_y}{\partial x} & \dfrac{\partial L_y}{\partial y} & \dfrac{\partial L_y}{\partial z} \\ \dfrac{\partial L_z}{\partial x} & \dfrac{\partial L_z}{\partial y} & \dfrac{\partial L_z}{\partial z} \end{bmatrix} \begin{bmatrix} dx \\ dy \\ dz \end{bmatrix}. \tag{7.107b}$$

Considering that the point P_M is compared with the point P'_T, one obtains

$$P'_T P'_{T1} - P_M P_{M1} = P_{M1} P'_{T1} \tag{7.108a}$$

or, in terms of components

$$\begin{bmatrix} L_x + dL_x \\ L_y + dL_y \\ L_z + dL_z \end{bmatrix} - \begin{bmatrix} L'_x \\ L'_y \\ L'_z \end{bmatrix} = \begin{bmatrix} d\bar{L}_x \\ d\bar{L}_y \\ d\bar{L}_z \end{bmatrix}, \qquad (7.108b)$$

where $(d\bar{L}_x, d\bar{L}_y, d\bar{L}_z)$ denote the components of difference displacement vectors between the two compared points. From (7.106–108), one obtain

$$P_{M1}P'_{T1} = P_{M1}P_{T1} + (P_M P'_T \cdot \nabla)P_T P_{T1}. \qquad (7.109)$$

This is a general equation of the difference in displacement vectors of two bodies in the case where the deformations can be considered to be quite small, such that the ratios of the type $\partial L_j/L_j$ obey:

$$\frac{\partial L_j}{L_j} \ll 1.$$

Equation (7.109) explains the influence of a possible misalignment, which results in an erroneous comparison of non-corresponding points on the two deformed bodies.

7.5.2 Difference Holographic Interferometry

This method consists of making a pair of holograms, one each for the undeformed and deformed states of the master object. The two reconstructed waves obtained by interrogating the holograms by conjugate reference waves are then used to illuminate sequentially the object under test. A double exposure hologram is recorded; the first recording corresponds to the test object in an undeformed state and illuminated by the master wave in the same state. The second recording corresponds to the test object in a deformed state and illuminated by the master wave in the same state. The double exposed hologram upon reconstruction displays fringes corresponding to the difference in response of the two objects submitted to the same stress levels.

The interpretation of the obtained fringes is understood by means of Fig. 7.47. Let us first consider the recording of the master object. The object is illuminated in the direction K_e and observed along K_o. The phase difference at a point P on the master object is given by

$$\varphi_m = (K_e - K_o) \cdot L'. \qquad (7.110)$$

The master object is removed from the system and replaced by a test object. The master hologram is placed in the observation arm of the preceding interferometer. Observation is carried along the direction formerly used to record the master hologram. The test object is illuminated and observed along wavevectors $(-K_o, -K_e)$. The phase difference at a point P_T, corresponding to P_M on the master object, is given by

$$\varphi_t = (K_e - K_o) \cdot L, \qquad (7.111)$$

278 7. Techniques to Measure Displacements, Derivatives and Surface Shapes

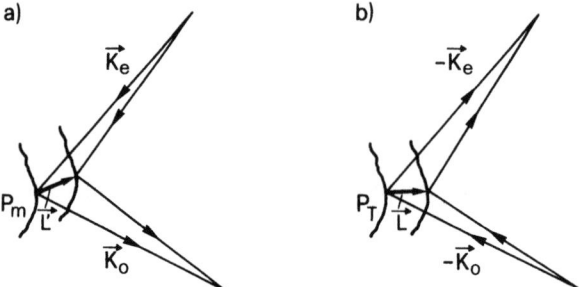

Fig. 7.47a, b. Fringe formation in difference holographic interferometry: (a) recording of the wavefronts corresponding to the master object; (b) recording of the difference holographic interferogram

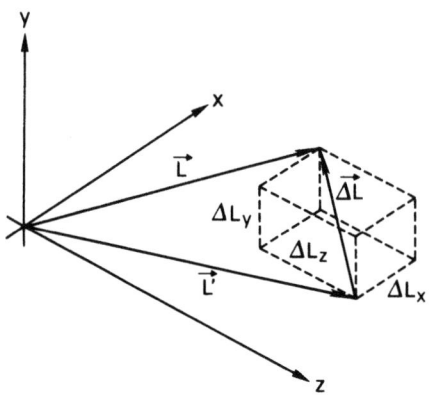

Fig. 7.48. Spatial representation of the displacement vector difference ΔL and its Cartesian components

where L is the displacement vector at the point P_T. The relative phase change at the point P_T is given by

$$\Delta\varphi = (\boldsymbol{K}_e - \boldsymbol{K}_o) \cdot \Delta \boldsymbol{L}, \tag{7.112a}$$

where ΔL is the difference in displacement vector between the test and master object displacements

$$\Delta \boldsymbol{L} = \boldsymbol{L} - \boldsymbol{L}'. \tag{7.112b}$$

The spatial representation of the vector ΔL and its components is shown in Fig. 7.48. Several experimental realizations of the method have been proposed. The optical layout of an experimental arrangement is shown in Fig. 7.49. Traced lines with shaded arrows indicate the beam path employed for exposing the master holographic plate twice, once each for the two different states, undeformed and deformed of the master object. The master object is replaced by the test specimen. Traced/dashed lines with unshaded arrows indicate the beam path for recording the difference hologram. The difference hologram is then

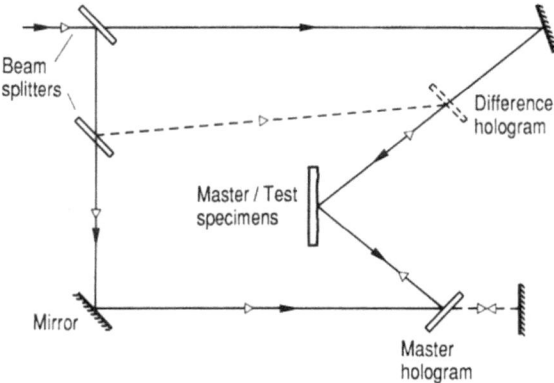

Fig. 7.49. Layout of the optical arrangement to record a difference hologram. Shaded arrows show the light path during the recording of the master hologram. Unshaded arrows show the corresponding path during the recording of the difference hologram

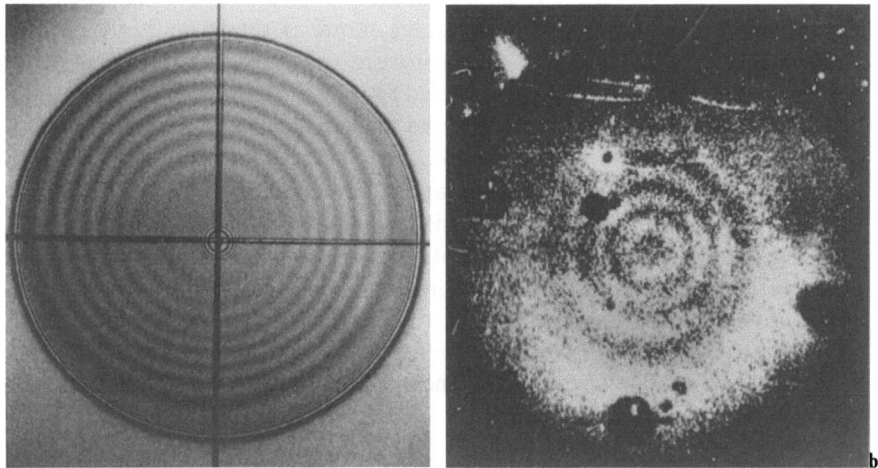

Fig. 7.50a, b. Examples of fringe patterns corresponding to (a) the difference in displacement component ΔL_z and (b) the topographic difference of two objects. (Courtesy of Z. Füzessy)

reconstructed to yield fringes due to the difference in displacements between the test and master components. The fringe contours clearly indicate the presence or absence of flaw in the test object. An example of fringe pattern is shown in Fig. 7.50a. The same object (pressure chamber) is used as both test and master objects by repainting the surface of the chamber after the recording of the master hologram. The fringe pattern corresponds to the difference in deflections obtained by subjecting the chambers to slightly different pressure levels.

Difference holographic interferometry has also been applied to detect differences in the shapes of two objects. Figure 7.50b displays the topographic difference of two objects obtained by two refractive index contouring.

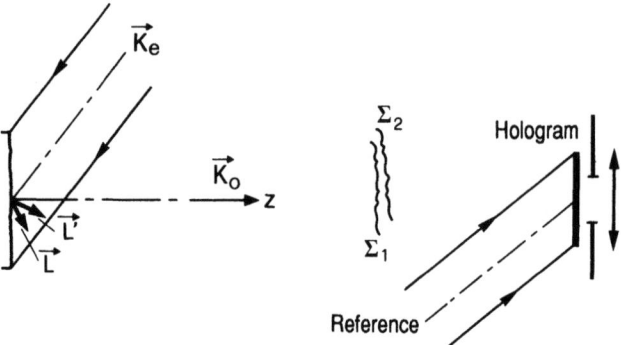

Fig. 7.51. Schematic diagram of the principle used in comparative holographic moiré

7.5.3 Comparative Holographic Moiré

The principle of comparative holographic moiré is shown in Fig. 7.51. The images of the master and test specimens are superimposed by means of an optical device. The wave field Σ_1 scattered toward the hologram is composed of two wave fronts diffracted by the test and master specimens, respectively. Σ_2 is the diffracted wave field after both specimens have been deformed. The specimens are supposed to be illuminated and observed along the wave vectors $\mathbf{K}_e$ and $\mathbf{K}_o$, respectively. The method consists of exposing the holographic plate to the wave field Σ_1. The holographic plate is processed and returned to its initial position. The test and master components are next subjected to the same stress levels. An observer looking through the hologram sees the appearance of a moiré interference pattern corresponding to the contours of the difference in displacements of the two specimens. The intensity in the reconstructed image is obtained as

$$\langle I(\mathrm{P}) \rangle = 4 \langle I_1 \rangle \left(1 - \cos \frac{\Delta \varphi}{2} \cos \psi \right)$$

where $\Delta\varphi$ is given by (7.112a). This equation describes the formation of a modulated fringe pattern related to the difference in phase $\Delta\varphi$. Since $\Delta\varphi$ appears in the form of a beat, a few conditions are imperative towards achieving a good interaction. The most important of these require the interference patterns to be of high density and localized on or in the neighbourhood of the specimens.

Equation (7.112) forms the basis of comparative holographic moiré: if the corresponding points P of the test and the master specimens are illuminated along the wave vector $\mathbf{K}_e$ and observed along the wave vector $\mathbf{K}_o$, and if $\mathbf{L}$ and $\mathbf{L}'$ are the displacement vectors of the point P for the test and the master specimens, the moiré fringe pattern provides information about the resolved part of the difference in the displacement $\Delta\mathbf{L}$ along the direction bisecting the illumination

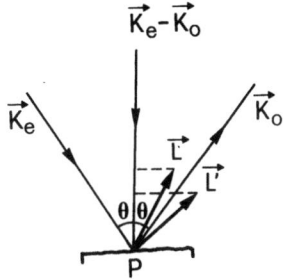

Fig. 7.52. Illustration of the measurement sensitivity in comparative holographic moiré

and the observation vector. This interpretation is schematically shown in Fig. 7.52. A point that needs emphasis is that the method can be applied to the testing of large size objects. In the case where relatively large models are to be studied, it becomes imperative to use a divergent wave to illuminate the objects. A disadvantage of using a spherical wave is that the interpretation of the interference pattern is no longer straightforward. However, this drawback is no means restrictive insofar as the main interest of the method resides in comparing the mechanical responses, that is to say, in discerning the presence and location of the relative phase differences between the master and the test components subjected to the same loading levels.

The testing operation in comparative holographic moiré can be classified into two types: those which require full-time access to the master object, and those which permit one to store the master "fingerprint" in the system memory. Whereas the first type of system is suitable for particular testing problems or trial studies to fix the testing modes, the latter is specially suited for routine testing. Evidently, in the latter case, the testing modes are supposed to be established beforehand using the first system. From the configuration point of view, the first system is simple and rapid to execute, and enables fringe-control techniques to be readily applied. An important feature of the system is its ability to work in real time which permits one to visualize the evolution of the fringes contouring the difference in displacements with increasing loads. On the other hand, and in order to avoid exposing the master object to repeated stresses, a few systems have been described [7.70] which include the capacity of storing the mechanical response of the master piece to predefined load levels. The stored displacement field serves as a signature with respect to which all the incoming test pieces are compared.

An example of fringe pattern depicting the out-of-plane difference displacement component ΔL_z is depicted in Fig. 7.53a. The interferogram compares deflections of two square aluminum plates clamped along the edges and submitted to centrally concentrated loads. Figure 7.53b shows an example of flaw detection. Two thick rubber plates of the same size and shape are machined from the same rubber sheet. In one rubber plate a defect is introduced by drilling a small hole on its rear surface. The other rubber plate is considered free of defects. The two plates are held in their respective loading frames, such that the

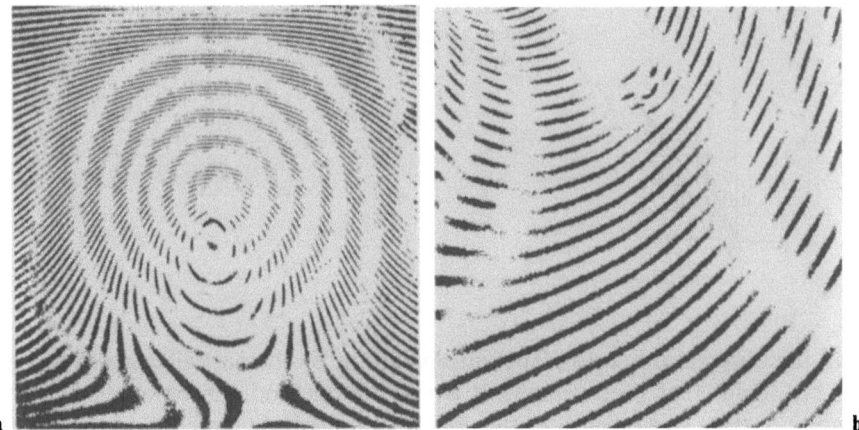

Fig. 7.53. (a) Example of fringe pattern corresponding to ΔL_z of two square plates clamped along the edges and subjected to centrally concentrated loads. (b) Comparative holographic moiré applied to nondestructive inspection: the moiré fringe pattern clearly displays and outlines the flaw. In both (a) and (b) an auxiliary system of fringes has been added to improve the quality of the moiré fringes

in-plane displacements at the edges are constrained. The presence of a closed set of moiré fringes clearly betrays the presence and location of the flaw. Auxiliary displacement fringes are added to obtain a good formation of the moiré fringes.

7.5.4 Formation of the Modulated Pattern

The discussion on holographic moiré fringes has so far been restricted to the simplified case of equal intensities and visibilities of the two superposed patterns. Considering the number of holographic moiré techniques devised, we will briefly examine the form of the modulated pattern in holographic moiré obtained in conditions of unequal intensities and visibilities of the interacting patterns. In such a case, the equation of the modulated pattern can be written as

$$\langle E \rangle = \frac{\langle I(\mathrm{P}) \rangle}{(\langle I_1 \rangle + \langle I_2 \rangle)} = 1 - \mathrm{Re}\left\{ \exp\left(i\frac{\varphi_1 + \varphi_2}{2}\right) \left[\frac{\langle I_1 \rangle V_1}{\langle I_1 \rangle + \langle I_2 \rangle} \cdot \exp\left(-i\frac{\varphi_1 - \varphi_2}{2}\right) + \frac{\langle I_2 \rangle V_1'}{\langle I_1 \rangle + \langle I_2 \rangle} \exp\left(i\frac{\varphi_1 - \varphi_2}{2}\right) \right] \right\}, \qquad (7.113)$$

where I_1 and I_2 are intensities and V_1 and V_1' are the fringe visibilities in the two interacting interference images; and

$$V_1 = \frac{2\sqrt{\langle I_{11} \rangle \langle I_{12} \rangle}}{\langle I_{11} \rangle + \langle I_{12} \rangle} |\gamma_1|, \qquad V_1' = \frac{2\sqrt{\langle I_{21} \rangle \langle I_{22} \rangle}}{\langle I_{21} \rangle + \langle I_{22} \rangle} |\gamma_2|,$$

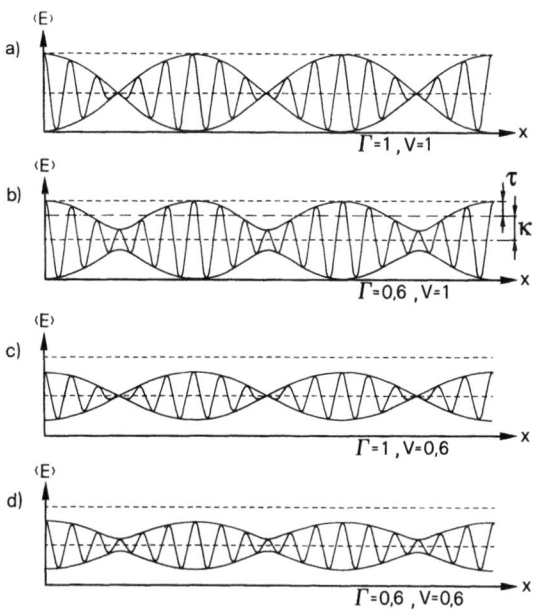

Fig. 7.54a–d. The influence of the depth of modulation and the carrier-fringe visibility on the formation of the moiré pattern

where ($\langle I_{11} \rangle$, $\langle I_{12} \rangle$) and ($\langle I_{21} \rangle$, $\langle I_{22} \rangle$) are the intensity pairs corresponding to the interfering wave fronts giving rise to the two interacting patterns; $|\gamma_{1,2}|$ are the degree of coherence between the respective interfering wave fronts. Relation (7.113) describes the formation of a slowly varying modulation superposed on the carrier fringes. The term representing the moiré fringes is not isolated. The observation of these fringes requires that the carrier fringes are resolved by the observation system. *Yoshina* and *Takasaki* [7.154] have demonstrated that the visibility of such a moiré is inherently null.

In practice, it is a difficult task to convene the conditions necessary to achieve an equation similar to (7.19). The quality of moiré (Fig. 7.54) starts deteriorating with the fall in the carrier-fringe visibilities under the influence of displacements and deformations of the specimens. The situation is aggravated in the presence of the basic ill-conditioning of the system specified by $\langle I_{11} \rangle \neq \langle I_{12} \rangle$, $\langle I_{21} \rangle \neq \langle I_{22} \rangle$ and $\langle I_1 \rangle \neq \langle I_2 \rangle$. An understanding of the variation of the quality of the moiré fringes is thus of great interest in improving the method's serviceability. This is accomplished by introducing two parameters, the depth of modulation and the fringe clarity, to describe the quality of the moiré fringes, of course, within the framework of zero-fringe visibility.

The depth of modulation is actually a means of expressing the degree to which the moiré fringes modulate the carrier pattern. The envelope of the modulated pattern varies in the interval from $\langle E_{\max} \rangle$ to $\langle E_{\min} \rangle$. The depth of modulation is defined as the ratio between τ and κ; τ and κ are shown in Fig. 7.54b. Their values are obtained as

$$\tau = \frac{\langle E_{max} \rangle - \langle E_{min} \rangle}{2}, \tag{7.114a}$$

$$\kappa = \frac{\langle E_{max} \rangle - \langle E_{min} \rangle}{2} - 1, \tag{7.114b}$$

whereupon the depth of modulation Γ follows as

$$\Gamma = \frac{\tau}{\kappa} = \frac{\langle E_{max} \rangle - \langle E_{min} \rangle}{\langle E_{max} \rangle + \langle E_{min} \rangle - 2}. \tag{7.115}$$

The effect of different amounts of modulation upon the carrier pattern – of unit visibility – is shown in Figs. 7.54a, b. Figure 7.54a is obtained for a unit modulation. It is generally desirable to operate with such a fully modulated carrier pattern. If the modulation is less than unity, the quality of the moiré is reduced (Fig. 7.54b). The depth of modulation in itself is not sufficient to describe the quality of the moiré fringes. Figure 7.54c illustrates that even if Γ is kept at unity, the quality of the moiré fringes deteriorates with the decrease in the visibility of the carrier fringes. Figure 7.54d, drawn for $\Gamma = 0.6$, $V_1 = V_1' = 0.6$, shows the combined influence of the depth of modulation and visibility on the quality of the moiré fringes. Another useful parameter to define the modulated fringes is fringe clarity C

$$C = \frac{\langle E_{max} \rangle - \langle E_{min} \rangle}{\langle E_{max} \rangle + \langle E_{min} \rangle}. \tag{7.116}$$

The maximum value of C is 1/3. Figure 7.55 depicts a three-dimensional plot of the moiré fringe clarity, where its maximum value has been normalized to unity, with respect to the carrier fringe visibility V and depth of modulation Γ.

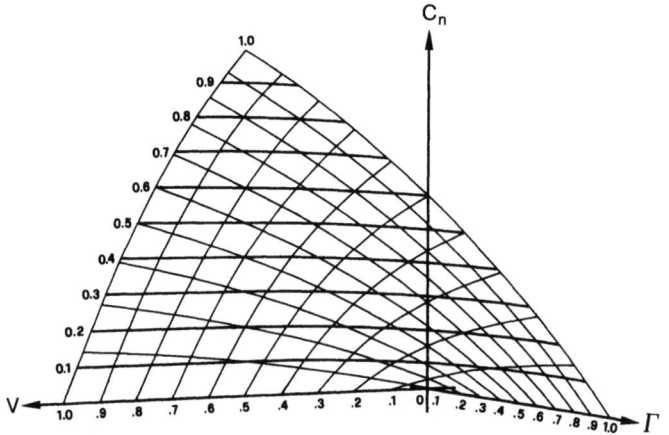

Fig. 7.55. A three dimensional plot of the variation of moiré fringe clarity as a function of the visibility and depth of modulation

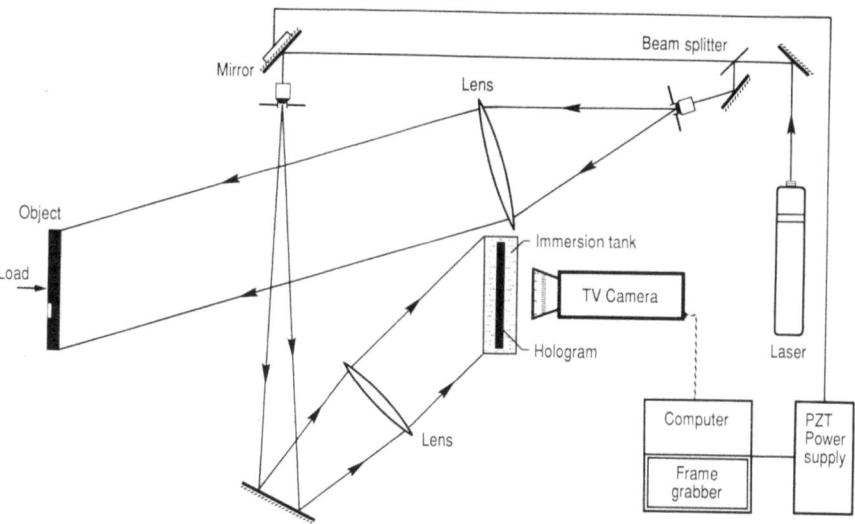

Fig. 7.56. Layout of the phase shifting comparative holographic moiré arrangement

7.5.5 Phase-Shifted Comparative Holography

Rastogi et al. [7.90] have combined phase shifting and computer image-processing facilities to holographic moiré with a view to obtaining a highly improved visualization of the phase difference corresponding to the difference in displacements of two nominally identical specimens. The schematic of the experimental set-up is depicted in Fig. 7.56. The test object, an aluminum plate, is identical to the master object except that a flaw is simulated by drilling a small untraversing hole on its rear surface. Both the plates are clamped along the edges and submitted to centrally concentrated loads. The moiré fringes resulting from the numerical subtraction of sawtooth fringe maps is depicted in Fig. 7.57a. Furthermore, and as shown in Fig. 7.58a, the most visible discordances between the absolute values of the difference $|\Phi_t - \Phi_m|$ and $\Delta\Phi$ occur toward the higher and lower gray levels. The quality of the display could thus subsequently be enhanced by varying the threshold in a manner to select the midrange gray levels only. The gray levels between the lower and upper thresholds are made equal to the maximum gray level, while the rest are made equal to zero. This is shown in Fig. 7.58b. The interferogram obtained by gray-level windowing of Fig. 7.57a is displayed in Fig. 7.57b.

The sawtooth fringe map related to the difference in displacements between the test and master objects and obtained in accordance to the procedure described in Sect. 7.2.4 is illustrated in Fig. 7.59a. The sawtooth fringe map shown in Fig. 7.59b is related to the difference in displacements of two nominally identical defect-free plates subjected to different loadings The sawtooth fringe map (Fig. 7.59c) is related to the difference in displacements of test and master

286   7. Techniques to Measure Displacements, Derivatives and Surface Shapes

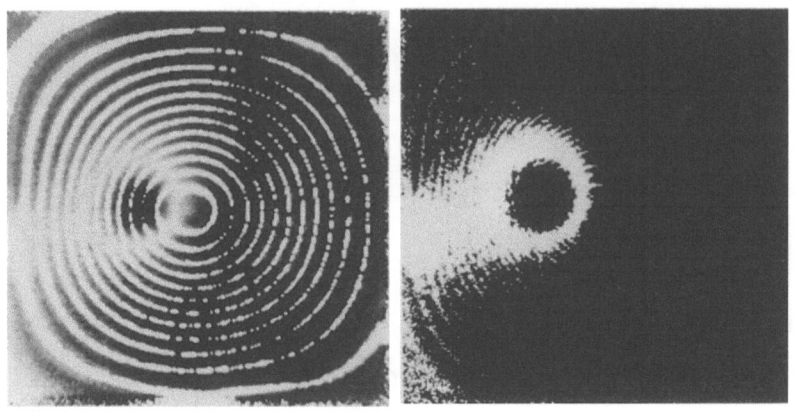

a   b

Fig. 7.57. (a) Moiré phase distribution resulting from the numerical subtraction of the sawtooth phase functions Φ_t and Φ_m. (b) Phase map obtained by gray-level windowing of (a)

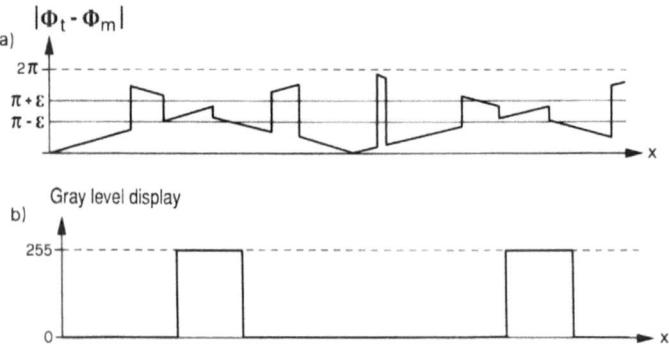

Fig. 7.58. (a) Display of the absolute value of $|\Phi_t - \Phi_m|$. The thin line indicates the setting of the threshold levels. (b) The resulting gray-level display

objects subjected to loadings similar to those to obtain the interferogram (Fig. 7.59b). The difference displacement fringes are modulated by the phase term arising from the difference in the mechanical response of the two plates resulting from the presence of the flaw. The plate containing the flaw is easily recognized. This result leads one to conclude that a flaw can still be detected by comparative holography even if the loading of the two specimens are not rigorously the same. As an example of data analysis, Fig. 7.60 shows a 3-D plot of the difference phase distribution obtained from the calculated phase map of Fig. 7.59b.

Rastogi [7.155–159], and *Simova* and *Stoev* [7.160–161] have recently developed phase shifting techniques inherently applicable to a range of interferometers using holographic moiré. A highlight of their approach is that it permits to provide access to information relative to both the difference and sum of phases existing in the two arms of a holographic moiré interferometer. This

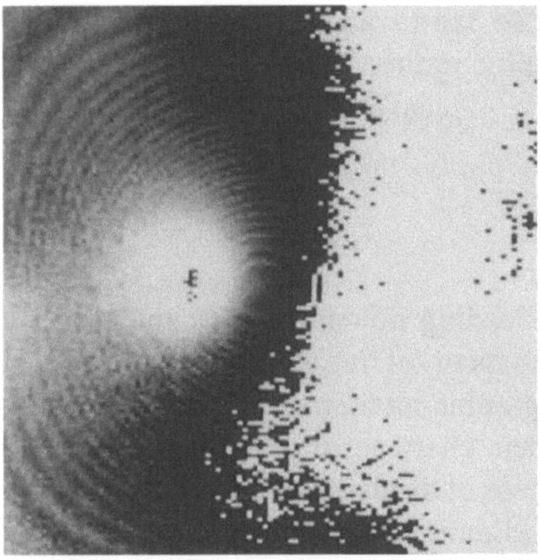

Fig. 7.59a–c. Interference phase distribution $\Delta\Phi$ corresponding to difference in displacements: **(a)** two plates, one flaw-free and the other with a flaw, subjected to identical loadings; **(b)** two identical flaw-free plates subjected to unequal loadings; and **(c)** two plates, one flaw-free and the other with a flaw, subjected to unequal loadings

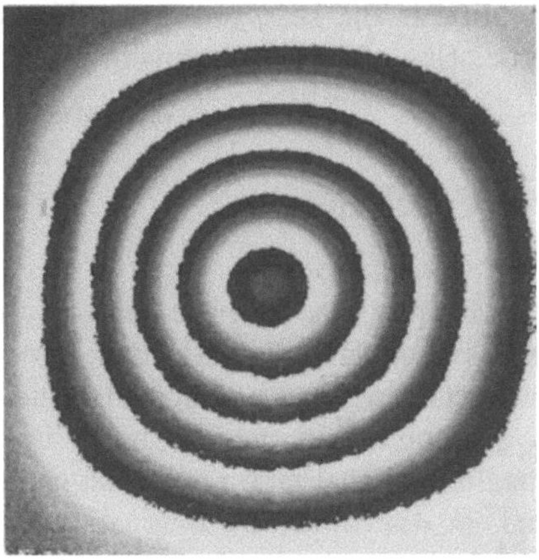

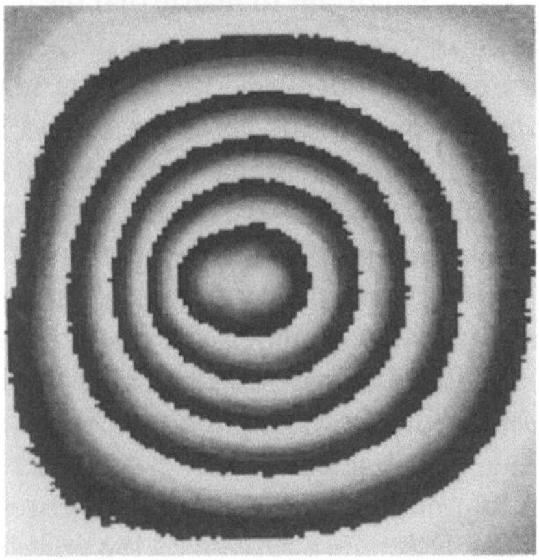

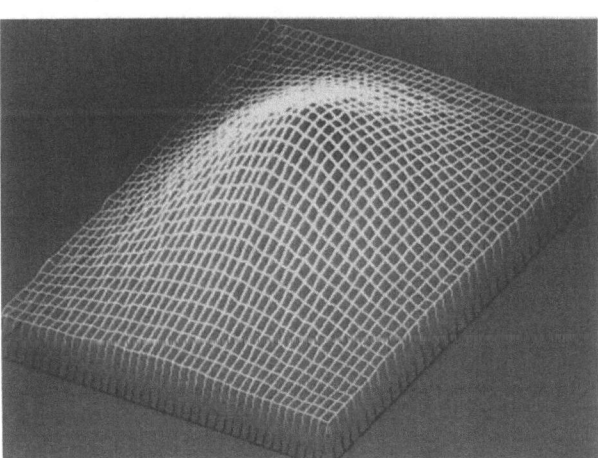

Fig. 7.60. Three dimensional plot of the difference phase distribution obtained in Fig. 7.59(b)

solution is the best, one has achieved so far to address fringe readout and processing of fringe patterns in holographic moiré.

7.6 Conclusions

This chapter has reviewed some of the leading holographic techniques in the deformation analysis and shape measurement of rough object surfaces. The successful functioning of most of the holographic-interferometry-based techniques is guaranteed, on the one hand, by their characteristic feature of providing a real-time visualization of the full-field maps of displacement and shape contours and, on the other, by their ability to eliminate the influence of rigid body movements on the displayed fringe patterns. It is important to realize that the type of loading and the mode of observation of the corresponding response both have a significant influence on the performance of the holographic methods in non-destructive inspection. For best results, the response of the specimen should be observed in terms of fringes characterizing the distribution of that displacement or derivative component in which the presence of that particular type of flaw is best distinguished. The development of comparative holographic interferometry in the context of flaw detection is an important step forward in non-destructive inspection. It is reasonable to expect that in the future the technique will find a broad base of applications in flaw detection, fatigue damage research, and in the comparative analysis of the mechanical behaviour of two neighbourly conceived engineering components. Another enticing development which has attracted considerable attention has been the consummation of the marriage between holographic interferometry and phase-shifting techniques. Driven by the need for automatic data analysis, more and more interferometers based on holography are incorporating phase shifting technique into their respective instrumentations. The integration of the two processes has led to a significant enhancement of the potentialities of holographic interferometry in non-destructive testing and in the measurement of deformations and shapes of diffuse object surfaces.

Some of the avenues in which the future of holographic interferometry looks to be most encouraging are in the conception of prototypes, in the evaluation of the performances of structural components and in the verification of finite-element computer codes. In the final count and notwithstanding the manifold merits of holographic interferometry, the popularity of the technique with the user will depend on his perception of its real utility and adaptability to the problem at hand.

References

7.1   K. Stetson, R.L. Powell: J. Opt. Soc. Am. **55**, 1694 (1965)
7.2   K. Stetson: Exp. Tech. **15**, 15 (1991)

7.3 R.K. Erf (ed.): *Holographic Non-Destructive Testing* (Academic, New York 1974)
7.4 C.M. Vest: *Holographic Interferometry* (Interscience, New York 1979)
7.5 W. Schumann, M. Dubas: *Holographic Interferometry*, Springer series in Optical Sciences 16 (Springer-Verlag, Heidelberg 1979)
7.6 Y.I. Ostrovsky, M.M. Butusov, G.V. Ostrovskaya: *Interferometry by Holography*, Springer series in Optical Sciences 20 (Springer, Verlag, Heidelberg 1980)
7.7 W. Schumann, J.P. Zürcher, D. Cuche: *Holography and Deformation Analysis*, Springer series in Optical Sciences 46 (Springer-Verlag, Heidelberg 1985)
7.8 Y.I. Ostrovsky, V.P. Schepinov, V.V. Yakovlev: *Holographic Interferometry in Experimental Mechanics*, Springer series in Optical Sciences 60 (Springer-Verlag, Heidelberg 1991)
7.9 C.M. Vest, Soc. Photo-Opt. Instr. Eng. **349**, 186 (1982)
7.10 G. Birnbaum, C.M. Vest: Int'l Adv. in Nondestructive Testing **9**, 257 (1983)
7.11 I. Yamaguchi, H. Saito: Jpn. J. Appl. Phys. **8**, 768 (1969)
7.12 S.P. Timashenko, J.N. Goodier: *Theory of Elasticity* (McGraw-Hill, New York 1970)
7.13 F.P. Kupper, C.A. Van Dijk: Optics Lasers Technol. **5**, 69 (1973)
7.14 P.D. Plotkowski, Y.Y. Hung, J.D. Hovanesian, G. Gerhart: Opt. Eng. **24**, 754 (1985)
7.15 D.C. Holloway, A.M. Patacca, W.L. Fourney: Appl. Opt. **17**, 1213 (1978)
7.16 P. Long, D. Hsu, B. Wang: Appl. Opt. **26**, 4282 (1987)
7.17 P. Long, D. Hsu, B. Wang: Opt. Eng. **27**, 867 (1988)
7.18 S. Toyooka: Appl. Opt. **16**, 1054 (1977)
7.19 T. Nomura, K. Yoshikawa, M. Hamada: Bull. Jpn Soc. Prec. Engg. **20**, 97 (1986)
7.20 T. Nomura, K. Yoshikawa, H. Tashiro, Y. Nagata, S. Kakunai: Bull Jpn. Soc. Prec. Engg. **21**, 239 (1987)
7.21 C. Froehly, J. Monneret, J. Pasteur, J.Ch. Viénot: Opt. Acta **16**, 343 (1969)
7.22 S. Walles: Opt. Acta **17**, 899 (1970)
7.23 S. Walles: Ark. Fys. **40**, 299 (1970)
7.24 J. Monneret: Etude théorique et expérimentale des phénomènes observables en interférométrie holographique, interprétation des interférogrammes et applications à la métrologie des microdéplacements. D.Sc. Dissertation, University of Besançon, Besançon (1973)
7.25 P.K. Rastogi, M. Spajer, J. Monneret: Opt. Lasers Eng. **2**, 79 (1981)
7.26 P.K. Rastogi, P. Jacquot, L. Pflug: Optica Acta **30**, 1067 (1983)
7.27 W. Schumann, M. Dubas: Optik **47**, 391 (1977)
7.28 E. Champagne, L. Kersch: J. Opt. Soc. Am. **59**, 1535 (1969)
7.29 L.A. Kersch: Mater. Eval. **29**, 125 (1971)
7.30 J.P. Waters: Appl. Opt. **11**, 630 (1972)
7.31 C. Shakher, R.S. Sirohi: Can. J. Phys. **57**, 2155 (1979)
7.32 C. Shakher, R.S. Sirohi: J. Phys. E. **13**, 284 (1980)
7.33 A. Stimpfling, P. Smigielski: Opt. Eng. **24**, 821 (1985)
7.34 J.B. Garcia, J.L. Fernandez, C. Lopez, A.F. Doval, M.P. Amor: Appl. Opt. **30**, 1588 (1991)
7.35 N. Abramson: Appl. Opt. **13**, 2019 (1974)
7.36 N. Abramson: Appl. Opt. **14**, 981 (1975)
7.37 H. Bjelkhagen: Appl. Opt. **16**, 1272 (1977)
7.38 N. Abramson, H. Bjelkhagen: Appl. Opt. **17**, 187 (1978)
7.39 N. Abramson, H. Bjelkhagen: Appl. Opt. **18**, 2870 (1979)
7.40 M. Dubas, W. Schumann: Opt. Acta **24**, 1193 (1977)
7.41 A. Dzubur, D. Vukicevic: Appl. Opt. **23**, 1474 (1984)
7.42 J. Monneret: Opt. Commun. **2**, 159 (1970)
7.43 P. Jacquot, P.K. Rastogi: Appl. Opt. **18**, 2022 (1979)
7.44 K.A. Stetson: J. Opt. Soc. Am. **64**, 1 (1974)
7.45 P. Jacquot: In Hologram Interferometry and Speckle Metrology – Tech. Digest (OSA, 1980) p. MA 4
7.46 D.B. Neumann, R.C. Penn: Exp. Mech. **15**, 241 (1975)
7.47 A.E. Ennos, M.S. Virdee: Exp. Mech. **22**, 202 (1982)

7.48 H. Kreitlow, T. Kreis, W. Jüptner: Appl. Opt. **26**, 4256 (1987)
7.49 E.B. Aleksandrov, A.M. Bonch-Bruevich: Sov. Phys. – Tech. Phys. **12**, 258 (1967)
7.50 A.E. Ennos: J. Sci. Instr. **1**, 731 (1968)
7.51 V. Fossati-Bellani, A. Sona: Appl. Opt. **13**, 1337 (1974)
7.52 P.M. Boone: Opt. Laser Technology **2**, 94 (1970)
7.53 J.N. Butters: *The Engineering Uses of Holography*, ed. by E.R. Robertson, J.M. Harvey (Cambridge Univ. Press, London 1970)
7.54 M. Schlüter, A. Nowatzyk: Opt. Acta **27**, 794 (1980)
7.55 J. Ebbeni: Proc. 5th Int'l Conf. Exp. Stress Anal. CISM, Udine, Italy (1974)
7.56 Y.Y. Hung, C.E. Taylor: J. Appl. Mech. **42**, 1 (1975)
7.57 C.A. Sciammarella, J.A. Gilbert: Exp. Mech. **16**, 215 (1976)
7.58 C.A. Sciammarella, J.A. Gilbert: Appl. Opt. **15**, 2176 (1976)
7.59 J. Monneret, P.K. Rastogi, M. Spajer, Proc. Soc Photo. Opt. Instr. Eng. **136**, 258 (1977)
7.60 C.A. Sciammarella, S.K. Chawla: Mech. Res. Commun. **4**, 333 (1977)
7.61 J.A. Gilbert, C.A. Sciammarella, S.K. Chawla: Exp. Mech. **18**, 321 (1978)
7.62 C.A. Sciammarella, S.K. Chawla: Exp. Mech. **18**, 373 (1978)
7.63 J.A. Gilbert: Exp. Mech. **18**, 436 (1978)
7.64 J.A. Gilbert, G.A. Exner: Exp. Mech. **18**, 382 (1978)
7.65 W.J. Beranek, A.J.A. Bruinsma: SESA Spring Meeting, San Francisco (1979)
7.66 P.K. Rastogi: Visualisation et mesure des déplacements tangentiels, et des déformations associées, par moirés holographiques et interférométrie speckle. Ph.D. Dissertation, Besançon, France (1979)
7.67 J.A. Gilbert, J.W. Herrick: Opt. Lasers Eng. **1**, 21 (1980)
7.68 S.K. Chawla, C.A. Sciammarella: Exp. Mech. **20**, 240 (1980)
7.69 C.A. Sciammarella: Opt. Eng. **21**, 447 (1982)
7.70 C.A. Sciammarella, P.K. Rastogi, P. Jacquot, R. Narayanan: Exp. Mech. **22**, 52 (1982)
7.71 Y. Katzir, I. Glaser: Appl. Opt. **21**, 678 (1982)
7.72 Y. Katzir, A.A. Friesem, I. Glaser: Opt. Lett. **8**, 163 (1983)
7.73 L. Pirodda: Appl. Opt. **28**, 1842 (1989)
7.74 L. Pirodda: *Practical Holography* IV, SPIE Proc. **1212**, 267 (1990)
7.75 L. Pirodda, L.J. Griffiths: Opt. Lasers Eng. **14**, 39 (1991)
7.76 P.K. Rastogi, E. Denarie: Appl. Opt. **31**, 2402 (1992)
7.77 P.K. Rastogi, Exp. Mech. **25**, 325 (1985)
7.78 J. Tsujiuchi, N. Takeya, K. Matsuda: Optica Acta **16**, 709 (1969)
7.79 T. Tsuruta, N. Shiotake, Y. Itoh: Optica Acta **16**, 723 (1969)
7.80 K.A. Stetson: Optik **29**, 386 (1969)
7.81 N.E. Molin, K.A. Stetson: Optik **31**, 3 (1970)
7.82 M. Dubas, W. Schumann: Optica Acta **22**, 807 (1975)
7.83 P. Hariharan, B.F. Oreb, N. Brown: Opt. Commun. **41**, 393 (1982)
7.84 P. Hariharan, B.F. Oreb, N. Brown: Appl. Opt. **22**, 876 (1983)
7.85 P. Hariharan: Opt. Eng. **24**, 632 (1985)
7.86 B. Breuckmann, W. Thieme: Appl. Opt. **24**, 2145 (1985)
7.87 K. Kinnstaetter, A.W. Lohmann, J. Schwider, N. Streibl: Appl. Opt. **27**, 5082 (1988)
7.88 K. Creath: *Progress in Optics* **26**, 349 (North-Holland, Amsterdam 1988)
7.89 K. Creath: Appl. Opt. **28**, 2170 (1989)
7.90 P.K. Rastogi, M. Barillot, G.H. Kaufmann: Appl. Opt. **30**, 722 (1991)
7.91 S. Timoshenko, S.W. Kreiger: *Theory of Plates and Shells*, 2nd edn. (McGraw-Hill, New York 1959)
7.92 K.A. Stetson: Opt. Laser Technol. **2**, 80 (1970)
7.93 P. Boone, R. Verbiest: Opt. Acta **16**, 555 (1969)
7.94 G. Cadoret: Annls Inst. Tech. Bâtim. **373**, 27 (1979)
7.95 P.K. Rastogi, Opt. Acta **31**, 159 (1984)
7.96 P.K. Rastogi, Opt. Commun. **58**, 1 (1986)

7.97 P.K. Rastogi, J. Mod. Optics **38**, 1251 (1991)
7.98 S. Toyooka: Opt. Acta **25**, 991 (1978)
7.99 M. Dubas, W. Schumann: Opt. Acta **21**, 547 (1974)
7.100 K.A Stetson: Appl. Opt. **14**, 272 (1975)
7.101 K.A. Stetson: Appl. Opt. **14**, 2256 (1975)
7.102 K.A. Stetson: J. Opt. Soc. Am. **66**, 627 (1976)
7.103 R.J. Pryputniewicz, K.A. Stetson: Appl. Opt. **15**, 725 (1976)
7.104 R.J. Pryputniewicz: Appl. Opt. **17**, 3613 (1978)
7.105 R.J. Pryputniewicz: Hologram Interferometry and Speckle Metrology – Tech. Digest (OSA 1980) p. MB 1
7.106 J.C. Charmet, F. Montel: Rev. Physique Appliquée **12**, 603 (1977)
7.107 J. Ebbeni, J.C. Charmet: Appl. Opt. **16**, 2543 (1977)
7.108 C.A. Sciammarella, J.A. Gilbert: Appl. Opt. **12**, 1951 (1973)
7.109 S.K. Dhir, J.P. Sikora: Exp. Mech. **12**, 323 (1972)
7.110 J.P. Sikora: Exp. Mech. **18**, 101 (1978)
7.111 S. Toyooka: Jpn. J. Appl. Phys. **18**, 1289 (1979)
7.112 J. Fang, F.L. Dai: Exp. Mech. **31**, 163 (1991)
7.113 J.R. Varner: In *Handbook of Optical Holography*, ed. by H.J. Caulfield (Academic, New York 1979) p. 595
7.114 F. Deschryver: Proc. Soc. Photo-Opt. Instr. Eng. **349**, 99 (1982)
7.115 B.P. Hildebrand, K.A. Haines: J. Opt. Soc. Am. **57**, 155 (1967)
7.116 J.S. Zelenka, J.R. Varner: Appl. Opt. **7**, 2107 (1968)
7.117 J.R. Varner: Appl. Opt. **10**, 212 (1971)
7.118 A.A. Friesem, U. Levy: Appl. Opt. **15**, 3009 (1976)
7.119 F.M. Küchel, H.J. Tiziani: Opt. Commun. **38**, 17 (1981)
7.120 D.E. Cuche: Proc. Soc. Photo-Opt. Instr. Eng. **599**, 88 (1985)
7.121 T. Tsuruta, N. Shiotake, J. Tsujiuchi, K. Matsuda: Jpn. J. Appl. Phys. **6**, 661 (1967)
7.122 J.S. Zelenka, J.R. Varner: Appl. Opt. **8**, 1431 (1969)
7.123 E.S. Marrone, W.B. Ribbens: Appl. Opt. **14**, 23 (1975)
7.124 N. Abramson: Appl. Opt. **15**, 200 (1976)
7.125 N. Abramson: Appl. Opt. **15**, 1018 (1976)
7.126 M. Yonemura: Appl. Opt. **21**, 3652 (1982)
7.127 R. Thalmann, R. Dändliker: Opt. Eng. **24**, 930 (1985)
7.128 P.K. Rastogi, L. Pflug: Appl. Opt. **30**, 1603 (1991)
7.129 P.K. Rastogi, L. Pflug: J. Mod. Optics **38**, 1673 (1991)
7.130 P.K. Rastogi, L. Pflug: Appl. Opt. **29**, 4392 (1990)
7.131 P.K. Rastogi, L. Pflug: J. Mod. Optics 37, 1233 (1990)
7.132 P.K. Rastogi, L. Pflug: Soc. Photo-Opt. Instr. Eng. **1554B**, 48 (1991)
7.133 X. Youren, C.M. Vest, E.J. Delp: Opt. Lett. **8**, 451 (1983)
7.134 D.B. Neumann: Hologram Interferometry and Speckle Metrology – Tech. Digest (OSA 1980) p. MB 2
7.135 D.B. Neumann: Opt. Eng. **24**, 625 (1985)
7.136 Z. Füzessy, F. Gyimesi: Soc. Photo-Opt. Instr. Eng. **398**, 240 (1983)
7.137 Z. Füzessy, F. Gyimesi: Opt. Eng. **23**, 780 (1984)
7.138 F. Gyimessi, Z. Füzessy: Soc. Photo-Opt. Instr. Eng. **473**, 65 (1984)
7.139 F. Gyimessi, Z. Füzessy: Opt. Commun. **53**, 17 (1985)
7.140 Z. Füzessy, F. Gyimesi: Opt. Commun. **57**, 31 (1986)
7.141 Z. Füzessy, F. Gyimesi, J. Kornis: Optics Laser Technol. **18**, 318 (1986)
7.142 F. Gyimesi, Z. Füzessy: J. Mod. Optics **35**, 1699 (1988)
7.143 Z. Füzessy, F. Gyimesi, I. Banyasz: Opt. Commun. **68**, 404 (1988)
7.144 F. Gyimesi, Z. Füzessy: Soc. Photo-Opt. Instr. Eng. **1183**, 338 (1989)
7.145 Z. Füzessy, F. Gyimesi: Soc. Photo-Opt. Instr. Eng. **1183**, 471 (1989)
7.146 Z. Füzessy, F. Gyimesi: Soc. Photo-Opt. Instr. Eng. **IS 8**, 194 (1990)

7.147 P.K. Rastogi: Appl. Opt. **23**, 924 (1984)
7.148 P.K. Rastogi: J. Phys. E: Sci-Instrum. **17**, 1094 (1984)
7.149 P.K. Rastogi: Proc. Soc. Photo-Opt. Instr. Eng. **661**, 16 (1986)
7.150 E. Simova, V. Sainov: Opt. Eng. **28**, 261 (1989)
7.151 E. Simova, V. Sainov: Opt. Eng. **28**, 550 (1989)
7.152 V. Sainov, E. Simova, E. Manoah: Opt. Lasers Eng. **11**, 15 (1989)
7.153 B. Breuckmann: Proc. Soc. Photo-Opt. Instr. Eng. **398**, 234 (1983)
7.154 Y. Yoshino, H. Takasaki: Appl. Opt. **15**, 1124 (1976)
7.155 P.K. Rastogi: Appl. Opt. **31**, 1680 (1992)
7.156 P.K. Rastogi: J. Mod. Optics **39**, 677 (1992)
7.157 P.K. Rastogi: Opt. Commun. **93**, 336 (1992)
7.158 P.K. Rastogi: Opt. Eng. **32**, 190 (1993)
7.159 P.K. Rastogi: Appl. Opt. **32**, 3669 (1993)
7.160 E.S. Simova, K.N. Stoev: Appl. Opt. **31**, 2405 (1992)
7.161 E.S. Simova, K.N. Stoev: Appl. Opt. **31**, 5965 (1992)

8. Study of Vibrations

C.S. Vikram

Center for Applied Optics, The University of Alabama in Huntsville, Huntsville, AL 35899, USA.

Holographic analysis of vibration of diffusively reflecting surface started with the pioneering work of *Powell* and *Stetson* [8.1]. They showed that the reconstructed image of a time-average hologram (exposure time much longer than the vibration period) of a sinusoidally vibrating object is modulated by a fringe pattern. The intensity modulation factor is mathematically known under certain conditions and can be related to the amplitude of vibration. For the real-time observation, *Stetson* and *Powell* [8.2] introduced recording the stationary object, reconstructing the image and interfering it with the wave from the vibrating object. Again, the live fringe pattern can be used to extract the information about the vibration amplitude. These introductory contributions resulted in enormous activity among researchers and application engineers. The technique itself has continuously been refined and extended to cases of importance. In this chapter, we shall discuss various methods available in holographic vibration analysis.

8.1 Time-Average Holographic Interferometry of Sinusoidal Vibration and Separable Motions

Suppose o is the complex amplitude $|o|\exp(i\phi)$ corresponding to the stationary object point. $|o|$ and ϕ are absolute amplitude and phase, respectively. To begin with, let us consider the case of sinusoidal vibration. Then the displacement $A(t)$ against time t can be written as

$$A(t) = A \sin \omega t, \tag{8.1}$$

where A and ω are the vector amplitude and the angular frequency, respectively, of the vibration. The complex object beam amplitude then becomes

$$o(t) = |o| \exp[i(\phi + \boldsymbol{K} \cdot \boldsymbol{A} \sin \omega t)], \tag{8.2}$$

where $\boldsymbol{K}$ is the sensitivity vector. For the linear recording and reconstruction, the complex amplitude of the reconstruction wave is proportional to the time average of $o(t)$. If T is the exposure time, the time average is

$$u = (1/T) \int_0^T |o| \exp[i(\phi + \boldsymbol{K} \cdot \boldsymbol{A} \sin \omega t)] \, dt = oM_T, \tag{8.3}$$

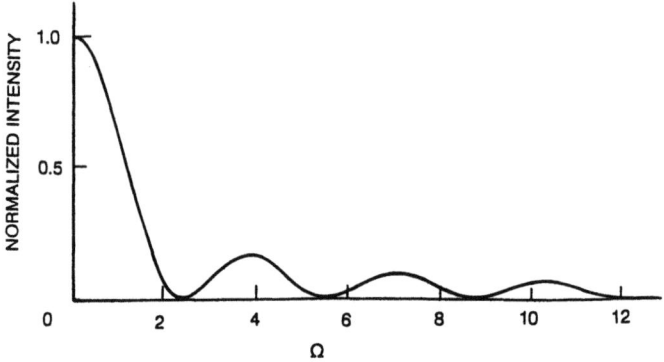

Fig. 8.1. Plot of the time-average fringe function $J_0^2(\Omega)$ of sinusoidal vibration

where M_T is called a *characteristic function*. For $T \gg 1/\omega$, the limiting case is

$$M_T = J_0(\mathbf{K}\cdot\mathbf{A}), \tag{8.4}$$

where J_0 is the zero-order Bessel function of the first kind. The normalized (as compared to that at rest) intensity of the reconstructed image is therefore

$$I = |M_T|^2 = J_0^2(\mathbf{K}\cdot\mathbf{A}). \tag{8.5}$$

Thus, the intensity at a point of the image will be corresponding to the local value of $\Omega = \mathbf{K}\cdot\mathbf{A}$. A fringe pattern is generally formed displaying the contours of equal vibration amplitude. The brightest regions correspond to the nodes. The function $|M_T|^2$ has been plotted against Ω in Fig. 8.1.

The fringes are of decreasing brightness against increased values of Ω. Therefore, only a limited number of fringes are generally observable. Special techniques of hologram/image processing to increase the brightness of higher-order fringes are available [8.3–6].

At this stage, a more general definition of the characteristic function can be considered for complicated motions. The obvious extension is *separable motions*. In this type, each point of the object surface moves with a common time function $f(t)$. The case of sinusoidal vibration discussed above is a particular example. The optical phase variation against time can then be represented as $\phi(t) = \Omega f(t)$ and the general *characteristic function* as

$$M_T(\Omega) = (1/T)\int_0^T \exp[i\Omega f(t)]\,dt. \tag{8.6}$$

Dealing with vibrations, detailed discussions of $M_T(\Omega)$ for damped harmonic vibrations [8.6] and nonlinear vibrations [8.7–10] are available. The interpretation of the characteristic function using the probability theory is also known [8.11].

8.2 Real-Time Interferometry of Vibration

In this approach [8.2], the hologram of the stationary object is recorded and the image is reconstructed. Also present is the actual object beam but now with the vibrating object. Two fields are allowed to mutually interfere. Suppose the object and the reconstructed waves are of equal irradiance and have the same polarization. From the negative hologram, the reconstructed wave phase is shifted by π rad. So, at a given time t, the normalized intensity is

$$I(t) = \tfrac{1}{4}(-1 + \exp[i\Delta\phi(t)])^2 = \tfrac{1}{2}\{1 - \cos[\Delta\phi(t)]\}, \tag{8.7}$$

where $\Delta\phi(t)$ is the phase change of the object beam from the steady state position. For the sinusoidal vibration case, $\Delta\phi(t)$ is $\Omega \sin \omega t$. Therefore,

$$I(t) = \tfrac{1}{2}[1 - \cos(\Omega \sin \omega t)]. \tag{8.8}$$

The response of the human eye is about 40 ms. Therefore, for usually high vibration frequencies, the *time-averaged* response is observed. The time-average of (8.8) yields the characteristic function in the present situation given by

$$|M_T|^2 = \left(\frac{1}{T}\right) \int_0^T [1 - \cos(\Omega \sin \omega t)] \, dt = 1 - J_0(\Omega). \tag{8.9}$$

This function has been plotted in Fig. 8.2. Notice that the real-time method has half as many fringes as compared to the time average method represented by (8.5).

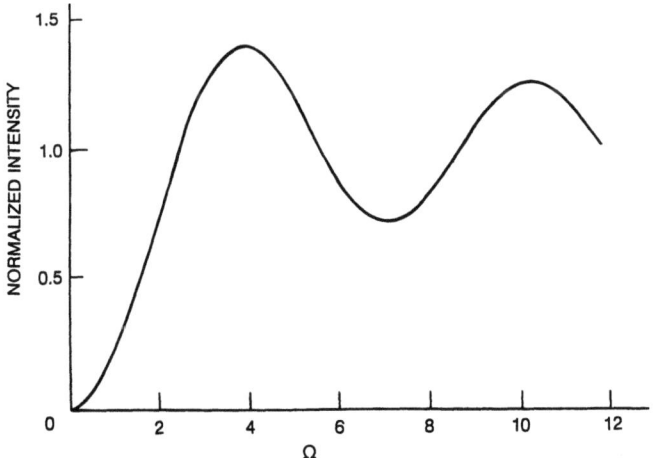

Fig. 8.2. Variation of the real-time fringe function $1 - J_0(\Omega)$ against Ω

In practice, it is difficult to obtain the same irradiance, polarization state, phase (other than the usual phase shift π between the waves), etc. [8.12]. Therefore, a more general form of (8.9) is

$$|M_T|^2 = 1 + C^2 + 2CJ_0(\Omega), \tag{8.10}$$

where C accounts for the reduced contrast. The desired value of C is ± 1. Anyway, the contrast will always be lower than that from the J_0^2 fringes.

8.3 Time-Average Holography of Nonseparable Motions and Multiple Modes

Several workers [8.13–21] addressed time-average holography of complex vibrations. These include nonseparable vibrations, multiple modes, and combination of vibration, drift and acceleration. The general theme is to evaluate [8.16] the *characteristic function* so that the experimental interferogram can be analyzed. The salient results are described in this section.

Equation (8.6) can be generalized to write the characteristic function of N separable motions as

$$M_T = (1/T) \int_0^T \exp\left[i \sum_{n=1}^{N} \Omega_n f_n(t)\right] dt, \tag{8.11}$$

where Ω_n and $f_n(t)$ represent the individual functions for the nth component. An example is the vibration simultaneously in N different modes. Combined displacement vector in that case is

$$A = \sum_{i=1}^{N} A_i \cos(\omega_i t + \theta_i), \tag{8.12}$$

where A_i, ω_i, and θ_i represent the individual vector amplitude, angular frequency, and phase, respectively. Another example is sinusoidal vibration superimposed with constant acceleration. The displacement vector in that case becomes

$$A = at^2 + bt + c + d \sin \omega t. \tag{8.13}$$

For combination of statistically independent and separable motions, the characteristic function of (8.11) becomes [8.14]

$$M_T = \prod_{n=1}^{N} M_T(n), \tag{8.14}$$

where $M_T(n)$ is the individual characteristic function due to the nth component of motion and can be obtained by deleting the summation operation in (8.11).

Due to the multiplicative nature of the combined characteristic function, a dark fringe is always obtained even if a single component yields a dark fringe.

8.3 Time-Average Holography of Nonseparable Motions and Multiple Modes

Another important result, experimentally demonstrated by *Molin* and *Stetson* [8.18], is that fringe patterns due to independent motion components are localized independently. The independent localization can be very useful in the analysis to separate different components of motion.

For temporally dependent motions, the characteristic function becomes complicated and analysis of results becomes difficult. A special case [8.13] of such a motion is vibration with different simultaneous modes at a single frequency ω but different geometrical structures. The optical phase variation against time of such a motion is

$$\phi(t) = \sum_{n=1}^{N} \Omega_n \cos(\omega t + \theta_n), \tag{8.15}$$

which can also be written as

$$\phi(t) = \Omega \cos(\omega t + \theta). \tag{8.16}$$

In (8.16) θ is a phase term and

$$\Omega = \left| \sum_{n=1}^{N} \Omega_n \right| \tag{8.17}$$

represents the *phasor sum* of the components. Ω_n is now represented in vector form where the angular orientations are represented by phase angles θ_n. The reconstructed pattern intensity then takes the well known form of (8.5), i.e. $J_0^2(\Omega)$. Again, in the presence of several modes, obtaining vibration parameters from the reconstruction is very difficult. However, some useful information is possible [8.19] if the number of modes are small, say two.

A slightly more complex case when the phase function is first two cosine terms of Fourier series, i.e.

$$\phi(t) = \Omega_1 \cos(\omega t - \theta_1) + \Omega_2 \cos(2\omega t - \theta_2). \tag{8.18}$$

Even in this simple but with rationally related frequency case, the characteristic function is as complex as [8.16]

$$M_T = \sum_{m=-\infty}^{+\infty} i^{-m} J_{-2m}(\Omega_1) J_m(\Omega_2) \exp[im(\theta_2 - 2\theta_1)]. \tag{8.19}$$

The observed fringe pattern is even more complex as it is proportional to $|M_T|^2$. The complicated characteristic functions can be evaluated using a digital computer [8.19] and experimentally verified [8.17]. However, the physical significance of such a characteristic function is difficult when an interferogram is to be analyzed for the vibration parameters.

The work of *Stetson* [8.20] was an important new development in this regard. The technique, called the method of *stationary phase* provides approximate solutions for the characteristic function. The method assumes the fact that for sufficiently large vibration amplitudes, the contribution to the fringe function is mainly from object positions where the velocity is zero. This process of the asymptotic analysis then becomes like multiple beam interferometry and hence easier interpretation.

For the simple sinusoidal-vibration case given by (8.1), the derivative is zero when $\omega t = \frac{1}{2}\pi, \frac{3}{2}\pi, \frac{5}{2}\pi$, and so on. The integrand then slows down and if the integration is performed near these points, the characteristic function becomes [8.12]

$$M_T = [2/(\pi\Omega)]^{1/2} \cos(\Omega - \pi/4), \tag{8.20}$$

which is also the approximate value (asymptotic) of $J_0(\Omega)$ for nonzero values of Ω. The approximation is very accurate except for very small values of Ω.

Stetson [8.20] considered the general multi-mode vibration whose frequencies are rationally related. He considered the case of two sinusoidal vibration with the frequency ratio of 2:1 along with experimental comparisons. The general conclusion was that, particularly for larger motions, the technique is a valuable tool in quantitative vibration analysis.

In the analysis of complex time-average holograms, cases exist when only one part in the characteristic function is dependent upon the exposure time. For example, in sinusoidal vibration with uniform slow drift superimposed, the vibration term in (8.14) is independent (time-averaged) of the exposure time. The slow drift term, which is a sinc function, contains the exposure time. Two or more interferograms with different exposure times can, in principle, separate out the individual characteristic functions [8.22, 23].

Tonin and *Bies* [8.24–26] considered the characteristic function for three-dimensional vibrations. When a time varying force with a constant frequency is applied, the general three-dimensional motion of the object point will form an ellipse as a locus. The general motion form thus have three orthogonal components. *Tonin* and *Bies* [8.24–26] introduced the concept of general characteristic function of three-dimensional vibrations with a single frequency. In some particular cases, the interpretation of the characteristic function is straightforward. However, the theory of generalized least squares is introduced to solve the non-linear characteristic equation for the component amplitude and phases of vibration. Several experimental results with vibrations of cylinders provide good applicability of the theory. Thus, 3-D orthogonal vibration (single frequency) including phase over a surface can be studied from the time-average reconstruction.

8.4 Stroboscopic Holographic Interferometry

Time-average holographic interferometry particularly of the sinusoidal vibration case is a very well established tool. This is because the experimental arrangement is in the simplest form. However, there are two technical drawbacks and could be important in particular situations. One is that the phase of

8.4 Stroboscopic Holographic Interferometry

the vibration across the object surface is not determined by the technique. Other is the loss of fringe intensity or the visibility with increasing the vibration amplitude (Fig. 8.1). Stroboscopic holographic interferometry [8.27–31], although experimentally more complex, solves these problems. The common procedure is to store the hologram using stroboscopy when the object surface is on its extreme positions of the vibration cycle. That means, according to the sinusoidal vibration case of (8.1), providing short pulses at $\omega t = \frac{1}{2}\pi, \frac{5}{2}\pi, \frac{9}{2}\pi, \ldots$ for one set and $\omega t = \frac{3}{2}\pi, \frac{7}{2}\pi, \frac{11}{2}\pi, \ldots$ for the other. This results in the intensity I and the characteristic function M_T given by the common double-exposure procedure as

$$I = |M_T|^2 = |\tfrac{1}{2}\exp(i\Omega) + \tfrac{1}{2}\exp(-i\Omega)|^2 = \cos^2\Omega. \tag{8.21}$$

Fringes in the real-time manner can also be observed for example by storing the static position, replacing the processed hologram (or in place processing), and then observing the interference between the reconstructed image and the field from stroboscopically illuminated object. The pulses at extreme positions of the vibration cycle are not necessary. Any two positions can be chosen. This is particularly important for large vibration amplitudes to reduce the sensitivity. Also, the example of sinusoidal vibration can easily be extended to any periodic separable motion. The method actually provides a difference between the two states of deformation.

Notice (8.21) that now the fringe intensity is not decreasing against increasing Ω. Nevertheless, this approximation is when the pulse widths can be neglected. In practice, particularly for higher frequency or amplitude, the object motion during the pulse may not be negligible. Increased pulse width are desirable so that the total exposure times are small for system stability. *Listovets* and *Ostrovskii* [8.32], and *Miler* [8.33] addressed the problem and obtained the characteristic functions for fine pulse width cases. If the pulse duration is Δ and the pulses appear symmetrically at times $\pm\tau$ in the sinusoidal vibration cycle, the first order approximation (valid for small Δ) becomes [8.12]

$$|M_T|^2 = \text{sinc}^2[\tfrac{1}{2}\Omega(\omega\cos\omega\tau)\Delta]\cos^2(\Omega\sin\omega\tau). \tag{8.22}$$

Thus, there is some loss of visibility as the pulse width increases. This was experimentally demonstrated by *Shajenko* and *Johnson* [8.30].

Double-pulse holography using a powerful laser solves the exposure time problem. Otherwise, when a continuous laser beam with chopping is used, a couple of techniques to reduce the time are available. One of the techniques [8.31] in stroboscopic holographic interferometry is to record the static surface with half of the required exposure. The remaining exposure is obtained with stroboscopic pulses at a given phase of the vibrating object. However, if more than half of the exposure time is used in the static situation, the time for the stroboscopic illumination can be reduced [8.34] but with reduced contrast of the fringe pattern. Two or more symmetrically located pulses in the vibration cycle can also be provided to reduce the total exposure time during the stroboscopic illumination [8.35–37]. Vibrating object may have different mean positions from that of the static object. The static object position as a reference can give errors. To solve this problem, time-average exposure of the vibrating object can act as a reference [8.38]. This approach is thus a combination of time-average and stroboscopic techniques.

Temporal changes in the vibration has always been of interest in the study of dynamic events. *Fryer* [8.39] described an arrangement where the reference beam scans across the hologram during the object's motion. The plate was already exposed with the object in the static or reference position. Scanning provides each region of the hologram with object deformation at different times. These early efforts of multiplexing for time-resolved holographic interferometry are continuing [8.40].

Another useful application [8.41] of stroboscopic holographic interferometry is when two rationally related sinusoidal modes are present. The object beam phase variation in such a case is

$$\phi(t) = \Omega_1 \sin(\omega t) + \Omega_2 \sin(n\omega t + \beta), \tag{8.23}$$

where Ω_1 and Ω_2 are dot products of the sensitivity vector to respective amplitude vectors. ω and $n\omega$ are the angular frequencies of vibration (n being a rational number). β represents the constant phase between the two modes. Sets of pulses can be provided at times such that $\sin(n\omega t + \beta)$ is constant. Such a hologram will yield information about Ω_1. Similarly, when pulses are provided for fixed values of $\sin \omega t$, information about Ω_1 can be obtained. Detailed descriptions are available in [8.41] about selection of the times of pulses.

One interesting aspect is when two modes have irrationally related frequencies. In that situation, pulses to keep the contribution of one mode fixed does not retain a fixed position in the cycle of the other. As the number of pulses increases, every position of the other mode is covered like that in the time-average method. Thus, when $\sin(n\omega t + \beta)$ is fixed by selecting values of t, we obtain $J_0^2(\Omega_1)$ in the reconstruction. Similarly, fixed $\sin \omega t$ would yield $J_0^2(\Omega_2)$.

Pulsed illumination can provide solutions to study the combinations of different types of motion [8.42–50]. The general theme is common. If the analytical form of motion is known, then a suitable number of double exposure holograms can give a set of equations. These equations can eventually be solved for the analysis. Several ways to obtain an independent set of equations are available. Different object illumination direction to obtain different sensitivity vectors is one approach [8.45, 47]. Methods proposed include damped oscillations [8.44, 46], superposition of two or more motions [8.45, 47], sinusoidal vibration with space variant phase [8.48], superposition of three sinusoidal vibrations [8.49], and many component motion [8.50]. As an example, let us consider the many component motion case [8.50]. The optical phase variation of the object beam in this case is

$$\phi(t) = \sum_{n=1}^{N} \Omega_n f_n(t), \tag{8.24}$$

where Ω_n determines the strength of the nth component whose law of motion is $f_n(t)$.

Now suppose we record a double-exposure hologram. For simplicity, let us assume one exposure with the static object and the other at the time t_k. The normalized reconstructed irradiance of such a hologram is

$$I = \cos^2\left[\frac{1}{2}\sum_{n=1}^{N} \Omega_n f_n(t_k)\right]. \tag{8.25}$$

Similarly, we can record N double-exposure holograms by changing the time of the recording ($k = 1, N$). Let us assume the kth recording gives the fringe order l_k (may be fractional). We now have N equations

$$\sum_{n=1}^{N} \Omega_n f_n(t_k) = 2\pi l_k, \tag{8.26}$$

where $k = 1, N$. Equation (8.26) can be written in the matrix form as

$$\boldsymbol{\Omega} f = 2\pi l \tag{8.27}$$

where $\boldsymbol{\Omega}$ and l are N-dimensional row vectors and f is $N \times N$ matrix whose nth row and the kth column is $f_n(t_k)$. Thus we obtain

$$\boldsymbol{\Omega} = 2\pi l f^{-1} \tag{8.28}$$

provided that $\det f \neq 0$. Thus, in principle, the strength Ω_n of different component can be determined. Some other stroboscopic techniques are described in conjunction with temporally modulated holography and phase shifting techniques in the next two sections.

8.5 Temporally Modulated Holography

In this form of holographic interferometry, object and/or reference beam are time dependent. The complete description of the technique has been presented by *Aleksoff* [8.51]. In usual off-axis holography, the reconstructed amplitude transmittance for the virtual image point is proportional to or^* where o and r are the amplitudes of the object and the reference beams, respectively. However, if during the exposure, suppose the object and the reference beam amplitudes are modulated by $f_o(t)$ and $f_R(t)$, respectively. Also, we represent the usual object motion by the additional phase factor in the object beam as $\exp[i\phi(t)]$. Then, the characteristic function is proportional to the time-average of the amplitude transmittance as

$$M_T = (1/T) \int_0^T f_o(t) f_R^*(t) \exp[i\phi(t)] \, dt. \tag{8.29}$$

Notice that in the absence of modulation $[f_o(t) = f_R(t) = 1]$, the characteristic function becomes the common one like that defined by (8.3 or 6). Some cases of particular interest are described below utilizing the modulation.

8.5.1 Frequency Translation

Frequency translation [8.51, 52] is generally applied to the sinusoidal vibration case $[\phi(t) = \Omega \sin \omega t]$. The reference beam is modulated so that the frequency of the reference beam is shifted by $n\omega$ where n is an integer and ω is the frequency of the vibrating object. Writing $f_o(t) = 1$ and $f_R(t) = \exp(in\omega t)$, (8.29) becomes, for $T \gg 2\pi/\omega$,

$$M_T = (1/T) \int_0^T \exp(-in\omega t) \exp(i\Omega \sin \omega t) \, dt = J_n(\Omega). \tag{8.30}$$

The intensity in the image is proportional to

$$I = |M_T|^2 = J_n^2(\Omega). \tag{8.31}$$

The advantages of the frequency translation can now be discussed. First, in the ordinary time-average holography ($n = 0$), for very small values of $\Omega \ll 1$, $J_0^2(\Omega)$ is nearly unity. The usually bright image does not show practical changes when Ω increases from zero. The result is that very small vibration amplitudes cannot be measured. By the frequency translation procedure, we can form $J_1^2(\Omega)$ fringes ($\approx \Omega^2/4$ as $\Omega \to 0$). Now, in the dark-field background, intensity changes are visible as Ω increases from zero. Vibration amplitudes as small as $\lambda/100$ can now be measured [8.51, 52]. *Ueda* et al. [8.53] estimated the detectable smallest amplitude as $2.7 \times 10^{-4} \lambda$. The role of the shift in the frequency slightly different from the frequency of the object's vibration and phase of the vibration has also been studied [8.54]. It has been found that these aspects can further enhance the sensitivity of the technique. Some applications of the frequency translation in the quantitative vibration analysis are described in Sect. 8.6.

Another application is the sensitivity reduction for the large amplitudes of the vibration. For large Ω, the approximate value of $J_n(\Omega)$ is $(2/\pi\Omega)^{1/2} \cos[\Omega - (\pi n/2) - (\pi/4)]$. The location of zeros thus increases nearly linearly with the order n. This aspect can be used to reduce the number of fringes. *Aleksoff* [8.51] showed this effect for a vibrating speaker with $n = 1, 9, 28$ and 75.

8.5.2 Amplitude Modulation

In this technique [8.55] the amplitude of the reference beam is modulated at the frequency of object's vibration but with phase difference Δ. So we have $f_o(t) = 1$ and $f_R(t) = \cos(\omega t - \Delta)$. Solving (8.29), for the sinusoidal vibration case:

$$I = |M_T|^2 = J_1^2(\Omega) \cos^2 \Delta. \tag{8.32}$$

Several holograms can be recorded with different values of Δ to obtain the phase variations over the object.

8.5.3 Laser Irradiance Modulation

Suppose the laser light output is modulated so that $f_o(t) = f_R(t)$. Stroboscopic holographic interferometry is an example. Stetson [8.56] provided general analysis of this type of modulation. The main feature of the irradiance modulation technique [8.56] is that the brightness of the high-order time-averaged vibration fringes increases although exposure times (as compared to common stroboscopy) remains low. If $f_i(t)$ is the irradiance modulation ($0 \leq f_i(t) \leq 1$), the characteristic function becomes

$$M_T = (1/C_0 T) \int_0^T f_i(t) \exp(i\Omega \cos \omega t) \, dt, \tag{8.33}$$

where

$$C_0 = (1/T) \int_0^T f_i(t) \, dt. \tag{8.34}$$

Now suppose $f_i(t)$ is periodic with period π/ω, and an even function. Also, suppose that the laser is completely turned on at the extreme positions of the object's vibration. Under these conditions and large amplitudes of vibration, Stetson [8.56] determined that

$$M_T^2 \approx [2/(\pi \Omega C_0^2)] \cos^2(\Omega - \pi/4). \tag{8.35}$$

Thus, the irradiance of the high order fringes increases by $1/C_0^2$.

The method can also be combined with holographic addition (a portion of exposure with the stationary object and the remaining in the time-average manner with vibrating object) [8.57].

8.5.4 Phase Modulation

Suppose the phase of the reference beam is modulated sinusoidally at the vibration frequency of the object. In this case $f_o(t) = 1$ and $f_R(t) = \exp[i\Omega_R \sin(\omega t)]$ where Ω_R is the *modulation depth*. The characteristic function (8.29) then becomes

$$M_T = (1/T) \int_0^T \exp(i\Omega_R \sin \omega t) \exp[i\Omega \sin(\omega t - \phi_o)] \, dt, \tag{8.36}$$

where ϕ_o is the phase of the vibration of the object point. The components can be added to obtain the form of (8.16) and under the assumption of $T \gg 1/\omega$, we obtain

$$M_T = J_0[(\Omega^2 + \Omega_R^2 - 2\Omega\Omega_R \cos \phi_o)^{1/2}]. \tag{8.37}$$

Several applications of the phase modulation technique are available. *Neumann et al.* [8.58] used the technique for the phase measurement by inspecting the zero-order fringe. Many interferograms are needed to cover the object surface unless real-time approach is used. *Aleksoff* [8.51, 59] applied the modulation for the vibration analysis of resonating ADP crystals. In the technique of *Metherell et al.* [8.60], very small vibrations are recorded as a linear function of image irradiance. The vibration amplitude and phase can then be determined. Time average as well as real-time approaches can be utilized. To understand the principle, let us consider the case when $\phi_o = 0$. Equation (8.37) gives the image irradiance as

$$I = M_T^2 = J_0^2(\Omega_R - \Omega). \tag{8.38}$$

If Ω is very small, then usual time average holography ($\Omega_R = 0$) yields practically uniform intensity ($I \approx 1 - \Omega^2/2$). However, in the case of finite Ω_R,

$$I \approx J_0^2(\Omega_R) + 2J_0(\Omega_R)J_1(\Omega_R)\Omega. \tag{8.39}$$

At the optimum phase modulation amplitude [8.60] of 65° or $\Omega_R \approx 1.1$, we have

$$I \approx 0.5179 + 0.6778\Omega. \tag{8.40}$$

Thus, the irradiance is a linear function of the amplitude rather than the square one. Tolerance limits in the modulation amplitude and phase have also been considered in detail [8.61].

Levitt and *Stetson* [8.62] have used the technique to obtain contours of constant phase. The phase term of (8.36) can be slightly modified to include the phase of the modulation ϕ_R. Then the dark fringes of the reconstruction are characterized by

$$\Omega^2 + \Omega_R^2 - 2\Omega\Omega_R \cos(\phi_o - \phi_R) = j_{0,n}^2, \tag{8.41}$$

where $j_{0,n}$ is the nth zero of the J_0 function.

Suppose two interferograms are obtained with the same modulation amplitude Ω_R but with different phases ϕ_{R1} and ϕ_{R2}. If two interferograms are superimposed then at the intersections at equal orders (n), (8.41) gives [8.62]

$$\phi_o = [(\phi_{R1} + \phi_{R2})/2] \pm n\pi, \tag{8.42}$$

where $n = 0, 1$. Thus, using N interferograms, $N(N-1)/2$ phase contours can be obtained [8.62]. The work was extended [8.63] to bright fringes are as well so that bright–bright and bright–dark combinations can also be used. Thus, if total number of fringes are small, the method could be used to locate phases intermediate to those located by *Levitt* and *Stetson*.

Phase modulation is useful in the study of large amplitudes also. If at a given region of the object, the amplitude Ω is large, then the usual time average pattern becomes dark. However, the *bias* Ω_R due to the modulation (8.38) will give the brightest fringe at $\Omega = \Omega_R$ and not at $\Omega = 0$. This fact has also been used for object motion compensation by reflecting the reference beam from an object point [8.64–66]. *Waters* [8.67] introduced the *speckle reference beam* approach for compensation of extraneous motion. The reference beam is derived by focusing the laser light on a spot of the object surface. Then, the motion relative to that of the spot is studied. *Kreitlow* et al. [8.68] presented several quantitative aspects of object motion modulated reference beam holographic interferometry. The modulation approach can also be applied to TV holography to study very small vibration amplitudes (Sect. 8.8).

8.5.5 Holographic Subtraction

Holographic subtraction has found applications to extend the measurement range [8.69–72]. Suppose two equal exposures are given, first the object at rest and the other in time-average manner while the object is vibrating but with phase shift of π in one of the beams. The reconstructed amplitudes are subtracted and the intensity is

$$I = [1 - J_0(\Omega)]^2. \tag{8.43}$$

Hariharan [8.69] used the approach to study very small vibration amplitudes. In usual time-average holography, the reconstructed intensity $J_0^2(\Omega)$ is about $1 - \Omega^2/2$. Therefore the changes in the intensity are small relative to the bright non-vibrating regions. In the subtraction procedure given by (8.43) the intensity is proportional to Ω^2. Therefore, as the vibration amplitude increases from the zero fringe or no-vibration region, changes in intensity in the dark background are readily seen. *Hariharan* [8.69] established the feasibility with experiments using the vibration of a circular diaphragm.

Sato et al. [8.71] introduced the weighted subtraction approach to generate contour lines of small vibration amplitudes. Rather than equal exposures, suppose ρth portion of the exposure is given with the stationary object while the remaining $(1 - \rho)$th portion of the object is vibrating. The reconstructed intensity then would be

$$I = [\rho - (1 - \rho)J_0(\Omega)]^2. \tag{8.44}$$

Now, the dark lines are the contours of

$$J_0(\Omega) = \rho/(1 - \rho). \tag{8.45}$$

In (8.43), or $\rho = 0.5$ case, the dark lines correspond to $\Omega = 0$. Thus, for several values of ρ, several contour lines of Ω can be generated. *Sato* et al. [8.71] presented experimental demonstrations of the weighted subtraction method and also considered the role of the phase error.

Holographic subtraction was primarily developed for small amplitudes of vibrations. However, the technique can increase the average reconstructed irradiance for large amplitudes ($\Omega \gg 1$) above noise level. Thus, the approach is useful for large vibration amplitudes also [8.70]. In this respect, the approach is similar to that of holographic addition [8.5].

8.5.6 Other Modulation Techniques

Besides the above basic modulation techniques, there have been several special applications. *Mottier* [8.73] used a triangular phase modulation of the reference beam to enhance the intensity of the higher-order fringes in time-average holography of sinusoidal vibration. Applications of generalized stroboscopy have been provided by *Mottier* [8.74, 75]. Measuring the vibration waveform (not necessarily the sinusoidal one) itself is the ultimate goal of the analysis. Modulation techniques can be used in this direction, as demonstrated by *Dallas* and *Lohmann* [8.76], and *Cutter* [8.77]. *Politch* [8.78] has developed *spectroscopic holography* to study random vibrations.

8.6 Fringe-Shifting and Quantitative Analysis Techniques

Quantitative analysis of the interferograms has been time consuming and the subject to considerable errors. The reason is that in the usual method, the quantitative information is available only at the fringe maxima and minima. Besides finding the locations of these positions, extrapolations are necessary for the analysis over the entire surface. The process thus becomes laborious and still inaccurate.

On the other hand, phase-stepping techniques in conjunction with digital data analysis can solve the speed and accuracy problems simultaneously. Detailed descriptions of this kind of phase measuring technique has been provided by *Creath* [8.79]. In this section we shall describe the applications concerning holographic vibration analysis.

The application of the phase shifting technique in holographic vibration analysis has been reported by several workers [8.80–84]. Stroboscopic holographic interferometry provides cosinusoidal fringes of the usual two-beam interference type. Accurate measurements using the conventional phase shifting technique is then possible [8.80, 82]. *Nakadate* et al. [8.82] provided the detailed description of the process, particularly the role of the pulse duration of the stroboscopic illumination. The vibration amplitude accuracy of $\lambda/100$ can be achieved by suitable parameter selection. The analysis was for the normal procedure of stroboscopic holography when the stationary object is first recorded with continuous illumination, processed *in situ* or replaced. Stroboscopic illumination is then provided for observing the real-time fringes. Accuracy is affected by the pulse width. However, a finite pulse width is required for the needed amount of light. *Hariharan* and *Oreb* [8.80] solved this problem by reversing the procedure. In their process, the hologram of the vibrating object is first recorded under stroboscopic illumination. Short pulse duration will simply increase the exposure time. The real time fringes can then be observed with stationary object and continuous illumination. Figure 8.3 shows time-average fringes and raw phase data obtained from the stroboscopic exposure at one

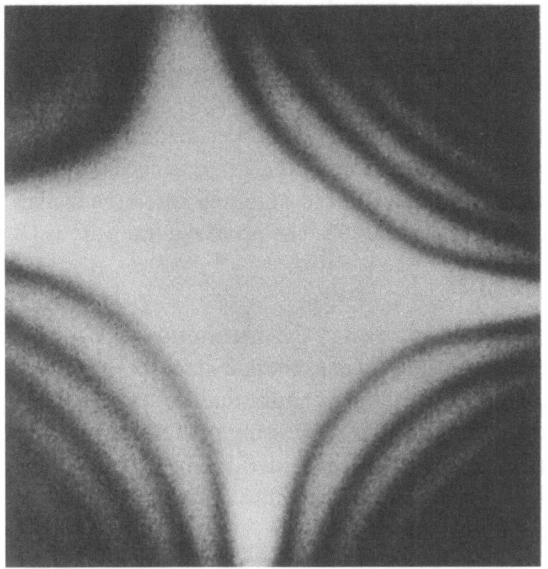

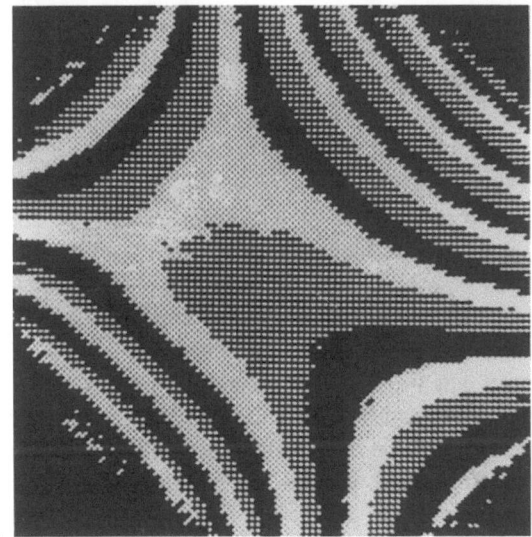

Fig. 8.3a, b. Holographic interferometry of a square metal plate vibrating at 231 Hz. (a) Time-average fringes and (b) raw phase data obtained from the stroboscopic exposure and the digital system [8.80]

extreme position. The object was a square metal plate vibrating at 231 Hz [8.80]. The digital data can be color coded and computer processed [8.80] for further analysis.

Hariharan et al. [8.83] applied the digital phase shifting technique to include the case when the motion direction is unknown or the motion is in two or three dimensions. Basically, the process is recording the holograms at different illumination directions to have different sensitivity vectors. The motion components along different directions can then be evaluated for the vector components.

Nakadate [8.81] presented a detailed theory and experimental results for the phase-shifting procedure using the time-average real-time fringes. In this process, the hologram of the stationary object stored and then real-time fringes are observed while the object is vibrating (Sect. 8.2). The observed intensity is [8.81]

$$I = \alpha + \beta \cos(\zeta + \Psi + \phi) J_0(\Omega), \tag{8.46}$$

where α and β can define the bias and contrast of the distribution. ζ, Ψ, and ϕ are the bias deformation of the vibration, phase change due to expansion or shrinking of the hologram plate, and the phase change due to the reference light respectively. Principle of the computer image processing with phase-shifting can be explained now. The period of the phase ϕ is divided into equal intervals ϕ_i ($i = 1, 2, 3, \ldots, N$) and the corresponding intensities I_i are determined. Mean values of a_0, a_1, and b_1 are then calculated:

$$a_0 = (1/N) \sum_{i=1}^{N} I_i = \alpha, \tag{8.47}$$

$$a_1 = (2/N) \sum_{i=1}^{N} I_i \cos \phi_i = \beta \cos(\zeta + \Psi) J_0(\Omega), \tag{8.48}$$

$$b_1 = (2/N) \sum_{i=1}^{N} I_i \sin \phi_i = \beta \sin(\zeta + \Psi) J_0(\Omega), \tag{8.49}$$

where $\phi_i = 2\pi i/N$ ($i = 1, 2, 3, \ldots, N$). A new fringe pattern can then be computed:

$$(a_1^2 + b_1^2)^{1/2}/a_0 = (\beta/\alpha)|J_0(\Omega)|. \tag{8.50}$$

The ratio β/α is the fringe contrast when the vibration amplitude is zero ($\Omega = 0$). However, the calculated interferogram can be normalized by the light intensity at the zero-vibration amplitude position to obtain the function $|J_0(\Omega)|$. Thus, the high-contrast fringe pattern is obtained without the effect of the unwanted background terms. This can be seen from the plot of Fig. 8.4 where the pattern is compared with real-time fringes. Also, the phase discontinuities at the zero of the function permit good measurement accuracy of the dark fringe position.

Nakadate [8.81] also introduced a phase-shifting double-exposure holographic interferometry technique for the vibration analysis. Although experimentally more complex, the approach does not require corrections due to the hologram deformation, the environmental disturbances are more tolerable, and the input fringes are of high contrast.

In the above mentioned methods, the vibration amplitude is still determined from the fringe orders. In the methods due to *Stetson* and *Brohinsky* [8.84], the vibration amplitude can be directly extracted from the irradiance measurements

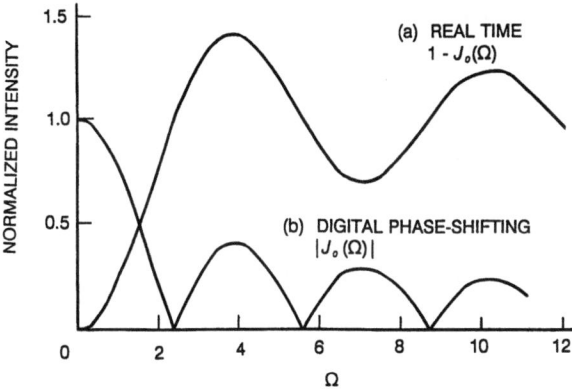

Fig. 8.4a, b. Plot of fringe functions of the vibrating object. (a) Digital phase-shifting fringes. (b) Real-time fringes

from the time-average as well as real-time fringes. A sinusoidal phase modulation (Sect. 8.5.4) is used to offset the argument of the Bessel-function pattern. Time-average interferograms with zero, plus, and minus offsets can be numerically processed like fringe patterns from phase-step interferometry. The irradiance of the phase modulated time-average reconstruction is

$$I = I_b + I_o J_0^2(\Omega - B), \tag{8.51}$$

where I_b and I_o represent the background and the object irradiances, respectively. B is the modulation depth. A number of holograms with different values of B but otherwise identical recording and processing conditions should now be made. A seed value of the unknown Ω function is now assumed. The set of equations from (8.51) can now be solved for I_b, I_o, the least-squared error, and then the actual squared error. The process is repeated with incremental seed values of Ω to estimate for the minimum squared error and hence the new seed value. The iterative process would ultimately yield the corrected value of Ω within a specified tolerance. The process requires a significant computation for every point considered. However, a simpler solution is possible [8.84, 85] using asymptotic form of the Bessel function.

Stetson and *Brohinsky* [8.84], and *Pryputniewicz* and *Stetson* [8.85] also introduced the parallel procedure for the real time (concomitant) time-average (or parallel electro-optic) holographic interferometry. First, different phase steps [8.82] are used to determine a factor

$$H = 4CI_o J_0(\Omega), \tag{8.52}$$

where C is the square root of the ratio of the irradiances of the reconstructed and transmitted fields. Different modulation depths can now be given to obtain H_1, H_2 and H_3 in terms of $J_0(\Omega - B)$, $J_0(\Omega)$ and $J_0(\Omega + B)$, respectively. With starting seed value of Ω and the iterative computation, the correct value of Ω can then be obtained. Again, the approximate periodic nature of the J_0 function can be used for reduced computation.

The methods proposed by *Stetson* and *Brohinsky* [8.84], and *Pryputniewicz* and *Stetson* [8.85] have the serious advantage of the analysis directly from the irradiance measurements. However, the amplitude and the phase of the bias vibration must be precisely known and several experimental aspects are critical [8.84].

In this connection, the role of frequency-shifting approach (Sect. 8.5.1) has been described recently [8.86, 87]. In one method [8.86], the well known recurrence relationship

$$J_{n-1}(\Omega) + J_{n+1}(\Omega) = 2nJ_{n+1}(\Omega)/\Omega \tag{8.53}$$

has been used. Suppose three frequency shifts corresponding to integers $n-1, n$ and $n+1$ are provided and time-average holography is performed. Then the irradiance data of (8.31) will provide maps proportional to $J_{n-1}(\Omega), J_n(\Omega)$ and $J_{n+1}(\Omega)$ except the signs. Keeping the sign ambiguity in mind, (8.53) yields

$$\Omega = \frac{2n|J_n(\Omega)|}{\||J_{n-1}(\Omega)| \pm |J_{n+1}(\Omega)|\|}. \tag{8.54}$$

Selection of the proper sign in (8.54) can be made [8.86]. Two values of Ω obtained from (8.54) can be tested in (8.53) to identify the proper Ω. Thus, the map of Ω over the object can be obtained using reconstructed irradiances from different frequency-translated time-average holograms. In the case of noise (8.31) takes the form

$$I = I_b + I_o J_n^2(\Omega), \tag{8.55}$$

where I_b and I_o represent the background and object irradiances, respectively. In the worst possible situation, I_b and I_o can vary over the image. Holograms with static object ($\Omega = 0$) and different n values can be recorded. $I_b + I_o$ is obtained by measuring I for $\Omega = n = 0$. Also, for $\Omega = 0$ and $n \neq 0$, I becomes I_b. Once I_b and I_o are known, $J_n^2(\Omega)$ information from (8.55) is available to use in (8.54).

In the second frequency-translation approach [8.87], the asymptotic approximation of $J_n(\Omega)$

$$J_n(\Omega) \approx (2/\pi\Omega)^{1/2} \cos(\Omega - n\pi/2 - \pi/4) \tag{8.56}$$

has been used to obtain

$$I = J_n^2(\Omega) + J_{n+1}^2(\Omega) \approx 2/(\pi\Omega). \tag{8.57}$$

Notice that the RHS of (8.57) is non-oscillatory. In theory, irradiances from the frequency-translated time-average holographic reconstructions can be added to obtain Ω. For lower values of Ω, care must be taken. $n = 0$ case is suitable for this purpose where the r.h.s. of (8.57) remains non-oscillatory [8.87] even in the lower value range of Ω. Anyway, once approximate value for Ω is obtained, precise analysis can be performed using $J_n^2(\Omega)$ and $J_{n+1}^2(\Omega)$ values [8.87]. In the presence of noise (8.55) is again useful.

8.7 Rotating Objects

Vibration of rotating components differ from that in the stationary situation. The analysis in the actual dynamic situation is therefore necessary. Early developments in this connection have been reported by *Vest* [8.12], and *Ostrovsky* et al. [8.88]. These efforts include matching the object position during the two exposures by synchronizing the pulses, image repositioning, sandwich holography, attaching the recording medium to the object itself, image derotating, etc. *Beeck* [8.89] has recently discussed the problems and solutions of pulsed holography for the vibration analysis on high-speed rotating objects. He demonstrated successful applications of these techniques for up to 15 000 rpm rotating objects under real operating conditions. Usual off axis holography of rotating objects creates two problems. First, vibration fringes due to the in-plane motion are modulated creating, analysis problems. Secondly, decorrelation of the object waves (speckle motion due to the rotation) occurs reducing (ultimately eliminating) the fringe contrast. Some relief is obtained by geometrical considerations where the fringe modulation due to the rotation can be avoided. The typical geometry insensitive to the in-plane motion is shown in Fig. 8.5. With the help of a beam splitter, illumination point source and the observation point are on the rotational axis. Holography and interferometry can be performed as the optical phase variations do not occur in this geometrical arrangement. Care must still be taken to restrict the lateral object motion

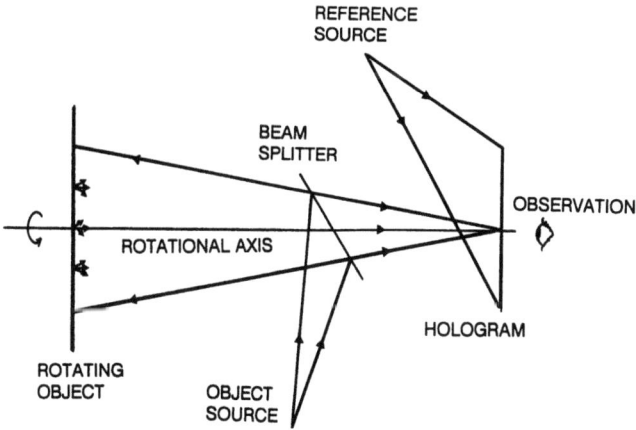

Fig. 8.5. Holographic recording arrangement insensitive to lateral rotation

(between double exposures) within half of the average speckle size. Otherwise poor fringe contrast will result due to decorrelation.

The approach is not very suitable in practical situations [8.89] because it is difficult to align the system perfectly. The following recording techniques are possible [8.89] to cope with the problem in real operating conditions.

8.7.1 Synchronized Triggering

In this procedure (preferably using the set up of Fig. 8.5), the double-exposure holography is performed at the same angular position of the object. However, the vibrational phase may be different allowing interferometry. The pulse separation is high as compared to the usual double-pulse holography to allow the rotation cycle to be completed. The high pulse separation may create additional fringes due to rigid body motions, contouring, etc.

8.7.2 Rotating Holographic Plate and Reference Beam

In this arrangement, the off-axis reference wave and the holographic plate both are rotated with the object speed. With reference to the recording plane, the system is virtually static. Usual double-pulse holographic interferometry will then contain only vibration related fringes.

8.7.3 Use of Image-Derotator

In this approach, the lateral rotational motion is compensated optically. The image of the rotating object observed through a prism appears stationary if the prism rotates at half the speed of the object. A typical arrangement with a reflective image derotation is shown in Fig. 8.6. Notice that the arrangement of Fig. 8.5 for collinear rotational, object illumination and rotational axes is used in conjunction with the image derotation. In Fig. 8.7, fluttering vibrations of an automobile engine cooling fan rotating at 2850 rpm are shown. Very low speckle noise is present indicating complete compensation of the rotatory motion.

8.8 Other Techniques and Applications

So far, we covered the basic methods of holographic vibration analysis. In this section we shall discuss some related developments and practical aspects. Holography with photographic emulsion involves wet chemical processing and is time consuming. Therefore, it has been important to develop methods for quasi-real time analysis. Time-average holographic interferometry using

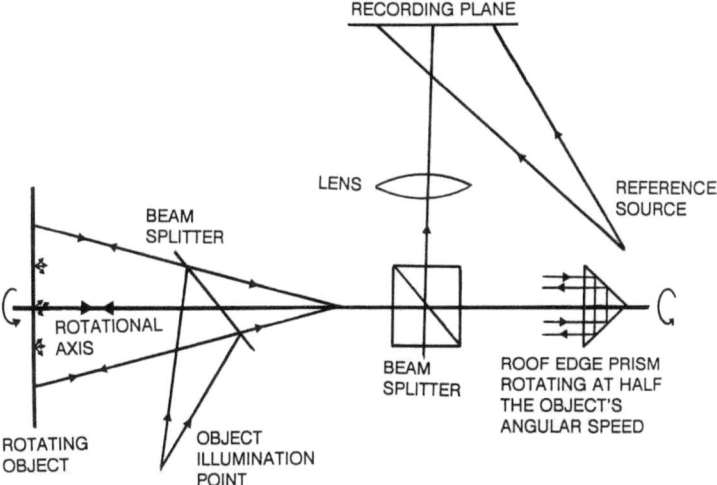

Fig. 8.6. Schematic diagram of image-derotated holography

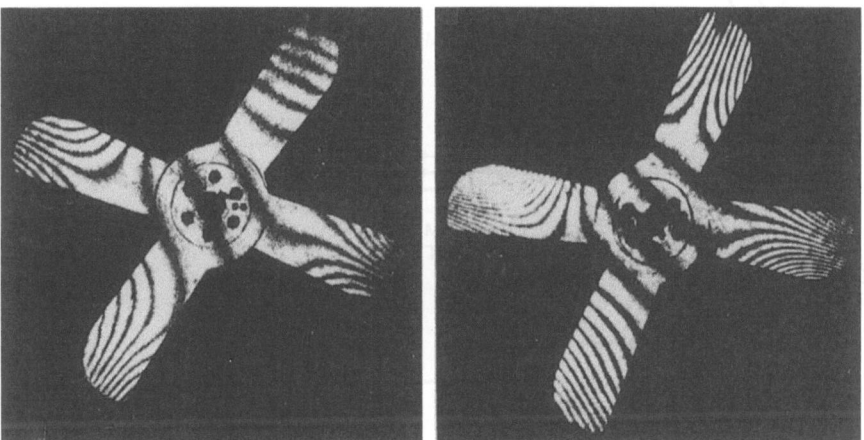

Fig. 8.7. Fluttering vibrations of a cooling fan rotating at 2850 rpm [8.89]

photorefractive crystals was an important development in this regard, as reported by *Huignard* et al. [8.90], and *Marrakchi* et al. [8.91]. Usually, the time constant (10 100 ms) [8.92] in building up the hologram is large as compared to the period of vibration. Time-average fringes can thus be observed in real-time. The basic arrangement is shown in Fig. 8.8. The formation of the hologram on the crystal surface is as usual with object and reference beams. However, the reference beam is reflected back using a mirror to act as a reconstruction beam. Consequently a real image is formed and is suitably placed using a beam splitter.

314   8. Study of Vibrations

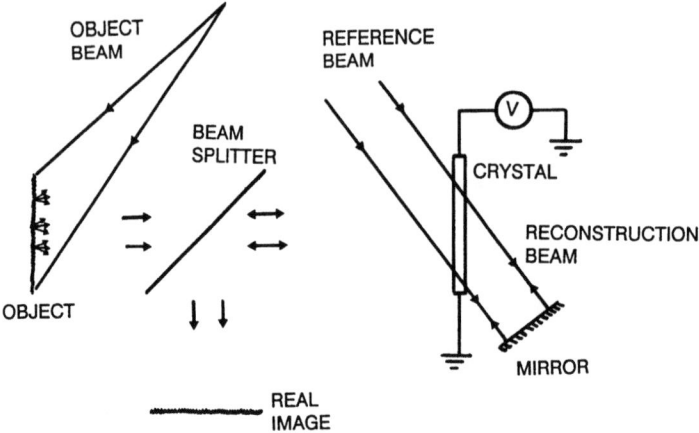

Fig. 8.8. Dynamic holography with phase conjugate imaging using a photo-refractive crystal

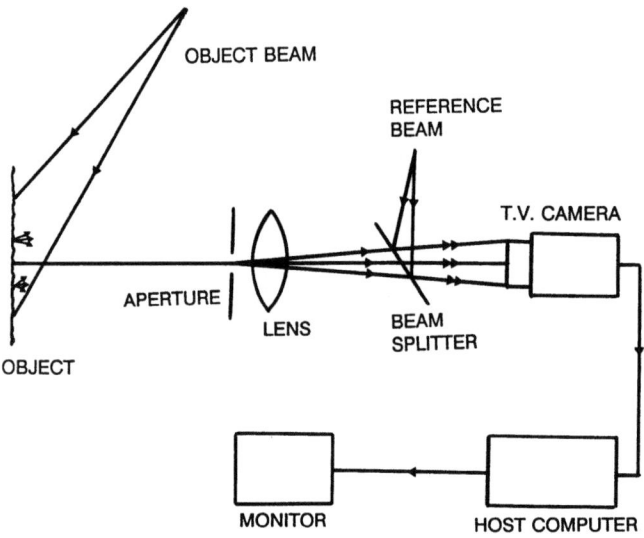

Fig. 8.9. Schematic diagram of a typical electronic speckle pattern interferometer

As stated earlier, time-average fringes are formed in quasi-real time manner. Vibration modes of a loudspeaker have been studied [8.91, 92]. Virtual image can also be used with different geometrical arrangement [8.92, 93].

Electronic Speckle Pattern Interferometry (ESPI) or sometimes called TV holography is another parallel development [8.94]. Basically, the hologram is video-recorded and reconstructed image is displayed by the electronic processing. The technique is well covered in optics literature and several extended discussions and reviews are available [8.94–97]. The basic procedure can be described by Fig. 8.9. The object is imaged by a low-aperture lens on the target of the camera. Also present is the coaxial reference beam. These things are essential so that the complex pattern (hologram) can be resolved by the TV system. The camera is connected to a host computer with a frame grabber. The processing can be done in almost real time.

One special case of vibration analysis is when the vibration period is small as compared to the scan time of the TV system. At a given time t, a point represents the intensity

$$I(t) = I_R + I_o + 2(I_R I_o)^{1/2} \cos(\phi_R - \phi_o - \Omega \sin \omega t), \qquad (8.58)$$

where I_R and I_o are the reference and object beam intensities, respectively. ϕ_R and ϕ_o represent the corresponding phases. The time-averaged value is

$$I_\tau = I_R + I_o + 2(I_R I_o)^{1/2} J_0(\Omega) \cos(\phi_R - \phi_o). \qquad (8.59)$$

I_τ averaged along many speckles is constant (average zero value of the cosine function). However, the speckle contrast varies as Ω changes. The camera signal can be passed through a band-pass filter and rectified. Areas of low contrast then appear nearly black. The contour lines of constant Ω can then be determined. Although noisy as compared to photographic time average fringes, the real-time approach is of serious advantage.

Nearly all of the techniques of usual holographic interferometry such as digital, phase shifting, stroboscopy, temporally modulated holography, have been applied to ESPI. Frame grabber, computer and digital image processing combination yields almost real-time results. Digital phase stepping techniques resulted in very accurate quantitative analysis. Efforts to improve different aspects of the technique are continuing [8.85, 96–105]. In the work due to *Ellingsrud* and *Rosvold* [8.105], the phase modulation techniques and digital image processing have been applied to TV holography to study very small vibration amplitudes ($\lambda/3000$) and phase of the vibration.

The optical set-up for holographic interferometry has been under constant modification for various applications. Use of fiber optics for remote applications and holoendoscope using a panoramic annular lens have been described by *Gilbert* and *Dudderar* [8.106]. Another interesting application is Substrate-Guided Wave (SGW) holo-interferometry including vibration analysis as introduced by *Huang* et al. [8.107]. In such holograms or other "edge-lit" holograms [8.108], the reference and reconstruction beams illuminate from the side. Such holograms have certain design advantages including system compactness.

Heterodyne holographic interferometry [8.109] commonly uses two reference (multiple reference beams are also possible) to store two wave fields independently. During the reconstruction, frequency of one of the beams is slightly shifted for the heterodyne readout. Basically, the time dependent phase variation is studied in the frequency space. Accuracy of 1/1000 of a fringe is common. In connection to vibration analysis, synchronized double-pulse holography and stroboscopic holographic interferometry are basically two-beam interference methods. Then concepts of heterodyne holographic interferometry can be used in holographic vibration analysis as well as for very accurate measurements and improved sensitivity [8.109].

Applications of holographic vibration analysis are continuing. We describe here some recent applications. Accelerometers are common in vibration analysis. Holographic interferometry [8.110] was used to study the effect of attaching the accelerometer on the vibration. It was found that even a microminiature accelerometer can alter the vibration pattern substantially. Holography also finds application in vibration testing of mirror concentrators for X-ray astronomy [8.111]. Ground vibrations from blasting [8.112] and earthquake engineering [8.113] can also be analysed with holography. Vibration analysis of a car body [8.114], other industrial applications [8.115] have also been reported. Reports on holographic vibration analysis of biological objects such as tissues [8.116] and tympanic membrane [8.117] are available. Study of flute vibrations [8.118] is another recent application. This brief description of some recent applications shows that holography continues to be a useful technique in vibration and related problems of engineering, biological and real world objects.

References

8.1 R.L. Powell, K.A. Stetson: J. Opt. Soc. Am. **55**, 1593–1598 (1965)
8.2 K.A. Stetson, R.L. Powell: J. Opt. Soc. Am. **55**, 1694–1695 (1965)
8.3 K.A. Stetson, K. Singh: Opt. Laser Technol. **3**, 104–108 (1971)
8.4 B.M. Watrasiewicz: Opt. Technol. **1**, 20–23 (1968)
8.5 M.R. Wall: Opt. Technol. **1**, 266–270 (1969)
8.6 J. Janta, M. Miler: Optik **36**, 185–195 (1972)
8.7 K.A. Stetson: J. Opt. Soc. Am. **62**, 297–298 (1972)
8.8 P.C. Gupta, K. Singh: Appl. Phys. **6**, 233–240 (1975)
8.9 P.C. Gupta, K. Singh: Appl. Opt. **14**, 129–133 (1975)
8.10 P.C. Gupta, K. Singh: Nouv. Rev. Opt. **7**, 95–100 (1976)
8.11 K.A. Stetson: J. Opt. Soc. Am. **61**, 1359–1362 (1971)
8.12 C.M. Vest: *Holographic Interferometry* (Wiley, New York 1979) Chap. 4
8.13 N.-E. Molin, K.A. Stetson: J. Phys. E. **2**, 609–612 (1969)
8.14 K.A. Stetson: J. Opt. Soc. Am. **60**, 1378–1388 (1970)
8.15 M. Zambuto, M. Lurie: Appl. Opt. **9**, 2066–2072 (1970)
8.16 A.D. Wilson: J. Opt. Soc. Am. **60**, 1068–1071 (1970)
8.17 A.D. Wilson, D.H. Strope: J. Opt. Soc. Am. **60**, 1162–1165 (1970)
8.18 N.-E. Molin, K.A. Stetson: Optik **33**, 399–422 (1971)
8.19 A.D. Wilson: J. Opt. Soc. Am. **61**, 924–929 (1971)
8.20 K.A. Stetson: Appl. Opt. **11**, 1725–1731 (1972)
8.21 C.S. Vikram: Opt. Commun. **8**, 355–357 (1973)
8.22 C.S. Vikram, R.S. Sirohi: Phys. Lett. **A35**, 460–461 (1971)
8.23 C.S. Vikram: Phys. Lett. **A45**, 426 (1973)
8.24 R. Tonin, D.A. Bies: J. Sound Vib. **52**, 315–323 (1977)
8.25 R. Tonin, D.A. Bies: Appl. Opt. **23**, 3713–3721 (1978)
8.26 R. Tonin, D.A. Bies: J. Opt. Soc. Am: **68**, 924–931 (1978)
8.27 E. Archbold, A.E. Ennos: Nature **217**, 942–943 (1968)

8.28   A.E. Ennos, E. Archbold: Laser Focus **4**, 58–59 (October 1968)
8.29   J.T. La Macchia: J. Appl. Phys. **39**, 5340–5341 (1968)
8.30   P. Sajenko, C.D. Johnson: Appl. Phys. **13**, 22–24 (1968)
8.31   B.M. Watrasiewicz, P. Spicer: Nature **217**, 1142–1143 (1968)
8.32   V.S. Listovets, Yu.I. Ostrovskii: Sov. Phys. Tech. Phys. **19**, 847–860 (1975)
8.33   M. Miler: Opt. Commun. **14**, 406–408 (1975)
8.34   C.S. Vikram: Nouv. Rev. Opt. **4**, 147–149 (1973)
8.35   C.S. Vikram: Opt. Commun. **6**, 295–296 (1972)
8.36   C.S. Vikram: Opt. Commun. **7**, 347–348 (1973)
8.37   K.N. Chopra, G.S. Bhatnagar: Appl. Opt. **13**, 2468–2470 (1974)
8.38   C.S. Vikram, G. Bose, J.N. Maggo: Nouv. Rev. Opt. Appl. **6**, 55–59 (1975)
8.39   P.A. Fryer: Appl. Opt. **9**, 1216 (1970)
8.40   T.E. Carlsson, B. Nilsson, J. Gustafsson, N. Abramson: Opt. Eng. **30**, 1017–1022 (1991)
8.41   C.S. Vikram: Opt. Commun. **11**, 360–364 (1974)
8.42   C.S. Vikram: Opt. Commun. **10**, 290–291 (1974)
8.43   C.S. Vikram: J. Opt. (India) **3**, 26–27 (1974)
8.44   C.S. Vikram: Optik **42**, 361–366 (1975)
8.45   C.S. Vikram: Optik **43**, 65–70 (1975)
8.46   C.S. Vikram, G. Bose: Optik **43**, 253–258 (1975)
8.47   C.S. Vikram: Optik **45**, 55–64 (1976)
8.48   C.S. Vikram: Pramana **8**, 420–426 (1977)
8.49   C.S. Vikram: Pramana **8**, 541–544 (1977)
8.50   C.S. Vikram: Opt. Quantum Electron. **10**, 527–528 (1978)
8.51   C.C. Aleksoff: Appl. Opt. **10**, 1329–1341 (1971)
8.52   M.H. Zambuto, W.K. Fischer: Appl. Opt. **12**, 1651–1655 (1973)
8.53   M. Ueda, S. Miida, T. Sato: Appl. Opt. **15**, 2690–2694 (1976)
8.54   C.S. Vikram: Optik **50**, 251–254 (1978)
8.55   N. Takai, M. Yamada, T. Idogawa: Opt. Laser Technol. **8**, 21–23 (1976)
8.56   K.A. Stetson: J. Opt. Soc. Am. **62**, 698–700 (1972)
8.57   C.S. Vikram: Appl. Opt. **12**, 2808 (1973)
8.58   D.B. Neumann, C.F. Jacobson, G.M. Brown: Appl. Opt. **9**, 1357–1362 (1970)
8.59   C.C. Aleksoff: Appl. Phys. Lett. **14**, 23–24 (1969)
8.60   A.F. Metherell, S. Spinak, E.J. Pisa: J. Opt. Soc. Am. **59**, 1534 (1969)
8.61   C.S. Vikram, K. Vedam: Optik **56**, 51–58 (1980)
8.62   J.A. Levitt, K.A. Stetson: Appl. Opt. **16**, 195–199 (1976)
8.63   C.S. Vikram: Appl. Opt. **16**, 1140–1141 (1977)
8.64   V.J. Corcoran, R.W. Herron, Jr., J.G. Jaramillo: Appl. Opt. **5**, 668–669 (1966)
8.65   F.M. Mottier: Appl. Phys. Lett. **15**, 44–45 (1969)
8.66   H.J. Caulfield: Appl. Phys. Lett. **16**, 234–235 (1970)
8.67   J.P. Waters: Appl. Opt. **11**, 630–636 (1972)
8.68   H. Kreitlow, T. Kreis, W. Jüptner: Appl. Opt. **26**, 4256–4262 (1987)
8.69   P. Hariharan: Appl. Opt. **12**, 143–146 (1973)
8.70   C.S. Vikram, G.S. Bhatnagar: Appl. Opt. **12**, 2239–2240 (1973)
8.71   T. Sato, H. Ogawa, M. Ueda: Appl. Opt. **13**, 1280–1282 (1974)
8.72   P. Hariharan: Opt. Eng. **15**, 279 (1976)
8.73   F.M. Mottier: App. Phys. Lett. **15**, 278–285 (1969)
8.74   F.M. Mottier: Nouv. Rev. Opt. Appl. **1** (Suppl.), 12 (1970)
8.75   F.M. Mottier: Holographic vibration analysis by generalized stroboscopy. Report KLR-71-03, Brown Boveri Research Center, Baden, Switzerland (1971)
8.76   W.J. Dallas, A.W. Lohmann: Opt. Commun. **13**, 134–137 (1975)
8.77   D. Cutter: Measurement of vibration wavefronts using multiplex holography. Dissertation, Friedrich-Alexander University Erlangen-Nürnberg (1976)
8.78   J. Politch: J. Opt. Soc. Am. **A7**, 1355–1361 (1990)

8.79 K. Creath: *Progress in Optics* **26**, 349–393 (North-Holland, Amsterdam 1988)
8.80 P. Hariharan, B.F. Oreb: Opt. Commun. **59**, 83–86 (1986)
8.81 S. Nakadate: Appl. Opt. **25**, 4155–4161 (1986)
8.82 S. Nakadate, H. Saito, T. Nakajima: Optica Acta **33**, 1295–1309 (1986)
8.83 P. Hariharan, B.F. Oreb, C.H. Freund: Appl. Opt. **26**, 3899–3903 (1987)
8.84 K.A. Stetson, W.R. Brohinsky: J. Opt. Soc. Am. **A5**, 1472–1476 (1988)
8.85 R.J. Pryputniewicz, K.A. Stetson: Proc. SPIE **1162**, 456–467 (1990)
8.86 C.S. Vikram: J. Mod. Opt. **39**, 1987–1989 (1992)
8.87 C.S. Vikram: Optik **93**, 155–156 (1993)
8.88 Y.I. Ostrovsky, V.P. Shchepinov, V.V. Yakolev: *Holographic Interferometry in Experimental Mechanics*, Springer Ser. Opt. Sci., Vol. 60 (Springer-Verlag, Heidelberg 1991) Chap. 8
8.89 M.-A. Beek: Opt. Eng. **31**, 553–561 (1992)
8.90 J.P. Huignard, J.P. Herriau, T. Velentin: Appl. Opt. **16**, 2796–2798 (1977)
8.91 A. Marrakchi, J.P. Huignard, J.P. Herriau: Opt. Commun. **34**, 15–18 (1980)
8.92 J.-P. Huignard, P. Günter: *Photorefractive Materials and Their Applications II*, ed. by P. Günter, J.-P. Huignard. Topics Appl. Phys., Vol. 62 (Springer, Berlin, Heidelberg 1989) pp. 205–273
8.93 H.J. Tiziani: Optical and Quantum Electron. **21**, 253–282 (1989)
8.94 R. Jones, C. Wykes: *Holographic and Speckle Interferometry*, 2nd edn. (Cambridge Univ. Press, Cambridge 1989)
8.95 J.R. Tyrer: SPIE Proc. **604**, 95–109 (1986)
8.96 O.J. Løkberg: *Laser Holography in Geophysics*, ed. by S. Takemoto (Ellis Horwood, Chichester, UK 1989) pp. 168–198
8.97 C. Joenathan: Recent developments in electronic speckle pattern interferometry, in Proc. SEM Conf. in Experimental Mechanics, Milwaukee, Wisconsin, USA (1991) pp. 198–204
8.98 K.A. Stetson, W.R. Brohinsky: Opt. Eng. **26**, 1234–1239 (1987)
8.99 M. Santoyo, M.C. Shellabear, J.R. Tyrer: Appl. Opt. **30**, 717–721 (1991)
8.100 R.J. Pryputniewicz: SPIE Institute Series **IS8**, 215–246 (1991)
8.101 C. Joenathan: Appl. Opt. **30**, 4658–4665 (1991)
8.102 C. Joenathan, B.M. Khorana: Opt. Eng. **31**, 315–321 (1992)
8.103 C. Joenathan, B.M. Khorana: Appl. Opt. **31**, 1863–1870 (1992)
8.104 S. Peng, C. Joenathan, B.M. Khorana: Opt. Lett. **17**, 1040–1042 (1992)
8.105 S. Ellingsrud and G.O. Rosvold: J. Opt. Soc. Am. **A9**, 237–251 (1992)
8.106 J.A. Gilbert, T.D. Dudderar: SPIE Institute Series **IS8**, 146–159 (1991)
8.107 Q. Huang, J.A. Gilbert, H.J. Caulfield: SPIE Proc. **1667**, 172–181 (1992)
8.108 A.N. Putlin, V.N. Morozov, Q. Huang, H.J. Caulfield: Opt. Eng. **30**, 1615–1619 (1991)
8.109 R. Dändliker: *Progress in Optics* **17**, 1–83 (North-Holland, Amsterdam 1980)
8.110 C.S. Vikram, T.E. McDevitt: Optics and Lasers in Engineering **9**, 77–83 (1988)
8.111 P. Delvò, M.L. Rizzi: Opt. Eng. **27**, 1072–1077 (1988)
8.112 T.E. Carlsson, G. Bjarnholt, N. Abramson, D.C. Holloway: Opt. Eng. **27**, 923–927 (1988)
8.113 J.D. Trolinger, D.C. Weber, G.C. Pardoen, G.T. Gunnarsson, W.F. Fagan: Opt. Eng. **30**, 1315–1319 (1991)
8.114 J.T. Malmo, E. Vikhagen: Experimental Techniques **12**, 28–30 (1988)
8.115 R.J. Parker, D.G. Jones: Opt. Eng. **27**, 55–66 (1988)
8.116 M.A. Pathak, C.F. Stanley, G. Nuñez: Experimental Techniques **13**, 30–31 (1989)
8.117 M. Maeta, S. Kawakami, T. Ogawara, Y. Masuda: SPIE Proc. **1429**, 152–161 (1991)
8.118 P.T. Ajith Kumar, P.J. Thomas, C. Purushothaman: Appl. Opt. **29**, 2841–2842 (1990)

Subject Index

Abel transform 160, 207
Aberrations 115, 130, 242
Absolute fringe order 47
Absorption 17
Ac fringe modulation 119
Acceleration 296
Accelerometer 316
Acceptance band 165
Acousto-optic modulator 77, 90, 97, 98, 102, 114, 116, 175
Adaptive thresholds 174
Additive
  moiré 246
  noise 163
Adhesive bonding 205
ADP crystals 304
Aerodynamics 207
Aircraft components 2
Air turbulence 21, 22, 104, 107, 126–128, 130, 131, 137, 138
Airy function 84, 87, 90
Algebraic operations 154
Aliasing 117, 140
Alignment 142
Aluminium
  beam 242, 243
  plate 104, 215, 216, 243, 247, 281, 282, 285
Amplifier 115
Amplitude
  hologram 14, 22, 25
  modulation 302
  spectrum 185, 189, 191
  transmittance 9, 15, 22, 24, 230, 256
Analog-to-digital converter 114, 136
Angular selectivity 15, 18, 24
Annular lens 315
Antisymmetric matrix 34, 36
Aperture
  function 83, 84, 220
  masks 100
Arcs 196, 197
Array processor 128, 142, 144, 145

Asymptotes 217
Asymptotic approximation 310
Auto-correlation 83–85, 92, 219, 220
Auxiliary fringes 221, 224–228, 282
Average windowing 154

Back-projection filter 208
Band limited function 154
Band-pass filter 184–186, 193, 195, 200, 315
Beam weakening 256
Beat 239, 261, 264, 280
Beat frequency 77, 102, 175
Bending moment 238
Bessel
  fringes 39, 40
  function 15, 65–68, 157, 294, 309
Bezier polynomial 174
Bias 304, 308
  modulation 68–70, 144
Bilinear
  interpolation 170, 174
  mapping 172
  transformation 170–173
Binary filtering 174
Biological objects 316
Biomechanics 3
Birefringent plate 175
Bleach 25
Blind hole 247, 281, 285
Boundary
  element method 153, 206
  layers 160, 208
Bragg condition 15, 16, 18, 25
Brazed cooling panels 165
BSO 27

Cameras 139–141
Cantilever
  beam 162
  plate 69, 70
Car body 316

Carbonfiber
  composite 233
  reinforced plastic 183
Carré algorithm 125, 128, 132, 133–135, 146, 180
Carrier fringes 184, 185, 217, 218, 229, 246, 249, 250, 252, 259–262, 264, 265, 283, 284
Cavity losses 19
CCD 38, 60, 66, 95, 127, 153, 177, 209
CCIR 139
Cellular automata 197
Characteristic function 66, 294–299, 301–303
Chopping 299
Circular
  carrier 217
  frequency 86
  plate 250
Clock rates 139
Coherence 19
Coherence length 19, 20
Coherent imaging 83, 84
Color coded 307
Comparative
  holographic moiré 280–282
  holography 213, 274–288
Computer
  -aided evaluation 152, 155, 173, 233, 285, 308, 315
  simulations 153, 163, 164
  tomography 207, 208
Concentration analysis 207
Concentrations 160
Concrete shrinkage 3
Conjugate
  image 7, 10
  reference 167, 277
  -wave holography 222, 232–233
Connectivity 170
Continuation 161
Continuity 170, 195
Contour sensitivity 121, 122, 215, 223, 231, 242, 243, 249–251, 254, 257–259, 261, 262, 265, 267, 269, 271–273
Contouring 121, 122, 254–274
  plane orientation 255, 257, 259, 266, 267, 269, 274
Contrast
  reversal 87
  stretching 154
Controlled phase 259
Convex quadrangle 170
Convolution integral 83

Correlation cell 85
Corresponding points 213, 275, 276
Cosinusoidal fringes 39–41, 66, 67, 69, 156, 160, 306
Coupled-wave theory 16
Coupling 18
Crack 2, 181, 201, 206, 207
Cross reconstructions 79, 82, 90, 91
Cross talk 88–90, 98
Crystal growth 2
Curvature 238, 243–249
Curve fitting 201
Cut-off frequency 140, 186, 191, 200
Cylindrical carrier 217

Dark field background 302
Data
  frame 114, 127, 128
  image 64, 65
  reduction 165
  storage 65
Dc intensity 127
Dc term 184, 185, 192
Debonds 183
Debrazes 165
Decorrelation 81, 89–94, 104, 163, 218–220, 229, 311
Defects 2, 164, 165, 201–203, 205, 206, 221, 247, 248, 275, 279, 281, 282, 285, 286, 288
Degree of correlation 87, 283
Delocalization 88, 219, 220, 264
Demodulation algorithm 196, 197
Density function 208
Dentistry 3
Depth of modulation 283, 284
Derivation direction 242, 243
De-rotator 312
Detector
  area 85, 92, 93
  array 102, 103, 111, 117–119, 134, 147
  bandwidth 177
  dynamic range 120
  element 114, 117–119
  function 85
  nonlinearities 126, 130, 131, 134–136
  saturation level 120, 183
  shot noise 138
  signal quantization 130, 131, 136–137
  size 84
  thermal noise 138
Determinant 168
Development of holography 7, 8
Diametral compression 233

Difference
  displacement   213, 275–281, 285, 287
  hologram   278
  holography   277–279
  shape   275, 279
Diffraction
  efficiency   15–18, 25, 27
  order   249–251
Digital image processing   153–155
Digitizer   153
Dirac function   84
Displacement
  measurement   31, 39, 41–53, 59, 151, 214–236
  observed   51, 52
  vector   29, 41, 47, 49, 57, 86, 99–103, 110, 158, 214, 239, 275, 278
Displacement derivatives   104, 204, 236, 254
  in-plane   204, 246, 252–254
  out-of-plane   166, 204, 238–243, 246, 250
Distance modulo 2π 196
Distortions   155, 163, 200
Double diffraction   253
Double exposure   28, 29, 56, 60, 66, 67, 75, 78, 82, 88, 95, 96, 98, 102, 107, 155, 156, 169, 177, 218, 221, 238, 255, 277, 299–301, 308, 309
Double pulse holography   82, 95, 97, 99–101, 107, 299, 312, 315
Drift   201, 296, 298
DSPI   143
Dyadic   245
Dynamic
  evaluation   167
  load   39, 40, 59, 60, 64–70
  range   18, 191

Earthquake engineering   316
Edge lit   315
Elastic ribbon   161
Elastomechanics   204
Electronic
  fringe interpolation   75, 77, 79, 82
  holography   51, 59–71, 142–144
  noise   163
  phase measurement   75–82
Electro-optic
  holography   51, 59, 60, 142–144, 195
  modulator   114
Embryonic behaviour   2, 215
Emulsion shrinkage   25
Envelope   283
Equidistant planes   254
Error   44, 45, 50, 68, 88–95, 104, 131, 200

sensitivity   126
sources   130–138
vector   46
ESPI   143, 195, 209, 314, 315
Expert systems   165, 209
External fixation   230
Extinction ratio   90

False images   24
Feature extraction   155
FFT algorithm   175, 185, 189
Fiber reinforced concrete   236
Finite element
  method   70–72, 153, 205, 206
  model verification   3, 152, 205, 288
Fizeau fringes   193
Flaw   2, 164, 165, 201–203, 205, 206, 221, 247, 248, 275, 279, 281, 282, 285, 286, 288
Flexural
  deformation   236
  moments   251
  rigidity   251
Flow visualization   159, 207
Flute vibrations   316
Fluttering vibrations   312
Focal spot   20
Four-fold correlation   84
Four-wave interferometer   145, 258, 263
Fourier
  series   131, 297
  transform   81, 83, 155, 220, 184–193, 197, 199–201, 208
Fracture
  mechanics   3, 206
  toughness   207
Frame grabbers   111, 139–141, 315
Frequency
  shift   175
  translation   302, 310
Friction   206
Fringe
  clarity   283, 284
  control   220, 221, 224, 281
  counting   48, 152, 173
  formation   86–88, 155–164, 219, 263, 264, 270
  interpolation   54, 79, 88
  localization   31, 54, 167, 219, 221, 224–229, 252, 260, 264, 269, 280, 297
  locus function   40, 41, 43, 46–49, 52, 56, 61, 63, 64, 66, 67, 169
  locus planes   54
  numbering   174

322 Subject Index

Fringe *(contd.)*
  order known 41–47
  order unknown 41, 47–51
  orientation 166, 229
  parallax 52
  shift 48, 50
  shifting in vibration 306–311
  skeletonizing 155, 173–175, 195, 197, 199, 200
  tensor 57
  tracking 173, 174
  vector 54, 57, 58, 81, 166, 169, 252
  visibility 19, 21, 76, 77, 81, 84, 87, 88, 89, 94, 97, 110, 112–114, 118, 123–125, 127, 137, 163, 218–221, 227, 228, 252, 259, 262, 265, 267, 272, 282–284, 308, 311
Frost resistance 2

Gain
  adjustment 134
  profile 19
Gaussian intensity profile 20, 163
Generalized phase shifting 122, 193
Generation of auxiliary fringes 224–226
Geometric
  operations 154
  sensor aberrations 154
Glass-plate shearing 241, 242
Global operations 154
Gradient operator 57, 174
Gravity 152
Gray-level windowing 285
Gray-levels 99, 101, 153, 173, 174, 187, 191, 285
Ground vibrations 316

Hadamard condition 168
Halfplanes 188, 189, 191, 192
Halo 10, 24
Harmonic vibrations 294
Heat transfer 3, 159, 205, 207
Hermitean distribution 185
Heterodyne 76, 77, 88, 90, 91, 93, 95, 102, 104, 106, 109, 175, 177, 184, 185, 199–201, 315
High byte image 66
High spatial density 71
Highpass filtering 154
High-speed shutter 110
Histogram 155
Holodiagram 30
Holoendoscope 315
Hologram
  amplitude 14, 22, 25
  diffusely reflecting objects 18
  imaging 7
  multiply exposed 18
  phase 14, 22, 24, 25
  recording 38, 39
  rotation 224, 227, 260, 312
  translation 225, 227, 260, 264
  types 14–18
  volume reflection 14, 15
  volume transmission 14, 16, 17
Holographic
  addition 303, 305
  moiré 145–147, 222–229, 233–236, 238–243, 252, 255, 258–267, 282, 285, 286, 288
  numerical analysis 40, 41, 47, 57
  subtraction 305
Holography development 7, 8
Homologous rays 213, 219, 225, 226, 228, 263, 267, 270
Honeycomb structure 183
Hooke's law 205
Host computer 67, 71
Hybrid evaluation 206

Identity matrix 35
Ill-conditioned matrix 168
Image
  coding 155
  compression 155
  derotation 311, 312
  enhancement 186, 195
  frame 66
  hologram 11, 225, 230, 260, 270
  restoration 155
Imaging parameters 229–230
Immersion
  method 254, 257, 258, 279
  tank 215
Impulse
  loading 152
  response 83, 84
In-plane displacements 84, 93, 100, 110, 145, 146, 204, 221–236
Integer-part function 234
Integrating bucket 76
Integration period 114
Intensity modulation 118–120
Interactivity etalon 20
Interferogram evaluation 164–173
Interferometric readout 39
Interlaced signals 139
Intermodulation 24

Interpretation
  multiple hologram 41, 51–53
  single hologram 41–51
Inverse
  Abel transform 207
  Fourier transform 185, 186, 208
  Radon transform 208
Isolated points 174
Isolation table 138
Isotropic bodies 205
Iterative process 205, 206, 309

J-integral 207

Lame-Navier equations 207
Large objects 216
Laser
  beam expansion 20–21
  irradiance modulation 303
  safety 22
  triangulation 169
Lateral magnification 12
Least-square error 47, 52
Least-squares technique 77, 113, 114, 125, 129, 166–168, 178, 201, 252, 298
Light sources 19–21
Line integral 207
Linear
  carrier 161, 217, 218
  filtering 154
  ramping 115
  recording 9
Liquid gate 25
Loading
  intensities 152
  types 152
Local operations 154
Logarithmic gray scale 191
Longitudinal
  modes 19, 20
  shift 256
Look-up table 64, 68, 144, 145
Loudspeaker 152, 314
Low byte image 66
Lowpass filtering 154, 155, 173, 195

Mach-Zehnder interferometer 159
Magnification change 256
Many component motion 300
Marginal pixels 187
Mass transfer 208
Master
  hologram 277, 278
  piece 275, 277–281, 285

wave 277
Matric product 35, 37, 57
Matrix
  evaluation 168
  rank 168
  transformations 33–37, 54, 57, 58
  transpose 37, 46, 47
Measurement errors 44–46, 50, 68, 88–95, 104, 131, 200
Mechanical loading 152
Michelson interferometer 96, 184, 185
Mirror concentrators 316
Misalignment 79, 88, 89, 93, 277
Mode shapes 66, 314
Modulated
  diffraction grating 249–254
  pattern formation 282–284
Modulation
  aspect factor 264
  depth 303, 309
  parameter 16, 17
  transfer function 23
Moiré 145, 146, 169, 222, 233, 234, 240, 241, 245–248, 252, 259, 261, 264, 265, 267, 269, 280, 282, 283
Moiré fringe quality 283, 284
Moments 251
Monobath 25
Motion compensation 220, 221
Moving surfaces 152
Multidimensional signals 153
Multiple
  -exposure holography 18, 98, 99
  holograms 41, 51–53, 252
  modes 296, 298, 300
  -sources contouring 258, 268–274
Multiplexing 100, 300
Multiplicative
  moiré 252
  noise 163
Multiprocessor networks 209
Mutual intensity 87, 91, 94
Mutual interference 59

Nearest neighbour interpolation 171
Neural network 164, 165, 209
Nodes 65, 196, 197, 294
Non-collinear directions 51, 52
Noncoplanar
  directions 41, 54, 170
  sensitivity vectors 43
Non-corresponding points 276, 277
Non-destructive
  inspection 248, 275, 288

Non-destructive *(contd.)*
  testing 2, 3, 5, 151, 164, 274, 275, 288
Non-interlaced signals 139, 141
Non-invasive 151
Non-linear
  filtering 154
  phase shifter errors 132, 133
  recording 24, 89, 90
  response 187
  vibrations 294
Non-oscillatory 310
Non-overlapping reconstructions 91–94
Nonseparable motions 296–298
Non-singular matrix 43
Normal projection 33, 35, 37
Normalized errors 131, 132
Number of speckles 85, 86, 91, 93
Numerical analysis holography 40, 41, 47, 59
Nyquist
  limit 116, 140
  sampling theorem 117, 118

Object-tilt contouring 266, 267
Oblique
  planes 274
  projection 33, 35, 37
Observed displacement vector 51, 52
Off-axis
  hologram 9–11
  reference beam 8, 11
Offset angle 10
Optical
  delay line 100, 107
  filtering 229, 249, 252, 253
  power 85, 86, 91
  slicing 274
  subtraction 259
Orthogonally polarized beams 96, 115
Orthoscopic image 10
Oscillating mirror 167
Outliers 180, 181
Out-of-plane displacements 110, 121, 145, 146, 166, 204, 214–217, 238
Overlapping reconstructions 88, 90, 94, 102

Parallax 51
Parallel procedure 309
Partial spectra 184
Pass band 188, 189
Path-dependent demodulation 196
Path-independent demodulation 196–198
Pattern segmentation 155

Peak-to-valley phase error 132–135
Performance evaluation 264–266
Perspective
  correction 58, 170–173
  distortion 154, 166, 170–173
  variation 54, 57, 100, 204, 205
Phase
  ambiguities 116, 217
  balance 271
  demodulation 195–197, 200, 267, 268, 271
  difference amplification 218, 264
  error 89–95, 101, 104, 131–138, 196, 200
  hologram 14, 22, 24, 25
  lock 195
  management 255, 260–262, 264, 265, 267
  measurement algorithms (2 + 1) frames 126
  measurement algorithms (N + 1) frames 126
  measurement algorithms five frames 124, 128, 132, 133–136, 179
  measurement algorithms four frames 123, 132, 133–136, 179
  measurement algorithms seven frames 126, 135
  measurement algorithms three frames 122, 132, 133–135, 138, 179, 235
  measurement Carré 125, 128, 132, 133–135, 146, 180
  measurement random phase shifts 127–128
  measurement scanning phase shifts 127–128
  measurement spatial 109, 184, 185
  measurement synchronous detection 125
  measurement temporal 109, 112, 122, 175
  modulation 15, 24, 66, 115, 116, 195, 303–305, 309, 315
  modulo 2π 68, 77, 114, 128, 143, 144, 180–182, 186, 191, 193, 195, 200, 234–236, 247, 248
  modulo π 68, 114
  neutralization 269–271
  objects 95
  organization 255
  reallocation 271
  sampling 154, 177–183, 199–201, 203
  saturation 267
  unwrapping 116, 140, 143, 144, 147, 161, 183, 195, 197
Phase shift 59–61, 76, 90, 95–97, 100, 109, 112–115, 119, 122–129, 132, 137, 141, 144, 180–183, 192–194, 246, 306–310
  shift histogram skewness 128
  shift histogram spread 128
  shift holographic moiré 145–147, 233–236, 238, 285–288

shift ramping 141
shift speckle interferometry 143
shift spread 128
shifted comparative holography 285–288
Phase shifter
  calibration 128–129, 136
  errors 124, 126, 130–133
Phase shifters 111, 114, 115
Phasemeter 77, 102, 175
Phasor
  bias 66
  sum 297
Photo-conductive sensors 153
Photodetectors 78, 91, 175
Photographic emulsions 25
Photomultiplier tube 102
Photorefractive crystals 26–28, 201, 313
Photosensitive medium 38
Photothermoplastics 25, 28
Phototransistors 153
Picture element 59, 76
Piecewise approximation 174
Piezo-electric transducer 96, 115, 235, 177, 195
Pinhole 20, 21
Pipeline processor 66
Pixels 66, 68, 95, 128, 139, 141, 153, 154, 166, 170, 171, 173, 174, 187, 191, 195, 200, 235
Plasma diagnostics 3, 159, 207
Plastic zone 207
Plate aspect ratio 68
Pockels cell 22, 90, 97, 100
Point operations 154
Point-symmetric 185
Poisson's ratio 217, 251
Polarization 13, 21, 142
Polarizing beam splitter 115
Position vectors 38, 56, 238, 244, 256, 267
Positive image 11
Precision 77, 78, 86, 88, 103, 104, 106, 130, 199–201, 248
Pressure
  chamber 152, 279
  loading 152
  vessels 104, 165, 203
Principle values 173
Projection matrices 33, 37, 41, 51, 52, 58, 59
Propagation vectors 29, 38, 100, 214, 215, 238, 244, 255, 259, 269
Prototype conception 288
Pseudocolor 236, 237
Pseudoscopic image 10
Pulse width 306

Pulsed
  holography 22, 79, 82, 90, 97, 201, 300, 311
  laser 22, 110
Pupil plane decorrelation 218–220, 229, 267

Q parameter 17, 18
Q switch 22, 97, 98, 100, 101, 107
Quadrangle 170
Quadratic errors 178
Quantization 130, 131, 136–137, 153, 187
Quarter-wave plate 116
Quasi real-time 312–315
Quasi-heterodyne 76, 77, 95, 107

Radially symmetric refractive index 160, 207
Radii of curvature illumination perspective 58
Radii of curvature observation perspective 58
Radon transform 208
Random phase shifts 127–128
Random vibrations 306
Real image 10
Real-time 27, 28, 59, 64, 66, 71, 76, 109, 143, 155–157, 178, 181, 191, 199, 229, 238, 255, 258, 266, 269, 270, 273–275, 281, 295, 296, 299, 304, 306, 308, 309, 313, 315
Reconstructed
  image 12–14
  reference wave 222, 230, 231
Recording
  materials 24–27
  medium 22–24
Reeds 3
Reference
  rotation 225, 260, 312
  scanning 167
Reflection hologram 11, 28, 221
Refractionless limit 208
Refractive index change 14, 16, 17, 151, 155, 159, 160, 207, 208, 258
Relief
  encoding 259, 266
  variations 254, 269
Repositioning sensitivity 79, 88, 89
Resolution 86, 106, 152, 199, 200
Resonance 40
Resonant frequency 19
Rigid-body movements 53, 57, 104, 161, 169, 170, 201, 220, 221
Ring 233
Ring pattern 165, 200

Rotating
  analyzer 116
  half-wave plate 116, 177
  mirror 167
  objects 311, 312
  radial grating 175, 177
Rotation
  matrix 58
  vector 81
Rotations 53, 54, 56, 58
RS-170 139
Rubber plates 281

Sampling
  requirement 116, 117
  theorem 154, 175
Sandwich hologram 29, 217, 220, 254, 255, 311
Saturation 120, 183, 187
Saw-tooth 195, 234, 285
Scalar product 34, 35, 41, 48
Scan time 103, 315
Scanning phase shift 127–128
Second moment 85
Second-order
  derivatives 243, 245, 246, 248, 250, 251
  moiré 245, 246
Secondary interference fringes 109, 110, 142
Seed values 309
Segmentation 173, 174
Self-reconstructions 79, 82
Sensing element 38
Sensitivity
  enhancement 217, 218, 224, 261, 262, 265, 271–273, 302
  reduction 302
  vector 29, 30, 38, 39, 41–44, 46–49, 54, 56–58, 99, 100, 110, 159, 162, 163, 168–170, 185, 204, 240, 241, 293, 308
Separable motions 294, 296, 299
Shading correction 154
Shadow 33–37
Shape measurement 121, 122, 254–274
Shear variation 242
Shearing 238, 240, 241, 243–246, 248–251
  interferometry 147
  strains 204
Sheet polarizer 21
Sign ambiguity 29, 77, 160, 161, 175, 186, 189, 192, 197, 217, 224, 310
Signal
  enhancement 181
  processors 154
  -to-noise ratio 14, 24, 174
Similarity property 208

Sinc function 113
Single hologram 41
Singular matrix 37, 168
Sinusoidal vibration 64, 65, 293–295, 298–300, 302, 306
Skeletonizing 155, 173–175, 195, 197, 199, 200
Slope change 238–243, 246–250
Slow-scan 139
Small
  deformations 217, 218
  vibrations 302, 304, 305
Solid-state sensors 153
Sources of errors 88–95
Space
  -charge pattern 26
  variant phase 300
  vectors 214
Sparse array sensor 144
Spatial
  coherence 19
  filtering 174, 186
  frequency domain 155
  light modulators 209
  noise 20
  resolution 78
  transform 170
Spatiotemporal 147
Specimen integrity 152
Speckle 13, 59, 81–83, 88, 90, 219, 223
  averaging 82, 86, 143
  fields 81–83
  interferometer 59, 143, 206
  metrology 59
  noise 85, 88, 90–93, 154, 163, 173, 174, 186–188, 312
  reduction 13
  reference beam 304
  shifts 220
  size 85–88, 90, 91, 94, 220, 265, 312
  statistics 83
Spectral contents 155, 186
Spectroscopic holography 306
Spherical
  perspective 56
  wavefront 56
Spline function 174
Split lens 243, 246
Spurious
  fringe patterns 88, 90
  reflections 130
  vibrations 181
Static
  evaluation 166
  load 39, 40, 59–64, 69

Stationary
  condition 226
  phase 297
Statistical error 82, 91
  features 154
  phase errors 88
  properties 83, 84
Steel
  beam 254
  plate 99
Strain-rotation matrix 54, 58
Strain-shear matrix 58
Strains 31, 53, 54, 57, 58, 102–106, 164, 201, 204, 205, 207, 219, 252–254
  homogeneous 54, 57
  inhomogeneous 57–58
Stress intensity factor 207
Stresses 204–206, 251
Stroboscopic holography 79, 156, 298–301, 303, 306, 315
Structural
  components evaluation 288
  failure 53
Substrate-guided wave 315
Subsurface void 164
Sum of phases 146, 147, 286
Surface of localization 226–228
Surface roughness 83, 84
Synchronized triggering 312
Synchronous detection 125, 126
System ill-conditioning 283
Systematic errors 88–90

Taylor series 56, 240, 249, 253
Telecentric imaging 142, 241, 245, 256, 258
TEM$_{00}$ mode 19, 20
Temporal filtering 154, 174
Temporally
  dependent motions 297
  modulated holography 157, 301–306, 315
Tensile
  stress 152
  test specimen 181, 201
Test piece 275, 277–281, 285
Thermal expansion 205
Thermal loading 152, 189, 190, 205
Thermoplastic plate 122
Thin gratings 14, 15
Threshold 191, 285
Threshold binarization 154
Tilted glass plate 177
Time average 60, 64, 66–69, 144–145, 157, 293–298, 300, 301, 304–306, 308, 310, 312–315

Time constant 313
Time function 294
Time-resolved holography 300
Tissues 316
Tolerance limits 304
Torsion 164
Torsional
  moments 251
  stress 152
Training samples 165
Transient phenomena 28
Transmission function 249, 250
Transparent objects 75, 88, 89, 206, 207
Transputer networks 154
Triangular phase modulation 306
Triple vector product 34, 35
True phase 236, 237
TV holography 51, 59, 127, 143, 209, 305, 314, 315
Twin-image 8
Twist 246, 251
Two angle contouring 122
Two reference beams 75, 78–82, 88–90, 95, 97, 98, 102, 107, 175, 177, 199, 315
Two-dimensional detectors 175
Two-frequency laser 175
Tympanic membranes 316
Types of holograms 14–18
Tyre 2

Ultrasonics 152, 208
Uniaxial tension 229, 230
Unknown phase steps 180–183

Vacuum chamber 152
Validation 3, 152, 205, 288
Valve 174
Variable sensitivity 255
Variance 14, 155
Vector product 34, 35, 169
Vibration 64–70, 97, 99, 101, 144, 145, 156, 157, 293–316
  effects 126, 128, 130, 131, 137–138, 141
  frequency 68, 69
  isolation 21, 127
  loading 152
  phase 315
Virtual image 10, 79, 83, 246
Volume
  gratings 15–18
  phase holograms 18
  reflection hologram 14, 15
  transmission hologram 14, 16

Wavelength
  changes   79, 81, 82
  -change contouring   254–257
  selectivity   15
Wear   2
Well-conditioned system   169
Wire frame
  grid   68
  plots   68
Wollaston prism   96, 97
Wrap-around pollution   187

X-ray   152, 208
  astronomy   316

Young's fringes   268

Zeeman laser   114, 116, 177
Zero-frequency   189
Zero-fringe visibility   283
Zero-order
  Bessel function   65–68, 157, 294
  fringe   41, 47, 65, 104
Zero-peak   185

Springer Series in Optical Sciences
Editorial Board: A. L. Schawlow K. Shimoda A. E. Siegman T. Tamir

1 **Solid-State Laser Engineering**
3rd Edition By W. Koechner

2 **Table of Laser Lines in Gases and Vapors**
3rd Edition
By R. Beck, W. Englisch, and K. Gürs

3 **Tunable Lasers and Applications**
Editors: A. Mooradian, T. Jaeger, and
P. Stokseth

4 **Nonlinear Laser Spectroscopy** 2nd Edition
By V. S. Letokhov and V. P. Chebotayev

5 **Optics and Lasers** Including Fibers and
Optical Waveguides 3rd Edition
By M. Young

6 **Photoelectron Statistics**
With Applications to Spectroscopy and
Optical Communication By B. Saleh

7 **Laser Spectroscopy III**
Editors: J. L. Hall and J. L. Carlsten

8 **Frontiers in Visual Science**
Editors: S. J. Cool and E. J. Smith III

9 **High-Power Lasers and Applications**
2nd Printing
Editors: K. -L. Kompa and H. Walther

10 **Detection of Optical and Infrared Radiation**
2nd Printing By R. H. Kingston

11 **Matrix Theory of Photoelasticity**
By P.S. Theocaris and E. E. Gdoutos

12 **The Monte Carlo Method in Atmospheric Optics**
By G. I. Marchuk, G. A. Mikhailov,
M. A. Nazaraliev, R. A. Darbinian,
B. A. Kargin, and B. S. Elepov

13 **Physiological Optics**
By Y. Le Grand and S. G. El Hage

14 **Laser Crystals** Physics and Properties
2nd Edition By A. A. Kaminskii

15 **X-Ray Spectroscopy**
2nd Edition By B. K. Agarwal

16 **Holographic Interferometry**
From the Scope of Deformation Analysis of
Opaque Bodies
By W. Schumann and M. Dubas

17 **Nonlinear Optics of Free Atoms and Molecules**
By D. C. Hanna, M. A. Yuratich, D. Cotter

18 **Holography in Medicine and Biology**
Editor: G. von Bally

19 **Color Theory and Its Application in Art and
Design** 2nd Edition By G. A. Agoston

20 **Interferometry by Holography**
By Yu.I. Ostrovsky, M. M. Butusov,
G. V. Ostrovskaya

21 **Laser Spectroscopy IV**
Editors: H. Walther, K. W. Rothe

22 **Lasers in Photomedicine and Photobiology**
Editors: R. Pratesi and C. A. Sacchi

23 **Vertebrate Photoreceptor Optics**
Editors: J. M. Enoch and F. L. Tobey, Jr.

24 **Optical Fiber Systems and Their Components**
An Introduction By A. B. Sharma,
S. J. Halme, and M. M. Butusov

25 **High Peak Power Nd : Glass Laser Systems**
By D. C. Brown

26 **Lasers and Applications**
Editors: W. O. N. Guimaraes, C. T. Lin,
and A. Mooradian

27 **Color Measurement** Theme and Variations
2nd Edition By D. L. MacAdam

28 **Modular Optical Design**
By O. N. Stavroudis

29 **Inverse Problems of Lidar Sensing of the
Atmosphere** By V. E. Zuev and
I. E. Naats

30 **Laser Spectroscopy V**
Editors: A. R. W. McKellar, T. Oka, and
B. P. Stoicheff

31 **Optics in Biomedical Sciences**
Editors: G. von Bally and P. Greguss

32 **Fiber-Optic Rotation Sensors**
and Related Technologies
Editors: S. Ezekiel and H. J. Arditty

33 **Integrated Optics: Theory and Technology**
3rd Edition By R. G. Hunsperger

34 **The High-Power Iodine Laser**
By G. Brederlow, E. Fill, and K. J. Witte

35 **Engineering Optics** 2nd Edition
By K. Iizuka

36 **Transmission Electron Microscopy** Physics of
Image Formation and Microanalysis
3rd Edition By L. Reimer

37 **Opto-Acoustic Molecular Spectroscopy**
By V. S. Letokhov and V. P. Zharov

38 **Photon Correlation Techniques**
Editor: E. O. Schulz-DuBois

39 **Optical and Laser Remote Sensing**
Editors: D. K. Killinger and
A. Mooradian

40 **Laser Spectroscopy VI**
Editors: H. P. Weber and W. Lüthy

41 **Advances in Diagnostic Visual Optics**
Editors: G. M. Breinin and I. M. Siegel

MIX
Papier aus verantwortungsvollen Quellen
Paper from responsible sources
FSC® C105338

If you have any concerns about our products,
you can contact us on
ProductSafety@springernature.com

In case Publisher is established outside the EU,
the EU authorized representative is:
**Springer Nature Customer Service Center GmbH
Europaplatz 3, 69115 Heidelberg, Germany**

Printed by Libri Plureos GmbH
in Hamburg, Germany